T0136972

Advances in Intelligent Systems and Computing

Volume 800

Series editor

Janusz Kacprzyk, Polish Academy of Sciences, Warsaw, Poland
e-mail: kacprzyk@ibspan.waw.pl

The series "Advances in Intelligent Systems and Computing" contains publications on theory, applications, and design methods of Intelligent Systems and Intelligent Computing. Virtually all disciplines such as engineering, natural sciences, computer and information science, ICT, economics, business, e-commerce, environment, healthcare, life science are covered. The list of topics spans all the areas of modern intelligent systems and computing such as: computational intelligence, soft computing including neural networks, fuzzy systems, evolutionary computing and the fusion of these paradigms, social intelligence, ambient intelligence, computational neuroscience, artificial life, virtual worlds and society, cognitive science and systems, Perception and Vision, DNA and immune based systems, self-organizing and adaptive systems, e-Learning and teaching, human-centered and human-centric computing, recommender systems, intelligent control, robotics and mechatronics including human-machine teaming, knowledge-based paradigms, learning paradigms, machine ethics, intelligent data analysis, knowledge management, intelligent agents, intelligent decision making and support, intelligent network security, trust management, interactive entertainment, Web intelligence and multimedia.

The publications within "Advances in Intelligent Systems and Computing" are primarily proceedings of important conferences, symposia and congresses. They cover significant recent developments in the field, both of a foundational and applicable character. An important characteristic feature of the series is the short publication time and world-wide distribution. This permits a rapid and broad dissemination of research results.

More information about this series at http://www.springer.com/series/11156

Fernando De La Prieta · Sigeru Omatu
Antonio Fernández-Caballero
Editors

Distributed Computing and Artificial Intelligence, 15th International Conference

 Springer

Editors
Fernando De La Prieta
BISITE Digital Innovation Hub. Edificio
    Multimedia I+D+i
University of Salamanca
Salamanca, Spain

Antonio Fernández-Caballero
Department of Computing Systems,
    School of Industrial Engineers at Albacete
University of Castilla-La Mancha
Albacete, Spain

Sigeru Omatu
Faculty of Engineering,
    Department of Electronics,
    Information and Communication
Osaka Institute of Technology
Osaka, Osaka, Japan

ISSN 2194-5357   ISSN 2194-5365   (electronic)
Advances in Intelligent Systems and Computing
ISBN 978-3-319-94648-1   ISBN 978-3-319-94649-8   (eBook)
https://doi.org/10.1007/978-3-319-94649-8

Library of Congress Control Number: 2018947362

Printed on acid-free paper

This Springer imprint is published by the registered company Springer International Publishing AG
part of Springer Nature
The registered company address is: Gewerbestrasse 11, 6330 Cham, Switzerland

Preface

Nowadays, most computing systems from personal laptops/computers to cluster/grid/cloud computing systems are available for parallel and distributed computing. Distributed computing performs an increasingly important role in modern signal/data processing, information fusion, and electronics engineering (e.g., electronic commerce, mobile communications, and wireless devices). Particularly, applying artificial intelligence in distributed environments is becoming an element of high added value and economic potential. Research on Intelligent Distributed Systems has matured during the last decade, and many effective applications are now deployed. The artificial intelligence is changing our society. Its application in distributed environments such as the Internet, electronic commerce, mobile communications, wireless devices, distributed computing is increasing and is becoming an element of high added value and economic potential, both industrial and research. These technologies are changing constantly as a result of the large research and technical effort being undertaken in both universities and businesses.

The 15th International Symposium on Distributed Computing and Artificial Intelligence 2018 (DCAI 2018) is a forum to present applications of innovative techniques for solving complex problems in these areas. This year's technical program will present both high quality and diversity, with contributions in well-established and evolving areas of research. Specifically, 62 papers were submitted to main track from over 20 different countries (Algeria, Angola, Austria, Brazil, Colombia, France, Germany, India, Italy, Japan, Netherlands, Oman, Poland, Portugal, South Korea, Spain, Thailand, Tunisia, UK, and USA), representing a truly "wide area network" of research activity. The DCAI'18 technical program has selected 42 papers, and, as in past editions, it will be special issues in JCR-ranked journals such as Neurocomputing, and International Journal of Knowledge and Information Systems. These special issues will cover extended versions of the most highly regarded works.

This symposium is organized by the University of Castilla-La Mancha, the Osaka Institute of Technology, and the University of Salamanca. The present edition was held in Toledo, Spain, from June 20 to 22, 2018.

We thank the sponsors (IBM, Indra, IEEE Systems Man, and Cybernetics Society Spain) and the funding supporting of the Junta de Castilla y León (Spain) with the project *"Moviurban: Máquina Social para la Gestión sostenible de Ciudades Inteligentes: Movilidad Urbana, Datos abiertos, Sensores Móviles"* (Id. SA070U16 - Project co-financed with FEDER funds), and finally, the Local Organization members and the Program Committee members for their hard work, which was essential for the success of DCAI'18.

June 2018   Fernando De la Prieta
                                                                                                 Sigeru Omatu
                                                                           Antonio Fernández-Caballero

Organization

Honorary Chairman

Masataka Inoue   President of Osaka Institute of Technology, Japan

Program Committee Chairs

Sigeru Omatu   Osaka Institute of Technology, Japan
Sara Rodríguez   University of Salamanca, Spain
Fernando De la Prieta   University of Salamanca, Spain

Local Committee Chair

Antonio Fernández   University of Castilla-La Mancha, Spain
Elena Navarro   University of Castilla-La Mancha, Spain
Pascual González   University of Castilla-La Mancha, Spain

Scientific Committee

Silvana Aciar   Instituto de Informática, Universidad Nacional
          de San Juan, Argentina
Naoufel Khayati   COSMOS Laboratory - ENSI, Tunisia
Miguel A. Vega-Rodríguez   University of Extremadura, Spain
Zita Vale   GECAD - ISEP/IPP, Portugal
Paulo Mourao   University of Minho, Portugal
Egons Lavendelis   Riga Technical University, Latvia

Mina Sheikhalishahi	Consiglio Nazionale delle Ricerche, Italy
Patricia Jiménez	Universidad de Huelva, Spain
Volodymyr Turchenko	Research Institute for Intelligent Computing Systems, Ternopil National Economic University, Ucrania
Olfa Belkahla Driss	University of Manouba, Tunisia, Tunisia
Amel Borgi	ISI/LIPAH, Université de Tunis El Manar, Tunisia
Heman Mohabeer	Charles Telfair Institute, Mauritius
Toru Fujinaka	Hiroshima University, Japan
Reza Abrishambaf	Miami University, EE.UU.
Luiz Romao	Univille, Brazil
Abdallah Ghourabi	Higher School of Telecommunications SupCom, Tunisia, Tunisia
Susana Muñoz Hernández	Universidad Politécnica de Madrid, Spain
Fabio Marques	University of Aveiro, Portugal
Ramdane Maamri	LIRE laboratory UC Constantine 2 Abdelhamid Mehri Algeria, Algeria
Julio Cesar Nievola	Pontifícia Universidade Católica do Paraná - PUCPR Programa de Pós Graduação em Informática Aplicada, Brazil
Reyes Pavón	University of Vigo, Spain
Raffaele Dell'Aversana	Research Center for Evaluation and Socio-Economic Development, Italy
Daniel López-Sánchez	BISITE, Spain
Emilio Serrano	Universidad Politécnica de Madrid, Spain
Amin Khan	INESC-ID. Instituto Superior Técnico. Universidade de Lisboa, Portugal
Daniel Hernández de La Iglesia	Universidad de Salamanca, Spain
Alberto López Barriuso	USAL, Spain
Eleni Mangina	UCD, Ireland
Diego Hernán Peluffo-Ordoñez	Yachay Tech - Ecuador, Ecuador
Alberto Fernandez	CETINIA. University Rey Juan Carlos, Spain
Angel Martin Del Rey	Department of Applied Mathematics, Universidad de Salamanca, Spain
María Navarro	BISITE, Spain
Muhammad Marwan Muhammad Fuad	Aarhus University, Denmark
Ichiro Satoh	National Institute of Informatics, Japan
Horacio Gonzalez-Velez	National College of Ireland, Ireland
Svitlana Galeshchuk	Nova Southeastern University, EE.UU.
Michal Wozniak	Wroclaw University of Technology, Poland
Florentino Fdez-Riverola	University of Vigo, Spain

Mariano Raboso Mateos	Facultad de Informática - Universidad Pontificia de Salamanca, Spain
Fábio Silva	University of Minho, Portugal
Li Weigang	University of Brasilia, Brazil
Nadia Nouali-Taboudjemat	CERIST, Algeria
Bozena Wozna-Szczesniak	Institute Of Mathematics and Computer Science, Jan Dlugosz University in Czestochowa, Poland
Javier Bajo	Universidad Politécnica de Madrid, Spain
Gustavo Santos-Garcia	Universidad de Salamanca, Spain
Jose-Luis Poza-Luján	Universitat Politècnica de València, Spain
Jacopo Mauro	University of Oslo, Norway
Worawan Diaz Carballo	Thammasat University, Thailand
Paulo Novais	University of Minho, Portugal
Peter Forbrig	University of Rostock, Germany
Nuno Silva	DEI & GECAD - ISEP - IPP, Portugal
Rafael Valencia-Garcia	Departamento de Informática y Sistemas. Universidad de Murcia, Spain
Yann Secq	Université Lille I, France
Tiago Oliveira	National Institute of Informatics, Japan
Rene Meier	Lucerne University of Applied Sciences, Switzerland
Aurélie Hurault	IRIT - ENSEEIHT, France
Fidel Aznar	Universidad de Alicante, Spain
Paulo Cortez	University of Minho, Portugal
Leandro Tortosa	University of Alicante, Spain
Rosalia Laza	Universidad de Vigo, Spain
Francisco A. Pujol	Specialized Processor Architectures Lab, DTIC, EPS, University of Alicante, Spain
Ângelo Costa	University of Minho, Portugal
Carlos Carrascosa	GTI-IA DSIC Universidad Politecnica de Valencia, Spain
Ana Almeida	ISEP-IPP, Portugal
Luis Antunes	GUESS/LabMAg/Univ. Lisboa, Portugal
Ester Martinez-Martin	Universitat Jaume I, Spain
José Ramón Villar	University of Oviedo, Spain
Faraón Llorens-Largo	Universidad de Alicante, Spain
Fernando Diaz	University of Valladolid, Spain
Jesus Martin-Vaquero	University of Salamanca, Spain
Maria João Viamonte	Instituto Superior de Engenharia do Porto, Portugal
Cesar Analide	University of Minho, Portugal
Pierre Borne	Ecole Centrale de Lille, France
Johan Lilius	Abo Akademi University, Finland
Camelia Chira	Technical University of Cluj-Napoca, Romania

Contents

Effects of Switching Costs in Distributed Problem-Solving Systems

Friederike Wall[(✉)]

Alpen-Adria-Universitaet Klagenfurt, 9020 Klagenfurt, Austria
friederike.wall@aau.at
http://www.aau.at/csu

Abstract. In many situations, changing the status quo may induce particular extra costs. Such switching costs are assumed to cause inertia and reduce performance. This paper studies the effects of switching costs in distributed problem-solving systems and, for this, employs an agent-based simulation based on NK fitness landscapes. The results indicate that the complexity of the problem to be solved considerably shapes the effects of switching costs. Depending on the period of time in the search for superior solutions, switching costs may even have beneficial effects in terms of stabilizing the search and increasing the system's performance.

Keywords: Agent-based simulation · Complexity · Effort
NK fitness landscapes · Switching costs

1 Introduction

Various domains related to systems with distributed problem-solving agents take note of the phenomenon that changing the status quo may come along with notable extra-costs in terms of, for example, additional consumption of resources or evoking particular cognitive efforts. Though the foci differ across domains, the particular costs for changing the status quo in favor of an alternative usually are named switching costs (e.g., [1–3]).

For example, organizational changes may cause costs for reorganization as well as for handling resistance of certain stakeholders [4] and using a new information system requires some learning efforts on the users' site and may cause costs for dealing with users' resistance [5,6]. Choosing another supplier could result in high costs for technical conversions [1,2] or switching to another cryptocurrency may result in losses due to switching costs and network effects [7]; changing the direction of a robot's movement in a multi-robot system could cost some extra time [8–10] and altering the task assignment in the course of job scheduling may rise some extra-costs of, for example, learning – in case of human as well as "artificial" agents [11,12]. These examples indicate on the wide range of manifestations and the relevance of switching costs in the course of searching for, broadly speaking, a superior course of action.

© Springer International Publishing AG, part of Springer Nature 2019
F. De La Prieta et al. (Eds.): DCAI 2018, AISC 800, pp. 1–9, 2019.
https://doi.org/10.1007/978-3-319-94649-8_1

A reasonable conjecture is that switching costs reduce the propensity that new and, potentially, superior solutions are implemented and, thus, could be an obstacle to enhancing the performance of a system employing distributed problem-solvers. However, one may also argue that switching costs could prevent a distributed problem-solving system (DPSS) from too many alterations, in particular, when the outcomes of the alterations are not known precisely in advance.

Against this background, the paper seeks to contribute to findings answers on two interrelated research questions:

1. Which effects on the performance of a DPSS result when switching is costly?
2. When do switching costs have less detrimental or, may be, even beneficial effects on a DPSS's performance?

The first research question is intended to capture the spectrum of possible effects of switching costs, while the second research question is directed to the conditions under which certain effects occur. For finding some answers on these questions, the paper makes use of an agent-based simulation where the task of the simulated DPSSs is modeled according to the framework of NK fitness landscapes [13,14] which was originally introduced in the domain of evolutionary biology and, since then, broadly employed [15]. A key feature of the NK framework is that it allows to easily control for the complexity of the task to be performed by the DPSS [16]. In the remainder of the paper, after the simulation model is outlined subsequently, Sect. 3 gives an overview over the simulation experiments, and in Sect. 4 results are introduced and discussed.

2 Outline of the Simulation Model

2.1 Organizational Structure and Overall Task of the DPSS

In the simulations, DPSSs are observed while searching for superior solutions for an N-dimensional binary problem $\mathbf{d_t} = (d_{1t}, ..., d_{Nt})$ with $d_{it} \in \{0, 1\}$, $i = 1, ...N$. The N-dimensional problem is decomposed into M disjoint sub-problems indexed by $r = 1, ..., M$ of equal size, i.e., each sub-problem r is of size $N^r = N/M$. Each of the M sub-problems is exclusively assigned to one agent $r = 1, ..., M$, and each agent searches for superior solutions for its N^r-dimensional sub-task without the intervention of any central decision-making authority.

Following the framework of NK-fitness landscapes, in every time step t, each of the two states of choice $d_{it} \in \{0, 1\}$, $i = 1, ...N$ of the N-dimensional binary search problem $\mathbf{d_t}$ contributes with C_{it} to the overall performance V_t of the DPSS. The overall performance V_t achieved in period t results as normalized sum of contributions C_{it} from

$$V_t = V(\mathbf{d_t}) = \frac{1}{N} \sum_{i=1}^{N} C_{it} \qquad (1)$$

A key feature of the NK framework is that the contributions C_{it} can be subject to interactions among the N single choices, and, thus, capture the overall task's complexity denoted by parameter K:

Parameter K (with $0 \leq K \leq N - 1$) reflects the number of those choices d_{jt}, $j \neq i$ which also affect the performance contribution C_{it} of choice d_{it}. With this, contribution C_{it} is not only affected by the single choice d_{it} but may also depend on K other choices:

$$C_{it} = f_i(d_{it}; d_{i_1 t}, ... d_{i_K t}), \qquad (2)$$

with $\{i_1, ..., i_K\} \subset \{1, ..., i - 1, i + 1, ..., N\}$. Hence, if choices do not affect each other, K equals 0; when every single choice i affects the performance contribution of each other choice $j \neq i$, interactions are at maximum with $K = N - 1$. C_{it} is randomly drawn from a uniform distribution with $0 \leq C_{it} \leq 1$. (Hence, from a more "technical" point of view, NK landscapes are stochastically generated pseudo-boolean functions with N bits, i.e., $F : \{0, 1\}^N \to \mathbb{R}^+$, [16].)

2.2   Agents' Options and Switching Costs

In every time step t, each agent has to decide which configuration $\mathbf{d_t^r}$ of the sub-problem assigned to that particular agent is to be implemented. In particular, an agent has three options to choose from: first, the status quo $\mathbf{d_{t-1}^{r*}}$, i.e., the option chosen in the previous period, could be kept; second, each agent discovers randomly an alternative $\mathbf{d_t^{r,a1}}$ that differs in one of the bits of the status quo-configuration; third, each agent finds an option $\mathbf{d_t^{r,a2}}$ where two single choices are modified compared to the status quo.

To capture the switching costs, a function g is introduced which gives the number of dimensions (bits) in which configuration $\mathbf{d_t^r}$ differs from the status quo $\mathbf{d_{t-1}^{r*}}$ (i.e., the Hamming distance):

$$g(\mathbf{d_t^r}) = \begin{cases} 0 \text{ if } \mathbf{d_t^r} = \mathbf{d_{t-1}^{r*}} \\ 1 \text{ if } \mathbf{d_t^r} = \mathbf{d_t^{r,a1}} \\ 2 \text{ if } \mathbf{d_t^r} = \mathbf{d_t^{r,a2}} \end{cases} \qquad (3)$$

The intuition behind is that the number of bits flipped could be regarded as the effort to be taken by agent r. The switching costs $S^r(\mathbf{d_t^r})$ of agent r are quadratically increasing with g, i.e.,

$$S_t^r(\mathbf{d_t^r}) = s^r \cdot (g(\mathbf{d_t^r}))^2 \qquad (4)$$

where s^r is a cost coefficient which, for the sake of simplicity, in the simulations is the same for all agents r.

2.3   Agents' Objectives and Information

Though the problem-solving agents are not hostile against each other, each pursues its "own" objective – related to its particular sub-problem. In every time step t, each agent seeks to identify that option out of the three $\mathbf{d_{t-1}^{r*}}$, $\mathbf{d_t^{r,a1}}$ and

$\mathbf{d_t^{r,a2}}$ which promises the highest net performance A^r, i.e., the highest difference between partial performance P_t^r – resulting from the contributions C_{it} of those particular single choices assigned to agent r – and the related switching costs:

$$A_t^r(\mathbf{d_t^r}) = P_t^r(\mathbf{d_t^r}) - S_t^r(\mathbf{d_t^r}) \qquad (5)$$

where the partial performance P_t^r is given by

$$P_t^r(\mathbf{d_t^r}) = \frac{1}{N} \sum_{i=1+(r-1)N^r}^{rN^r} C_{it} \qquad (6)$$

with $N^r = N/M \forall r$. However, when assessing performance P_t^r of options and, thus, forming preferences, each agent r suffers from two kinds of imperfect information: First, agent r it not informed about the intentions of the other agents $q \neq r$ and, thus, assumes that they will stay with their previous choices. Second, agent r is not able to precisely assess the consequences of its options on partial performance P^r. In particular, distortions from the true consequences of options are depicted as a relative errors z^r ([17], for other functions see [18]) where the z^r follow a Gaussian distribution $N(0; \sigma)$ with expected values 0 and standard deviations σ^r. For simplicity's sake, σ^r is the same for each agent r; errors are assumed to be independent from each other. With this, agent r, in fact, maximizes the *perceived* net performance so that Eq. 5 is modified to

$$\widetilde{A_t^r}(\mathbf{d_t^r}) = \widetilde{P_t^r}(\mathbf{d_t^r}) - S_t^r(\mathbf{d_t^r}) \qquad (7)$$

where

$$\widetilde{P_t^r}(\mathbf{d_t^r}) = P_t^r(\mathbf{d_t^r}) + z^r(\mathbf{d_t^r}) \qquad (8)$$

3   Simulation Experiments

The simulation experiments are intended to provide some findings on the effects of switching costs on the overall performance achieved by DPSSs in differently complex task environments. Therefore, the experiments (Table 1) distinguish for different levels of switching costs and different levels of complexity of the underlying search problem. In the experiments, after a landscape reflecting the task environment is generated, the DPSSs are "thrown" randomly into the landscape and observed over 250 periods while searching for superior solutions.

For depicting the complexity of the DPSSs' task as well as the interactions among the $M = 4$ agents, two parameters are used: Parameter K captures the complexity (see Sect. 2.1), and K^* denotes the level of interactions *across sub-problems* and, with that, also reflects the interactions among the agents' choices. The simulation experiments are obtained for three different interaction structures characterized by K and K^* for $N = 12$ (Table 1): (1) In the perfectly decomposable structure the overall search problem is decomposed into $M = 4$ disjoint parts with maximal intense intra-sub-problem interactions, but no cross-sub-problem interactions (i.e., $K^* = 0$) and this captures a block-diagonal structure [19]. (2) In the nearly decomposable structure with $K^* = 1$ only slight

Table 1. Parameter settings

Parameter	Values/Types
Observation period	$T = 250$
Number of choices	$N = 12$
Number of agents	$M = 4$ with $\mathbf{d^1} = (d_1, d_2, d_3)$, $\mathbf{d^2} = (d_4, d_5, d_6)$, $\mathbf{d^3} = (d_7, d_8, d_9)$, $\mathbf{d^4} = (d_{10}, d_{11}, d_{12})$
Interaction structures	Decomposable: $K = 2; K^* = 0$
	Nearly decomposable: $K = 3; K^* = 1$
	Full interdependent: $K = 11; K^* = 9$
Precision of ex-ante evaluation	$\sigma^r = 0.1 \forall r$
Cost coefficient	$s^r \in \{0; 0.001; 0.0055; 0.01; 0.02; 0.03; 0.055\}$

cross-sub-problem interactions occur in that every performance contribution C_i in primary control of agent r is affected by only one choice made by another agent $q \neq r$. (3) In the full interdependent case ($K = 11$, $K^* = 9$) all single options d_i affect the performance contributions of all other choices and, hence, the agents affect each other's choices at maximum.

4 Results and Discussion

The results of the simulation experiments are introduced in two steps. First, results obtained for selected levels of switching costs are analyzed more into detail (Sect. 4.1) before the sensitivity of results is studied for a broader range of switching costs (Sect. 4.2).

4.1 Analysis of Baseline Scenarios

For the interaction structures under investigation, Table 2 displays condensed results, for costless switching, a low and a high level of switching costs (column 1). In particular, the performance enhancement obtained by the DPSS in the first 10 periods (2), the final performance (3) and the performance achieved on average over the 250 periods (4) give some indication on the effectiveness of the search processes. The ratio of periods in which a new configuration $\mathbf{d_t}$ is implemented (5) and the relative frequency of periods where the status quo was altered in favor of false positive options from system's perspective (6) as well as the effort made by the M agents and averaged over the 250 periods (7) characterize the adaptive search processes more into detail.

Figure 1.a plots that ratio of the $N = 12$-dimensional configuration $\mathbf{d_t}$ that is altered less when switching is costly ($s = 0.0055$) than compared to costless alterations; Fig. 1.b displays the differences in performance obtained with costly against costless effort.

Table 2. Condensed results

Cost coeff. s^r (1)	$V_{t=10}$ $-V_{t=0}$ (2)	$V_{t=250}$ and CI* (3)	$\overline{V}_{(0;250)}$ and CI* (4)	Freq. New $\mathbf{d_t}$ (5)	Freq. False-Pos (6)	Av. Effort $\overline{g}_{(0;250)}$ (7)
Decomposable structure, $K = 2; K^* = 0$						
0	0.3024	0.9875 ±0.0012	0.9857 ±0.0010	21.16%	10.03%	0.1253
0.0055	0.3012	0.9824 ±0.0016	0.9809 ±0.0015	6.58%	2.69%	0.0292
0.02	0.2484	0.9345 ±0.0033	0.9331 ±0.0033	1.60%	0.29%	0.0072
Nearly decomposable structure, $K = 3; K^* = 1$						
0	0.2392	0.9308 ±0.0043	0.9228 ±0.0031	30.05%	14.93%	0.2301
0.0055	0.2619	0.9417 ±0.0033	0.9373 ±0.0031	7.42%	3.29%	0.0395
0.02	0.2293	0.8990 ±0.0044	0.8970 ±0.0043	1.84%	0.49%	0.0084
Full interdependent structure, $K = 11; K^* = 9$						
0	0.0312	0.8185 ±0.0110	0.7108 ±0.0090	51.25%	25.50%	0.6619
0.0055	0.0538	0.8487 ±0.0076	0.7650 ±0.0075	31.47%	15.56%	0.3203
0.02	0.0972	0.7910 ±0.0084	0.7635 ±0.0077	9.46%	4.47%	0.0581

*Confidence intervals at a confidence level of 0.99.

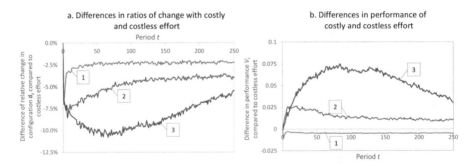

Fig. 1. Differences in (a) ratio of alterations in $\mathbf{d_t}$ and (b) performance V_t obtained for (1) decomposable, (2) nearly decomposable and (3) full interdependent interaction structures for costly ($s = 0.0055$) compared to costless switching. Each line represents the differences between the averages of 2,500 adaptive walks, i.e., 250 distinct fitness landscapes with 10 adaptive walks on each. For further parameter settings see Table 1.

The results provide broad support for the conjecture that costly switching reduces alterations from the status quo – and that the reduction of alterations is the higher the higher switching costs. However, the results *do not universally support* the conjecture that switching costs are detrimental. On the contrary, it appears that switching costs may have their beneficial effects:

This is most obvious for the *full interdependent structure* in Fig. 1.b where the performance excess obtained with costly alterations goes up to nearly 7.5 points of percentage in periods 50 to 100. Moreover, as can be seen in Table 2, in this structure and for $s = 0.0055$, the initial performance enhancement and the final as well as the average performance are remarkably higher than with costless switching; even with a high level of switching costs ($s = 0.02$) the initial performance enhancement and the average performance remain at a remarkably higher level than with costless switching. The *nearly decomposable* structure shows similar results like the full interdependent structure, though not that pronounced. In the *perfectly decomposable structure*, performance appears to be rather robust against low switching costs (see columns 2 to 4 Table 2), while, with high costs, the performance measures notably drop down. An interesting question is, of course, what may cause the effects found which is discussed subsequently.

4.2 Sensitivity to the Level of Switching Costs

In order to gain a deeper understanding of the effects described so far, simulations for further levels of switching costs were run. Since the effects are most pronounced (though not qualitatively different) for the full interdependent structure, the respective results are displayed in Fig. 2. The figure shows the performance differences obtained for different cost levels $s > 0$ to costless switching.

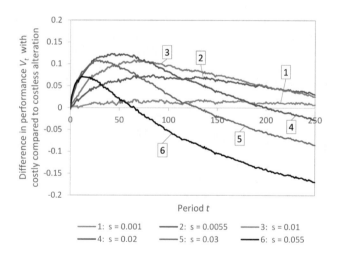

Fig. 2. Differences of performance V_t with costly to costless switching for full interdependent interactions structures. Each line represents the difference between the averages of 2,500 adaptive walks, i.e., 250 distinct fitness landscapes with 10 adaptive walks on each. For parameter settings see Table 1

Obviously, with increasing switching costs the periods of time where switching costs are beneficial become shorter, concentrating more in the beginning of the search processes. Moreover, with increasing switching costs up to a level of $s = 0.02$ the performance excess goes up to 12 points of percentage at around $t = 50$; for even higher switching costs maximum performance excess decreases.

An explanation may lie in the subtle interference of switching costs with the imperfect information at the agents' site in cases of cross-agent interactions ($K^* > 0$): agent r – forming its preferences in t without knowing about the intentions of the fellow agents (Sect. 2.3) – in $t + 1$, may not only be surprised by the actual performance P^r achieved in t (see Eq. 6 vs. 8) but also by the other agents' choices which, due to $K^* > 0$, affect r's performance too. Thus, this eventually lets agent r adapt the "own" choices in $t + 1$ and so forth – leading to *frequent time-delayed mutual adjustments*, particularly in the early stages of search when performance gains are more likely to occur. However, whether agent r leaves the status quo is shaped by the switching costs and, in particular, switching costs apparently cause that only worthwhile alterations are made, i.e., alterations whose perceived performance gains exceed the costs of effort. In later stages of search (usually promising lower performance gains) and in case of higher switching costs the detrimental effects of switching costs (i.e., making alterations less attractive and, in the extreme, inducing inertia) predominate the benefit of increasing the efficiency of search.

5   Conclusion

The results of this simulation study can be summarized in the three hypotheses: (1) The complexity of the problem to be solved by a DPSS shapes the effects of switching costs, (2) switching costs may eventually be beneficial in terms of increasing the overall performance obtained by a DPSS, and (3) the effects of switching costs vary with the phase in the search process for superior solutions.

These findings may be of particular relevance for the understanding and design of DPSS. For example, switching costs may not only be exogenously given (depending on the context, for example, by technical features, human nature or economic constraints) but may also be subject to deliberate choices: Well-known examples are the switching costs charged in the telecommunication sector or by energy suppliers relevant, for example, in the domain of smart grids. In particular, since according to the results obtained in this study switching costs are beneficial in complex environments, it might be even of interest to deliberately introduce some switching costs in order to stabilize the search processes of decentralized agents.

Moreover, this study suggests that switching costs may be particularly beneficial in the early stages of search and, hence, with frequent external shocks setting a DPSS again and again back to the start of search, purposefully shaped switching costs may be an appropriate means to reduce "hyperactivity" of the problem-solving agents in a distributed system.

Hence, this study calls for further research to study switching costs in the presence of environmental turbulence. Other promising fields would be to analyze switching costs in the context of more sophisticated coordination mechanisms than employed in this study or to study the interference with imperfect agents' information more into detail.

References

1. Basaure, A., Suomi, H., Hämmäinen, H.: Transaction vs. switching costs - comparison of three core mechanisms for mobile markets. Telecommun. Policy **40**, 545–566 (2016)
2. Thomas, A.B., Judy, K.F., Vijay, M.: Consumer switching costs: a typology, antecedents, and consequences. J. Acad. Mark. Sci. **31**, 109–126 (2003)
3. Blatt, J.M.: Optimal control with a cost of switching control. J. Aust. Math. Soc. Ser. B Appl. Math. **19**, 316–332 (1976)
4. Hannan, M.T., Freeman, J.: Structural inertia and organizational change. Am. Sociol. Rev. **49**(2), 149–164 (1984)
5. Kim, H.W.: The effects of switching costs on user resistance to enterprise systems implementation. IEEE Trans. Eng. Manag. **58**(3), 471–482 (2011)
6. Polites, G.L., Karahanna, E.: Shackled to the status quo: the inhibiting effects of incumbent system habit, switching costs, and inertia on new system acceptance. MIS Q. **36**(1), 21–42 (2012)
7. Luther, W.J.: Cryptocurrencies, network effects, and switching costs. Contemp. Econ. Policy **34**(3), 553–571 (2016)
8. Agmon, N., Kraus, S., Kaminka, G.A.: Multi-robot perimeter patrol in adversarial settings. In: IEEE International Conference on Robotics and Automation, 19–23 May 2008, pp. 2339–2345 (2008)
9. Le Ny, J., Dahleh, M., Feron, E.: Multi-agent task assignment in the bandit framework. In: 2006 IEEE Conference on Decision and Control, pp. 5281–5286. IEEE (2006)
10. Yu-Han, L., Devin, B.: Optimal trajectories for kinematic planar rigid bodies with switching costs. Int. J. Rob. Res. **35**, 454–475 (2015)
11. Ho, K.I.J., Sum, J.: Scheduling jobs with multitasking and asymmetric switching costs. In: IEEE International Conference on Systems, Man, and Cybernetics (SMC), 5–8 October 2017, pp 2927–2932 (2017)
12. Le, T., Szepesvári, C., Zheng, R.: Sequential learning for multi-channel wireless network monitoring with channel switching costs. IEEE Trans. Sig. Process. **62**, 5919–5929 (2014)
13. Kauffman, S.A., Levin, S.: Towards a general theory of adaptive walks on rugged landscapes. J. Theor. Biol. **128**, 11–45 (1993)
14. Kauffman, S.A.: The Origins of Order: Self-organization and Selection in Evolution. Oxford University Press, Oxford (1993)
15. Wall, F.: Agent-based modeling in managerial science: an illustrative survey and study. RMS **10**, 135–193 (2016)
16. Li, R., Emmerich, M.M., Eggermont, J., Bovenkamp, E.P., Bäck, T., Dijkstra, J., Reiber, J.C.: Mixed-integer NK landscapes. In: Parallel Problem Solving from Nature IX, vol. 4193, pp. 42–51, Springer, Berlin (2006)
17. Wall, F.: The (beneficial) role of informational imperfections in enhancing organisational performance. In: Lecture Notes in Economics and Mathematical Systems, vol. 645, pp. 115–126, Springer, Berlin (2010)
18. Levitan, B., Kauffman, S.A.: Adaptive walks with noisy fitness measurements. Mol. Diversity **1**(1), 53–68 (1995)
19. Rivkin, R.W., Siggelkow, N.: Patterned interactions in complex systems: implications for exploration. Manag. Sci. **53**, 1068–1085 (2007)

Fire Detection Using DCNN for Assisting Visually Impaired People in IoT Service Environment

Borasy Kong[✉], Kuoysuong Lim, and Jangwoo Kwon

Department of Computer Engineering, Inha University,
Incheon 22201, South Korea
borasykong@gmail.com, limkuoysuong@gmail.com,
jwkwon@inha.ac.kr

Abstract. In an emergency, such as fire in a building, visually impaired people are prone to danger more than non-impaired people, for they cannot be aware of it quickly. Current fire detection methods such as smoke detector is very slow and unreliable. But by using vision sensor instead, fire can be proven to be detected much faster as shown in our experiments. Previous studies have applied various image processing and machine learning techniques to detect fire, but they usually don't generalize well because those techniques use hand-crafted features. With the recent advancements in the field of deep learning, this research can be conducted to help solve the problem by using deep learning-based object detector to detect fire. Such approach can learn features automatically, so they can usually generalize well to various scenes. We introduced **two** object detection models (R1 and R2) with slightly different model's complexity. R1 can detect fire at 90% average precision and 85% recall at 33 FPS, while R2 has 90% average precision and 61% recall at 50 FPS. The reason why we introduced two models is because we want to have a benchmark comparison as no other research on fire detection with similar techniques exists. We also want to give two model choices when we wish to integrate the model into an IoT platform.

Keywords: Object detection · Deep convolutional neural network
Tensorflow · Darkflow

1 Introduction

According to the World Health Organization in 2014, an estimated 285 million people are visually impaired worldwide. Among those people, 39 million are completely blind. Life for these people can be very tough, that is why various technologies have been innovated specifically for them. Recently, the advancement of artificial intelligence (AI), computer vision and image recognition technology have helped the visually impaired to 'see' in a whole new way. Wearable technologies such as smart glass can describe the scene in front of the person in real-time. But these technologies are created to help them in their daily life basis. What still lacks for them is the event of an emergency. During such situation, blind people have a much higher risk then a non-impaired person. On top of the fact that they cannot identify the emergency quickly, it is difficult for them to navigate their way off the

© Springer International Publishing AG, part of Springer Nature 2019
F. De La Prieta et al. (Eds.): DCAI 2018, AISC 800, pp. 10–17, 2019.
https://doi.org/10.1007/978-3-319-94649-8_2

danger zone to a safer zone. To help mitigate this problem, we conducted a research where we trained a deep convolutional neural network to detect fire quickly and reliably. With this model, we hope to integrate deep learning-based object detection into an IoT platform where our model acts as the "brain" that identify a dangerous situation and allows other applications to build on top of this information to eventually alert the visually impaired people and assist them in navigating away from the danger.

2 Related Work

2.1 Traditional Approach

Detecting fire for safety purposes is not a new phenomenon. Traditional methods date back to the early 1900s [1]. One of those technologies is smoke detection. It has been a sensible approach as smoke is produced much earlier than other fire signatures during the development of fire. The detection of smoke is done by sensing the interaction of smoke particles with a beam of light [5] or electromagnetic radiation [3, 4]. Smoke detectors can be very sensitive as they can sense smoke particles ranging from 0.01 to 1.0 micron and 1.0 to 10.0 micron for ionization chamber and optical smoke detector, respectively [2]. However, they are very sensitive to false alarm due to general cooking and the interference of dust particles.

A more modern approach in the early 2000s is the use of machine vision fire detection system (MVFDS) which is a combination of video cameras, computers, and AI techniques [6–8]. This technique uses the advancement of image processing techniques to constantly analyze images fed from the camera on the computer. Such image processing techniques as summarized by Wieser and Brupbacher include histogram, temporal and rule-based techniques [8]. Experiments showed that while the system can rapidly detect fire and smoke, it failed to generalize to a wide range of domain of fire as the features for detecting fire are manually hand-crafted. Additionally, hand-making these features can be a very difficult and tedious task.

2.2 Deep Learning Approach

Seeing the grand success of deep learning algorithm in the last few years, many researchers and practitioners are quick to apply it to various domains of science to solve their specific task, from a simple one such as cucumber classification to a more advance one such as cancer detection [9]. Machine learning approach such as deep learning doesn't require experts to handcraft features for an object. Instead those features are learnt during the training process of the network, thus, saving time and it can generalize well to a wide range of domain. However, applying deep CNN to detect fire is a relatively recent task, as there are not many in the literature. The only appropriate work that talks about a proper model and training phase is the work of Frizzi et al. [10] in 2016, in which they train a CNN model to classify video of fire and smoke. Although the result is quite promising, their training approach and architecture's components are quite old. They have a rather shallow model (9 layers) compared to ours. Thus, their architecture might not be able to scale well in the future. They also use SGD optimizer which has been proven to converge much slower compared to a newer optimizer such as Adam [11] as our model uses. The training resolution is also quite small at 64×64 pixels compared to ours at 416×416, which may cause their model to overfit quite easier than ours. Additionally, they use fully-connected layers in their model, while newer architectures such as ours use fully convolutional layers. Lastly, they use a sliding window technique to localizes the object on the last convolutional features. Such technique is very slow compared to our approach with YOLOv2.

3   Methodology

Based on the work Huang *et al.* [15], we decided to choose YOLOv2 [14] which is a detection without proposals technique. Instead of having, a RPN to propose some regions of interest (ROI) like [12], it divides the image into an odd number of grids. Each grid contains some number of anchors boxes which are responsible for regressing to a final box with 5 numbers: $(dx, dy, dh, dw, confidence)$. Each grid also predicts scores for a class that might reside in that grid. This technique allows for very fast detection speed. We also like the fact that YOLOv2 has stable and reliable Tensorflow implementation (Darkflow)[1]. Training on Tensorflow allows us to get access to the latest technology in the field, such as the Adam optimizer [11]. In addition, it is much easier to deploy our model with protobuf file that can be further quantized and deployed relatively easy.

3.1   Model Architecture

To conduct our experiments, we trained 2 different models (R1 and R2) which are based on the Darknet architecture. R1 (see Table 2) is only finetuned with the final layer. It's designed to be more accurate but slower. R2 (see Table 3), on the other hand, has more modified and finetuned layers. It's designed to be faster but less accurate. The summary of the design choices can be found in Table 1.

Table 1. From R2 to R1 design decisions

Design choice	R1	R2	Parameters
Subtract 2 layers (24th and 29th)	31 layers	29 layers	Reduced 21 million
Reduce filters' size from 3×3 to 1×1	23rd layer: 3×3 size with 1024 filters	23rd layer: 1×1 size with 512 filters	Reduced 8.9 million
Filters' size and depth change	26th layer: 1×1 size with 64 filters 27th reorg: 256 filters 28th route: 1280 filters	25th layer: 3×3 size with 128 filters 26th reorg: 512 filters 27th route: 1024 filters	Added 5 thousand

---

[1] Darkflow: darknet implementation in Tensorflow. Github repository: https://github.com/thtrieu/darkflow.

Table 2. R1 architecture.

Layer	Type	Filters	Size / Pad / Stride	Output
0	Convolutional	32	3 x 3 / 1 / 1	416 x 416
1	Maxpool		2 x 2 / 0 / 2	208 x 208
2	Convolutional	64	3 x 3 / 1 / 1	208 x 208
3	Maxpool		2 x 2 / 0 / 2	104 x 104
4	Convolutional	128	3 x 3 / 1 / 1	104 x 104
5	Convolutional	64	1 x 1 / 0 / 1	104 x 104
6	Convolutional	128	3 x 3 / 1 / 1	104 x 104
7	Maxpool		2 x 2 / 0 / 2	52 x 52
8	Convolutional	256	3 x 3 / 1 / 1	52 x 52
9	Convolutional	128	1 x 1 / 0 / 1	52 x 52
10	Convolutional	256	3 x 3 / 1 / 1	52 x 52
11	Maxpool		2 x 2 / 0 / 2	26 x 26
12	Convolutional	512	3 x 3 / 1 / 1	26 x 26
13	Convolutional	256	1 x 1 / 0 / 1	26 x 26
14	Convolutional	512	3 x 3 / 1 / 1	26 x 26
15	Convolutional	256	1 x 1 / 0 / 1	26 x 26
16	Convolutional	512	3 x 3 / 1 / 1	26 x 26
17	Maxpool		2 x 2 / 0 / 2	13 x 13
18	Convolutional	1024	3 x 3 / 1 / 1	13 x 13
19	Convolutional	512	1 x 1 / 0 / 1	13 x 13
20	Convolutional	1024	3 x 3 / 1 / 1	13 x 13
21	Convolutional	512	1 x 1 / 0 / 1	13 x 13
22	Convolutional	1024	3 x 3 / 1 / 1	13 x 13
23	Convolutional	1024	3 x 3 / 1 / 1	13 x 13
24	Convolutional	1024	3 x 3 / 1 / 1	13 x 13
25	Route [16]	512		26 x 26
26	Convolutional	64	1 x 1 / 0 / 1	26 x 26
27	Reorganize	256	2 x 2 / 0 / 2	13 x 13
28	Route [27][24]	1280		13 x 13
29	Convolutional	1024	3 x 3 / 1 / 1	13 x 13
30	Convolutional	30	1 x 1 / 0 / 1	13 x 13

Table 3. R2 architecture

Layer	Type	Filters	Size / Pad / Stride	Output
0	Convolutional	32	3 x 3 / 1 / 1	416 x 416
1	Maxpool		2 x 2 / 0 / 2	208 x 208
2	Convolutional	64	3 x 3 / 1 / 1	208 x 208
3	Maxpool		2 x 2 / 0 / 2	104 x 104
4	Convolutional	128	3 x 3 / 1 / 1	104 x 104
5	Convolutional	64	1 x 1 / 0 / 1	104 x 104
6	Convolutional	128	3 x 3 / 1 / 1	104 x 104
7	Maxpool		2 x 2 / 0 / 2	52 x 52
8	Convolutional	256	3 x 3 / 1 / 1	52 x 52
9	Convolutional	128	1 x 1 / 0 / 1	52 x 52
10	Convolutional	256	3 x 3 / 1 / 1	52 x 52
11	Maxpool		2 x 2 / 0 / 2	26 x 26
12	Convolutional	512	3 x 3 / 1 / 1	26 x 26
13	Convolutional	256	1 x 1 / 0 / 1	26 x 26
14	Convolutional	512	3 x 3 / 1 / 1	26 x 26
15	Convolutional	256	1 x 1 / 0 / 1	26 x 26
16	Convolutional	512	3 x 3 / 1 / 1	26 x 26
17	Maxpool		2 x 2 / 0 / 2	13 x 13
18	Convolutional	1024	3 x 3 / 1 / 1	13 x 13
19	Convolutional	512	1 x 1 / 0 / 1	13 x 13
20	Convolutional	1024	3 x 3 / 1 / 1	13 x 13
21	Convolutional	512	1 x 1 / 0 / 1	13 x 13
22	Convolutional	1024	3 x 3 / 1 / 1	13 x 13
23	Convolutional	512	1 x 1 / 0 / 1	13 x 13
24	Route [16]	512		26 x 26
25	Convolutional	128	3 x 3 / 1 / 1	26 x 26
26	Reorganize	512	2 x 2 / 0 / 2	13 x 13
27	Route [26][23]	1024		13 x 13
28	Convolutional	30	1 x 1 / 0 / 1	13 x 13

4   Experiments

4.1   Dataset Collection Phase

There are no official datasets for fire detection task. Some fire datasets in[2] and[3] are of forest fire, which isn't what we want. This leads us to collect our own dataset. On top of using it as training data for our network, the collected data can also be contributed to the research community and those who are interested in indoor fire detection research. We collected 1200+ images from ImageNet[4], Google Image, Flickr and Pixalbay, and annotated them using LabelImg[5].

4.2   Training Phase

We train R1 for 140 epochs and R2 for 300 epochs[6] with a starting learning rate of $1e^{-5}$ and $1e^{-4}$ for R1 and R2, respectively. The learning rate is divided by 10 at 70 epochs for R1 and at 180 epochs for R2. We validate the training process at various

---

[2] Forest Fires Data Set: http://archive.ics.uci.edu/ml/datasets/forest+fires.

[3] Fire Detection Dataset: http://mivia.unisa.it/datasets/video-analysis-datasets/fire-detection-dataset.

[4] ImageNet: http://www.image-net.org/.

[5] LabelImg: https://github.com/tzutalin/labelImg.

[6] R2 has to be trained longer as more of its layers are being re-initialized.

epochs with around 100 images. While the original YOLOv2 uses SGD to train their network, we use Adam [11] instead. We train the network using a GTX Titan X.

Through our validation process, we observed that the networks no longer improve after 140th epoch for R1 and 300th epoch for R2. In fact, the network's performance got slighter worse which indicate overfitting. We decided to conclude our final networks at 140th and 300th epoch for R1 and R2, respectively. Figure 1 shows R1's and R2's training loss. Both models' loss values show a smooth downward trend which indicates a good learning rate.

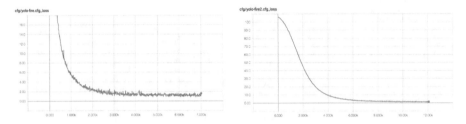

Fig. 1. R1 and R2 loss value

5   Results

After we have trained and validate our network and got the converged networks that we desired, we finally tested the network against 100 plus images of fire from various scenarios and lighting conditions to ensure a maximum result. The result is quite promising as R1 has 90% average precision and 85% recall while R2 has 90% average precision and 61% recall. In order to calculate the precision and recall, we use the formulas (2) and (3) for precision and recall, respectively.

$$AP = \frac{\sum_{i=1}^{n} TP/(TP+FP)}{n} \tag{2}$$

$$RC = \frac{\sum_{i=1}^{n} TP/(TP+FN)}{n} \tag{3}$$

AP denotes average precision, while *TP* and *FP* denotes true positives and false positives, respectively. The total number of testing dataset is represented with *n*. Through our thorough analysis we discovered that the *TP* to *FP* ratio is around 9 to 1 which gave us the 90% average precision. *RC* denotes recall, while *FN* denotes false negatives. Although our network can precisely detect fire images, it has a slightly worse accuracy when it comes to generalization, as it fails to detect some fire images as our experiments show.

Our test also shows that with an image of size 416 × 416, the R1 processes an image at 0.03 s (33 FPS) and R2 at 0.02 s (50 FPS). This means that, both models can be used in real-time in surveillance CCTV cameras well. Although, R2's 61% recall seems like a bad drop in accuracy compared to R1's, if we think about the fact that CCTV cameras usually have around 30 FPS, a 61% of detecting fire among those 30 FPS is enough for identifying if there is fire or not. In other words, not all 30 frames in a single second have to be detected. The probability of over half is very well satisfiable. In Fig. 2 we show some of the tested data which can be associated with real-life scenarios where fire could occur.

Fig. 2. Our model detects multiple fire with various sizes in different scenarios and lighting

5.1 Applications of IoT Service

In this section we show a hypothetical yet realistic approach to deploying our model onto an IoT platform where our fire detection model stands as a brain in the system for inferencing. As shown in Fig. 3 there are 3 levels of hierarchy connection between IoT devices and the cloud platform. In order for IoT devices to connect to cloud platform they need to go through the IoT gateway such as a Router or Switch. We mentioned earlier already that the aim of this research is not only for fire detection but also using this analytics component and incorporate it into the a rich IoT cloud platform where different IoT devices may gain access to this analytics knowledge. This shared knowledge allows smartwatch or smartphone devices to perform various functions such as escorting the blind people away from the fire. An easy to understand scenario of this IoT platform would be the following:

1. CCTV cameras deployed at various locations in an infrastructure such as a house constantly feed images at 30 to 60 FPS to the cloud platform through edge devices using REST API protocol.
2. Edge devices collect packets from all IoT devices that wish to connect with the cloud platform using TCP/IP connection.
3. The IoT cloud platform manage the resources of various IoT devices that are connected to the platform and constantly read data fed by them. In our case here, the data is images from the CCTV cameras.
4. Our fire detection model (R1 or R2) is in daemon mode awaiting to process the images fed by the CCTV cameras. The IoT platform should have a multitasking computing technique to feed images to the model.
5. When an image containing fire is detected, the image is trackbacked to the originated camera whose location is revealed to the platform.

6. At this point, the IoT platform may perform a few different tasks. One of which to determine if there is a human present nearby using a connected smartwatch, smartphone and sensors. If true, the IoT platform will initiate an extracting protocol where the smartwatch and smartphone are working together to escort the visually impaired person away from the fire and off the compound. Sound and voice direction may be used.

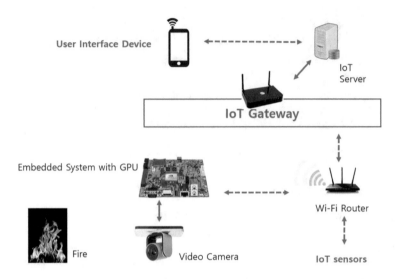

Fig. 3. Block diagram of IoT service

6 Conclusion and Future Work

We introduced two deep learning-based models (R1 & R2) that can detect fire with promising speed and accuracy in a household environment. R1 and R2 can detect fire at 90% average precision and 85% and 61% recall at 33 and 50 FPS, respectively. With these models, we help make detecting fire much more accurate and faster compared to traditional methods. In future work we will build the IoT platform and deploy our model in order to see the detection in real-life environment and for further testing and experiments.

Acknowledgment. This research was supported by the MSIT (Ministry of Science and ICT), Korea, under the ITRC (Information Technology Research Center) support program (IITP-2017-0-01642) supervised by the IITP(Institute for Information & communications Technology Promotion). This research was, also, supported by Basic Science Research Program through the National Research Foundation of Korea (NRF) funded by the Ministry of Education (2010-0020163). This paper is an extension to our past paper [13].

References

1. Milke, J.: The History of Smoke Detection: A Profile of How the Technology and Role of Smoke Detection Has Changed. SUPDET (2011)
2. Liu, Z., Kim, A.: Review of recent developments in fire detection technologies. J. Fire Protect. Eng. **13**, 129–151 (2003)
3. Litton, C.D.: The Two Faces of Smoke. Chapter 10, Mine Health and Safety
4. Morgan, A.: Automatic fire detection friend or foe? Fire Eng. J. (1999)
5. Morgan, A.: Automatic fire detection let there be light. Fire Eng. J. (1999)
6. Jacobson, E.: Finding Novel Fire Detection Technologies for the Offshore Industry, p. 26, March 2000
7. Lloyd, D.: Video Smoke Detection (VSD-8), Fire Safety, January 2000
8. Wieser, D., Brupbacher, T.: Smoke detection in tunnels using video images. In: 12th International Conference on Automatic Fire Detection, Gaithersburg, USA, March 2001
9. Liu, Y., Gadepalli, K., Norouzi, M., Dahl, G.E., Kohlberger, T., Boyko, A., Venugopalan, S., Timofeev, A., Nelson, P.Q., Corrado, G.S., Hipp, J.D., Peng, L., Stumpe, M.C.: Detecting Cancer Metastases on Gigapixel Pathology Images. arXiv preprint arXiv:1703.02442
10. Frizzi, S., Kaabi, R., Bouchouicha, M., Ginoux, J.M., Moreau, E., Fnaiech, F.: Convolutional neural network for video fire and smoke detection. In: IECON 2016 - 42nd Annual Conference of the IEEE Industrial Electronics Society, 23–26 October 2016, Florence, Italy (2016)
11. Kingma, D.P., Ba, J.: Adam: a method for stochastic optimization. In: International Conference on Learning Representations (ICLR) (2015)
12. Ren, S., He, K., Girshick, R., Sun, J.: Faster R-CNN: Towards Real-Time Object Detection with Region Proposal Networks. arXiv preprint arXiv:1506.01497
13. Kong, B., Won, I., Woo, J.: Fire detection using deep convolutional neural networks for assisting people with visual impairments in an emergency situation. J. Rehabil. Res. **21**, 129–146 (2017)
14. Redmon, J., Farhadi, A.: YOLO9000: Better, Faster, Stronger. arXiv preprint arXiv:1612.08242
15. Huang, J., Rathod, V., Sun, C., Zhu, M., Korattikara, A., Fathi, A., Fischer, I., Wojna, Z., Song, Y., Guadarrama, S., Murphy, K.: Speed/accuracy trade-offs for modern convolutional object detectors. In: Conference on Computer Vision and Pattern Recognition 2017. (Accepted)

Development of Agent Predicting Werewolf with Deep Learning

Manami Kondoh[✉], Keinosuke Matsumoto, and Naoki Mori

Osaka Prefecture University, 1-1 Gakuencho, Nakaku, Sakai, Osaka 599-8531, Japan
kondoh@ss.cs.osakafu-u.ac.jp

Abstract. In recent years, the development of AI that plays Werewolf attracts attention. This study researches on Werewolf, which is an incomplete information game. In order to create a good game agent, we tried to get unknown information that makes us advantageous in the game. Since Werewolf is communication game, we assumed that there are common strategies or features. For learning such something from enormous game logs, we proposed using LSTM that is a kind of deep learning.

Keywords: Deep learning · Game agent · AI agent

1 Introduction

Werewolf that is an incomplete information game is one of the most interesting topics in the artificial intelligence (AI) field. Several studies of applying AI to game environments have been reported [1]. We aim to develop a strong game agent that plays Werewolf well for overcoming the challenges of an incomplete information game. One way to play the game advantageously is to get unknown information, and how to get it is important. In machine learning fields Deep Learning has gained importance because of its high performance. The gaming environments have been reported as the new target of deep learning [2]. In this research, it was used to predict unknown information. By learning and capturing the characteristics of agent, we get useful information to win the game.

Several studies about AI Werewolf have been reported [3,4], however, research that adopts long short-term memory (LSTM) [5] to estimate agents inner states is only by us [6]. Although that was an experiment to predict in a simple environment, this paper deals further complicates environment, makes another prediction, and applies it to a game agent.

2 Werewolf

Werewolf is a party game that models a conflict between a minority "Werewolf" and a majority "Villager" [7]. Initially, each player is secretly assigned a role affiliated with one of these teams. The team of villagers is the majority, and its

F. De La Prieta et al. (Eds.): DCAI 2018, AISC 800, pp. 18–26, 2019.
https://doi.org/10.1007/978-3-319-94649-8_3

objective is to search for the werewolves and to kill them. The team of werewolves is the minority, and its objective is to behave like villagers and to kill them. Generally, there are three phases in Werewolf: "Talk", alive players talk each other, exchange information and discuss elimination by voting. "Vote", players vote for a player whom they want to execute. The most voted is executed, and the executed player will die. "Action", players who have an ability exercise it. Their result is not disclosed to others.

2.1 Role

All players have their roles, which are allocated randomly. Players are divided into two teams, according to their roles. The villager team has to penetrate werewolf through conversation. The werewolf team has to pretend to be cooperative with villagers. Except for werewolves, who can know each other, a player can't know the role of other players. For winning, the villager team has to eliminate all werewolves; the werewolf team has to eliminate humans such that humans become equal to or fewer in number than the werewolves.

In this paper, there are two roles each team: "VILLAGER" and "SEER" in Villager team, "WEREWOLF" and "POSSESSED" in Werewolf team. Villager and possessed have no ability. Seer can "divine", that makes him know whether a player is a werewolf or a human by selecting a player (possessed is a human). Werewolf can "attack". By selecting a player, he can attack the player and the attacked player is eliminated from the game.

2.2 Setting

In this research, we consider the game to be played by AI agents [8]. AI Wolf Server was developed by Toriumi et al. [9], and "Artificial Intelligence based Werewolf" [10] published on the platform ver 0.4.9. This server can be freely accessed for playing Werewolf with AI agents. We used the setting of this platform. First, the action of "divine" is carried out. Next, the Talk and the Vote phase are performed in this order. A game ends if villagers meet the winning condition, and if not, the Action phase starts. A game ends if werewolves meet the winning condition, and if not, it returns to the Talk phase. In this platform, the content that players can speak is limited. We show the types of remarks that are used in this research in Table 1.

3 Proposed Learning Method

3.1 Outline

In order to learn and estimate the behavior or thinking of agents, deep learning is employed. Generally, in Werewolf, the value and amount of information and the environment change with time, and thus, the state of the game has time series. Therefore, a network that takes time series property into account must be used.

Table 1. Talk contents

ESTIMATE	Express an estimation of others
COMINGOUT	Assert a role
DIVINATION	Declaring the target of the divine
DIVINED	Give the divined result
VOTE	Declare the voting destination
AGREE/DISAGREE	Agree/Disagree with someone's statement
Skip	Wait to listen to more talks and want to continue the discussion
Over	Agree to finish the discussion on the day
REQUEST	Request a specific action from other players

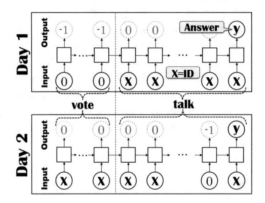

Fig. 1. Learning process of one game

LSTM, which is a kind of Recurrent Neural Network (RNN) [11], can take time series characteristics into consideration over a longer period than normal RNN. In Werewolf, since it uses not only information on a day but also past days (that may take a long time), LSTM is considered to be suitable as a learning method. In our paper [6], we did the voting destination prediction in games with only a single agent and proved that LSTM is useful for finding features that lead to voting. In this research, we did voting destination prediction and werewolf prediction in the game with multiple agents, more complex environment.

3.2 Learning Process

We use game logs as the data for learning. Logs are the 3rd AIWolf Competition Protocol Department published site [10]. The 3rd AIWolf Competition Protocol Department follows the platform 0.4.9. Those logs are games with five agents randomly selected from fifteen finalists. In a log, all the information such as the name of the agent, alive and death information, conversation in the game, voting

Table 2. Experimental setting

Input dimension	574
Node	800
Optimization	Adam
Error function	Soft_max_entropy

destination, etc. are described. Considering the application to the game, only the known information can be used for the input information. Therefore, only the information actually available for the player in the game is used from the logs: Specifically, the information of the conversation in the Talk phase, and that of who voted for whom in the Vote phase are used. As the actual input values, use the ID given to all patterns of remarks and votes. For example, "No. 3 estimate No. 4 werewolf" is ID 1, "No. 1 comingout No. 1 seer" is ID 2. However, when "AGREE" or "DISAGREE", if the remark is "No. 4 agree day: 1 index: 5", the ID is given not itself but the content of remarks which did at fifth in day 1.

We prepare models to predict voting destination of each player or werewolf. On the first day, only the conversation, on the second day or later, the conversation and votes of the previous day are inputted. Data is inputted one by one, and when all data of a day are inputted, learning is advanced by comparing the prediction value and the actual value. When the length of the data is less, it is adjusted with padding. The model is updated at the end of a day, and this process is repeated until the game is over. The state of the model is reset when the game is over. Learning process of one game is shown in Fig. 1.

4 Experiments

In this research, we use LSTM for learning from game logs and predicting voting destinations and werewolf player. In addition, we apply the prediction model obtained by learning to the game agent "JuN1Ro".

The game logs used in these experiments consisted of five players: two villagers, one seer, one werewolf and one possessed. In this setting, a game is sure to end on the first or second day. Logs mentioned in 2.2, which are 12727 games in all, were divided into two groups: 8000 (100 games 80 sets) for training and 4728 for evaluation. One set per epoch is randomly selected when training. IDs in this usage log are numbered 1 to 572. Padding and number 573 for assigning to a statement not in the registration pattern are added to IDs, so the number of IDs is 574. Common parameters of the experiments are shown in Table 2.

4.1 Exp. 1: Voting Destination Prediction

In this experiment, the voting destination, the subject of the vote, of a specific player on a day was used as teaching information. The baseline of the voting destination prediction uses the frequency with which an agent makes voting

	Table 3. Exp. 1: Result				Table 4. Exp. 2: Result		

Table 3. Exp. 1: Result   **Table 4.** Exp. 2: Result

	Number of test	Accuracy [%]	Baseline [%]
Day 1	23640	90.04	84.79
Day 2	8196	90.91	
All	31836	90.27	

	Number of test	Accuracy [%]	Baseline [%]
Day 1	4728	59.58	33.97
Day 2	2732	74.45	52.37
All	7460	65.03	40.70

remarks and actually votes in accordance with them. Comparing the baseline with the LSTM predictions, we can decide which it is better to follow: the voting declaration of the agent or the LSTM predictions. After predicting, eliminated players are excluded from the future prediction candidates.

Results of using the learned model to evaluate games is shown in Table 3.

4.2 Exp. 2: Werewolf Prediction

In this experiment, a werewolf index was used as teaching information. The baseline of the werewolf prediction uses the probability that werewolf will be positive when trusting either the revelation that someone is a werewolf or the result of the divine by the seer. For example, if there is a remark "divined X werewolf", it assumes X is a werewolf and if there is a remark of "divined Y werewolf", assumes that someone other than Y is a werewolf. As in Experiment 1, eliminated players are excluded from the prediction candidates (because when a werewolf is eliminated, the game ends).

Results of evaluations are shown in Table 4.

4.3 Application to Agent

We proposed to apply the learned model to an agent. Ignoring the agent's basic strategies, we use the prediction of the model for decision-making.

Agent "JuN1Ro". The agent applying this model is "JuN1Ro", which is developed by us. In our previous paper [6], we used "JuN1Ro" but it was the version developed in 2016. This paper's "JuN1Ro" was developed in 2017, and was ranked 6th in the 3rd AIWolf Competition Protocol Department. The following difference should be noted.

The algorithm JuN1Ro for the game of five players is shown briefly. However, rules that are obvious have been omitted from the description; for example, "A seer who gives a wrong result is a fake.", "When my role is a seer, a player who says that he is a seer is a fake.".

– Common
  When two players each claim to be a seer, JuN1Ro assumes them "seer and possessed", and votes for others. When three players make a claim, JuN1Ro assumes them "seer, possessed and werewolf" and votes for one of them.

- VILLAGER
  The Villager believes the seers: If a seer says "divined X werewolf", the Villager votes X. If a seer says "divined Y human", he does not vote Y.
- SEER
  The Seer randomly selects a target of "divine". When a werewolf is known to, he votes for it. When a human is known to, he does not vote for it.
- WEREWOLF
  If the Werewolf is declared as a werewolf by the seer, he thinks the seer is real. If the result of the declared divine was wrong, he thinks the seer is fake. If he knows a real seer, he attacks it. If not, he randomly selects someone from amongst the seers for attack.
- POSSESSED
  The Possessed claims to be a seer and randomly selects from players who do not claim to be a seer. He says that selected player is a werewolf.

All of the above and also more-detailed conditional rule branches are written manually. If the predicting the werewolf becomes possible, various patterns and the description task are not considered.

Application to "JuN1Ro". The prediction model was incorporated into JuN1Ro. When activating a game agent, the model is loaded. When trying to use the voting destination prediction model for the other four players, a timeout error occurs. If a response time of 100 ms or more is required to respond to a request from the game server, it becomes a regulation violation of the AIWolf Competition. In this research, we incorporated only the werewolf prediction model.

When updating each turn or phase, any new information from all previous updates is inputted to the model. After obtaining the result of werewolf prediction, Jun1Ro uses it to decide on a remark or an action. However, the following first-person perspective information is used as the top priority.

5   Analysis

5.1   Analysis of Agent

The results of Experiment 2 are analyzed and discussed.

The winning-rate when the role is werewolf in evaluation games and the prediction accuracy of each agent are shown in Fig. 2. For investigating whether there is a correlation between an agent's strength and prediction accuracy, Spearman's rank correlation coefficient was calculated using (1).

$$r = 1 - \frac{6 \times \sum_{i=1}^{n} D^2}{N^3 - N} \tag{1}$$

Here, D is the difference between the ranking of the agent and prediction accuracy, N is the number of agents. The null hypothesis is "no correlation", and will be rejected at $|r| > 0.521$ when the significance level is 5%. In this experiment,

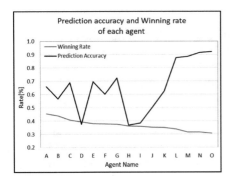
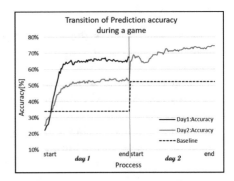

Fig. 2. Prediction accuracy and winning rate of each agent

Fig. 3. Transition of prediction accuracy during a game

r was -0.507, so the null hypothesis could not be rejected. It cannot be said really that a strong agent is hard to predict or a weak one is easy.

Agents for which prediction was particularly bad, "D", "H" and "I" are discussed. There are agents who say a werewolf strategically, but these did not do that. "H" and "I" had almost the same behavior as the villagers. "D" pretended to be a seer well, so it is considered to have been judged, not a werewolf. On the other hand, agents for which prediction was good, "N" and "O", also pretended to be seers. However, they committed characteristic mistakes, such as stating the result of a divine twice in one day. Therefore, it is considered that it was difficult to predict an agent that does not adopt special behavior well or could pretend to be another role well.

5.2   Analysis of Game

The transition of prediction accuracy during the game is shown in Fig. 3. The prediction accuracy is shown by black when the game ends in one day, and by ash when the game ends in two days. It is said that predictions exceeding the baseline were succeeded in about one-third of the game progress. It is inferred from Fig. 3 that the information obtained in the early stage is of high importance. As there are many Over and Skip, which do not have direct meaning in the second half of the day, this result is acceptable. By restricting input data and using this model, it will be possible to identify the remarks or the flow of the conversation that is important for guessing. This is a point to be noted for the future.

6   Consideration

From Tables 3 and 4, we can see that these results show good accuracy. Therefore, it can be said that predictions by LSTM have proved to be more useful than considering rule-based predictions from confirmed information. In particular, in werewolf prediction, it was possible to automatically make predictions without having to think or manually describe a theory or use logical reasoning in Werewolf. Therefore, the proposed method is useful for discovering unknown information in Werewolf.

We also achieved the application of the model to the agent. However, we failed to apply all models. Of course, considering actual games, it is not realistic to take a long time in loading and predicting. To be able to operate smoothly in the game environment, it is necessary to develop a model that is lightening-fast and approaches the original performance as closely as possible. In addition, no evaluation has been made of the created agent yet. It is necessary to quantitatively evaluate the advantages of utilizing a model, such as JuN1Ro before and after applying the agent in the same environment, and comparing the winning percentage.

7   Conclusion

This research shows that it is possible to predict unknown information game in multi-agent Werewolf environment using LSTM. In the analysis, we discussed from two viewpoints and showed that the model may be useful for game analysis. We also succeeded in incorporating prediction model into an actual agent. However, consideration of practicality is required.

As future tasks, game properties using models should be analyzed. Furthermore, an improvement of models considering practicality and consideration of the quantitative evaluation method of model application agents should be approached. It is also important to consider effective algorithms for using model.

A part of this work was also supported by JSPS KAKENHI Grant, Grant-in-Aid for Scientific Research(C), 26330282.

References

1. Suyama, J., Hashiyama, T., Tano, S.: Feature extraction from skilled player's logs in PuyoPuyo. In: 31st Fuzzy System Symposium (2015)
2. Firoiu, V., Whitney, W.F., Tenenbaum, J.B.: Beating the world's best at super smash bros. with deep reinforcement learning. Cornell University Library (2017)
3. Nakamura, N., Inaba, M., Takahashi, K., Shinoda, K.: Constructing a human-like agent for the werewolf game using a psychological model based multiple perspectives. In: The 2016 IEEE Symposium Series on Computational Intelligence (IEEE SSCI 2016) (2016)
4. Kaziwara, K., et al.: Development of AI wolf agent using SVM to detect werewolves. In: The 30th Annual Conference of the Japanese Society for Artificial Intelligence (2016)
5. Gers, F.A., Schmidhuber, J., Cummins, F.: Learning to forget: continual prediction with LSTM. Technical report IDSIA 01-99 (1999)
6. Kondoh, M., et al.: Agent of werewolf game applying deep learning predictions. In: The 6th Asian Conference on Information Systems (2017)
7. Toriumi, F., et al.: AI wolf contest-development of game AI using collective intelligence. In: Computer Games Workshop at IJCAI (2016)

8. Toriumi, F., et al.: Artificial Intelligence Based Werewolf Cheat, Penetrate, Persuade. Morikita Publishing Co., Ltd. (2016). (in Japanese)
9. Toriumi, F., et al.: Development of AI wolf server. In: The 19th Game Programming Workshop (2014)
10. Artificial intelligence based werewolf. http://aiwolf.org/
11. Zaremba, W.: Recurrent neural network regularization. In: Under Review as a Conference Paper at ICLR 2015 (2015)

A Comparative Study of Transfer Functions in Binary Evolutionary Algorithms for Single Objective Optimization

Ramit Sawhney[✉], Ravi Shankar, and Roopal Jain

Department of Computer Engineering, Netaji Subhas Institute of Technology,
New Delhi, India
{ramits.co,ravis.co,roopalj.co}@nsit.net.in

Abstract. Binary versions of evolutionary algorithms have emerged as alternatives to the state of the art methods for optimization in binary search spaces due to their simplicity and inexpensive computational cost. The adaption of such a binary version from an evolutionary algorithm is based on a transfer function that maps a continuous search space to a discrete search space. In an effort to identify the most efficient combination of transfer functions and algorithms, we investigate binary versions of Gravitational Search, Bat Algorithm, and Dragonfly Algorithm along with two families of transfer functions in unimodal and multimodal single objective optimization problems. The results indicate that the incorporation of the v-shaped family of transfer functions in the Binary Bat Algorithm significantly outperforms previous methods in this domain.

Keywords: Binary evolutionary algorithm · Bat Algorithm
Gravitational search · Transfer function · Single objective optimization

1 Introduction

Evolutionary algorithms (EAs) [1] are relatively new unconstrained heuristic search techniques which require little problem specific knowledge. EAs consist of nature-inspired stochastic optimization techniques that mimic social behavior of animal or natural phenomena known as Swarm Intelligence techniques [2]. EAs are superior to other techniques such as Hill-Climbing algorithms [3] that are easily deceived in multimodal problem spaces and often get stuck in some sub optima.

There are many optimization problems possessing intrinsic search spaces of a binary nature such as feature selection [4] and dimensionality reduction [5]. In this paper, we aim to compare the binary versions of 3 evolutionary algorithms with other high performance algorithms for unimodal and bimodal single objective optimization problems. We also investigate the effect of 2 different families of Transfer Functions (TFs) that are instrumental in mapping continuous variables

© Springer International Publishing AG, part of Springer Nature 2019
F. De La Prieta et al. (Eds.): DCAI 2018, AISC 800, pp. 27–35, 2019.
https://doi.org/10.1007/978-3-319-94649-8_4

into binary search spaces that allow us to apply Binary evolutionary algorithms (BEAs). The main contributions of our work can be summarized as:

1. We present a comparative study between three BEAs.
2. Our work focuses on global minimization of 4 standard benchmark functions across 4 evaluation metrics.
3. We study the variations in performance due to different TFs in BEAs.

The rest of the paper is organized as follows: Sects. 2 and 3 describe the various BEAs and their transfer functions. Section 4 lays down the experimental details for which the results are provided and discussed. Section 5 summarizes the ideas presented in this paper and outlines the directions for future work.

2   Binary Evolutionary Algorithms

Binary versions of EAs have been adapted from their regular versions that are employed in continuous search spaces. BEAs operate on search spaces that can be considered as hypercubes. In addition, problems with continuous real search space can be converted into binary problems by mapping continuous variables to vectors of binary variables. A brief presentation of the algorithms used for single objective optimization in this paper are shown in the following subsections.

2.1   Gravitational Search Algorithm

Gravitational Search algorithm (GSA) proposed by Rashedi et al. [6] is a nature inspired heuristic optimization algorithm based on the law of gravity and mass interactions. The position of each agent represents the solution of the problem. In GSA, the gravitational force causes a global movement where objects with a lighter mass move towards objects with heavier masses. The inertial mass that determines the position of an object is determined by the fitness function, and the position of the heaviest mass presents the optimal solution. The binary version of GSA, known as BGSA proposed by Rashedi et al. [7] is an efficient adaptation of GSA in binary search spaces.

2.2   Bat Algorithm

Bat Algorithm (BA) [8] is a meta-heuristic optimization algorithm based on the echolocation behavior of bats. BA aims to behave as a colony of bats that track their prey and food using their echolocation capabilities. The search process is intensified by a local random walk. The BA is a modification of Particle Swarm Optimization in which the position and velocity of virtual microbats are updated based on the frequency of their emitted pulses and loudness. The binary version of BA, [9] models the movement of bats (agents) across a hypercube.

2.3 Dragonfly Algorithm

Dragonfly Algorithm (DA) proposed by Mirjalili [10] is a novel swarm optimization technique originating from the swarming behavior of dragonflies in nature.
DA consists of two essential phases of optimization, exploitation and exploration
that are deigned by modeling the social interaction of dragonflies in navigating,
search for good, and avoiding enemies when swarming. The binary version of
DA, known as BDA maps the five parameters: cohesion, alignment, escaping
from enemies, separation and attraction towards food to be applicable for calculating the position of dragonflies (agents) in the binary search space.

3 Transfer Functions for Binary Evolutionary Algorithms

The continuous and binary versions of evolutionary algorithms differ by two
different components: a new transfer function and a different position updating
procedure. The transfer function maps a continuous search space to a binary
one, and the updating process is based on the value of the transfer function to
switch positions of particles between 0 and 1 in the hypercube. The agents of a
binary optimization problem can thus move to nearer and farther corners of the
hypercube by flipping bits based on the transfer function and position updating
procedure. Various transfer functions have been studied to transform all real
values of velocities to probability values in the interval $[0, 1]$.

In this paper, we investigate two families of transfer functions first presented
in [11], namely s-shaped transfer functions and v-shaped transfer functions.
Table 1 presents the s-shaped and v-shaped transfer functions that are in accordance with the concepts presented by Rashedi et al. [7].

Table 1. Transfer functions for s-shaped and v-shaped families.

s-shaped family		v-shaped family	
Name	Transfer function	Name	Transfer function
S_1	$S_1(x) = \frac{1}{1+e^{-2x}}$	V_1	$V_1(x) = \|erf(\frac{\sqrt{\pi}}{2}x)\| = \|\frac{\sqrt{\pi}}{2}\int_0^{\frac{\sqrt{\pi}x}{2}} e^{-t^2} dt\|$
S_2	$S_2(x) = \frac{1}{1+e^{-x}}$	V_2	$V_2(x) = \|tanh(x)\|$
S_3	$S_3(x) = \frac{1}{1+e^{-\frac{x}{2}}}$	V_3	$V_3(x) = \|\frac{x}{\sqrt{1+x^2}}\|$
S_4	$S_4(x) = \frac{1}{1+e^{-\frac{x}{3}}}$	V_4	$V_4(x) = \|\frac{2}{\pi}arctan(\frac{\pi}{2}x)\|$

The names s-shaped and v-shaped transfer functions arise from the shape
of the curves as depicted in Fig. 1 [11], and study the variations in performance
due to these.

Due to the drastic difference between s-shaped and v-shaped transfer functions, different position updating rules are required. For s-shaped transfer functions, Formula (1) is employed to update positions based on velocities.

$$x_i^k(t+1) = \begin{cases} 0 & \text{if } rand < T(v_i^k(t+1)) \\ 1 & \text{if } rand \geq T(v_i^k(t+1)) \end{cases} \tag{1}$$

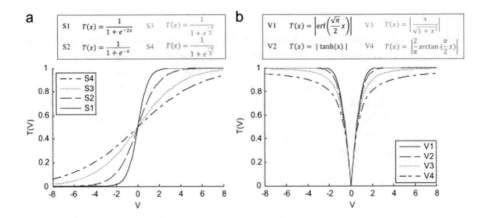

Fig. 1. (a) s-shaped and (b) v-shaped family of transfer functions

Formula (2) [11] is used to update positions for v-shaped transfer functions.

$$x_i^k(t+1) = \begin{cases} (x_i^k(t))^{-1} & \text{if } rand < T(v_i^k(t+1)) \\ x_i^k(t) & \text{if } rand \geq T(v_i^k(t+1)) \end{cases} \tag{2}$$

Formula (2) is significantly different from Formula (1) as it flips the value of a particle only when it has a high velocity as compared to simply forcing a value of 0 or 1. Algorithm 1 presents the basic steps of utilizing both s-shaped and v-shaped functions for updating the position of particles and finding the optima that can be generalized and used with any Binary Evolutionary Algorithm.

Algorithm 1. Generic Binary Evolutionary Algorithm

Input: Input: A binary vector, $V : \{x_1, x_2, ..., x_n \mid x_i = \{0, 1\}\ 1 \leq i \leq n\}$ that represents the binary search space.
Output: Output: Solution vector, S; Best value of fitness function.

1. Randomly initialize the agents $X_i (i = 1, 2, ..., n)$.
2. Compute velocity for each particle using velocity calculation equations for the corresponding BEA.
3. Calculate probabilities for changing positions of the agents by using the specified transfer function.
4. Update the positions for the agents based on formula (2) or formula (3) depending on the family of the transfer function used.
5. Repeat steps 2–4 for maximum number of iterations.

---

4 Experiment Settings and Results

4.1 Benchmark Functions and Evaluation Metrics

In order to compare the performance of the BEAs with both s-shaped and v-shaped transfer functions, four benchmark functions [12] are employed. The objective of the algorithms is to find the global minimum for each of these functions. Both unimodal (F_1, F_2) and multimodal (F_3, F_4) functions are used for performance evaluation, and are shown in Table 2. The Range of the function quantifies the boundary of the functions search space. The global minimum value for each of the functions is 0. The two dimensional versions of the functions are shown in Fig. 2. We use a 15 bit vector to map each continuous variable to a binary search space, where one bit is reserved for the sign of each variable. As the dimension of the benchmark function is 5, the dimension of each agent is 75. We use the following four metrics for evaluation and comparison.

1. Average Best So Far (ABSF) solution over 40 runs in the last iteration.
2. Median Best So Far (MBSF) solution over 40 runs in the last iteration.
3. Standard Deviation (STDV) of the best so far solution over 40 runs.
4. Best indicates the best solution over 40 runs in all iterations.

Table 2. Benchmark functions and their input ranges

Function	Range
$F_1(x_1 \ldots x_n) = \sum_{i=1}^{n} x_i^2$	$[-100, 100]^n$
$F_2(x_1 \ldots x_n) = \sum_{i=1}^{n-1}(100(x_i^2 - x_{i+1})^2 + (1 - x_i)^2)$	$[-30, 30]^n$
$F_3(x_1 \ldots x_n) = 10n + \sum_{i=1}^{n}(x_i^2 - 10cos(2\pi x_i))$	$[-5.12, 5.12]^n$
$F_4(x_1 \ldots x_n) = 1 + \frac{1}{4000}\sum_{i=1}^{n} x_i^2 - \prod_{i=1}^{n} cos(\frac{x_i}{\sqrt{i}})$	$[-600, 600]^n$

Table 3. Parameters for experiments

Parameter	Value
Number of agents	20
Maximum epochs	500
G_0 for BGSA	1
Loudness for BBA	0.25
Pulse rate, r for BBA	0.4
Minimum step frequency $q_{m}in$ for BBA	0
Maximum step frequency $q_{m}ax$ for BBA	2

4.2 Experiment Setting

The algorithms are run 40 times with a random seed on an Intel Core 2 Duo machine, 3.06 GHz CPU and 4 GB of RAM. Metrics are reported over multiple runs to reduce the effect of random variations keeping in mind the stochastic nature of the algorithms. Table 3 shows the parameters for simulation of the algorithms.

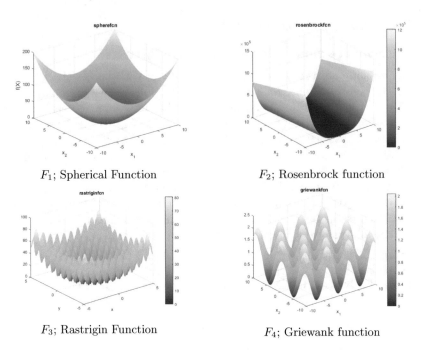

F_1; Spherical Function F_2; Rosenbrock function

F_3; Rastrigin Function F_4; Griewank function

Fig. 2. Performance comparison on evaluation metrics

4.3   Results and Discussion

As may be seen from the results presented in Tables 4 and 5 both BBA and BDA perform better than BGSA. We also observe that the v-shaped family of functions outperforms the s-shaped family and can significantly improve the ability of BEAs in avoiding local minima for these benchmark functions. It is also evident from the results, that BBA with v-shaped transfer functions also performs better than BDA in terms of most aspects. Both BBA and BDA with v-shaped transfer functions significantly outperform high performance algorithms such as CLPSO [13], FIPS [14], and DMS-PSO [15]. Despite relatively accurate results, a high standard deviation is observed which is attributed to the stochastic behavior of swarm optimization algorithms. The number of iterations drastically affects the ability of all 3 algorithms to converge to global minima easily in unimodal functions particularly F_2, characterized by an extremely deep valley that makes convergence at the minima extremely slow. Similarly, for multimodal functions, F_3, F_4, increasing the number of iterations combined with the effect of using v-shaped transfer functions allows a significant improvement in the accuracy as well as convergence rate when compared to s-shaped transfer functions.

Table 4. Minimization results of unimodal benchmark functions over 40 runs

F	TF	Algorithm											
		BGSA				BBA				BDA			
		ABSF	MBSF	STDV	Best	ABSF	MBSF	STDV	Best	ABSF	MBSF	STDV	Best
F_1	S_1	27.26	0.187	82.02	**0.0**	105.64	1.679	220.65	**0.0**	40.267	0.195	118.88	**0.0**
	S_2	11.656	1.402	21.994	**0.0**	37.970	1.0	70.28	**0.0**	9.159	0.208	27.359	**0.0**
	S_3	48.755	3.253	119.057	0.156	97.65	2.220	170.77	**0.0**	43.690	0.630	115.05	**0.0**
	S_4	87.249	3.409	121.509	1.230	71.96	6.650	205.30	**0.0**	72.530	1.980	105.35	0.062
	V_1	5.891	1.689	15.614	**0.0**	28.050	0.039	88.497	**0.0**	21.885	0.322	31.635	**0.0**
	V_2	3.4787	0.308	8.375	**0.0**	6.316	0.248	11.462	**0.0**	3.418	0.253	90.217	**0.0**
	V_3	3.116	1.343	7.956	**0.0**	7.212	0.001	16.223	**0.0**	11.160	0.625	22.934	0.003
	V_4	3.867	4.507	68.730	0.015	**0.442**	0.015	**1.255**	**0.0**	31.411	16.164	57.504	**0.0**
F_2	S_1	274.83	3.862	777.78	3.540	358.10	10.58	667.96	1.415	103.20	4.038	225.40	3.722
	S_2	49.748	3.951	140.90	2.682	125.70	13.69	263.66	3.446	176.25	3.920	55.49	3.340
	S_3	161.59	4.477	476.407	3.777	48.205	5.379	12.854	3.431	104.95	17.37	230.90	2.679
	S_4	479.73	9.411	150.497	3.975	54.201	26.27	111.12	2.684	106.01	31.632	32.263	3.989
	V_1	33.585	4.398	65.845	2.189	**3.352**	3.350	0.490	**0.015**	25.199	3.669	47.462	3.071
	V_2	32.822	3.924	91.075	3.138	3.579	3.598	**0.296**	3.151	21.310	6.329	0.915	0.755
	V_3	44.124	8.102	6.099	1.321	3.087	3.340	0.743	1.426	17.605	**2.937**	321.10	2.878
	V_4	59.869	6.634	227.035	0.981	46.991	3.866	97.131	3.292	64.804	19.545	126.14	3.189

Table 5. Minimization results of multimodal benchmark functions over 40 runs

F	TF	Algorithm											
		BGSA				BBA				BDA			
		ABSF	MBSF	STDV	Best	ABSF	MBSF	STDV	Best	ABSF	MBSF	STDV	Best
F_3	S_1	13.761	38.705	7.486	0.765	5.269	2.136	16.56	**0.0**	5.455	1.854	7.943	**0.0**
	S_2	16.372	13.04	14.96	**0.0**	15.393	24.84	17.56	1.640	5.523	5.263	**1.675**	**0.0**
	S_3	60.6	8.608	153.9	1.0	9.768	23.45	12.9	3.280	166.66	48.36	267.03	**0.0**
	S_4	198.825	210.87	379.893	**0.0**	65.997	53.24	106.5	3.280	61.5	28.65	98.276	26.06
	V_1	184.618	3.847	239.543	**0.0**	6.663	0.882	9.662	**0.0**	9.743	0.382	22.711	**0.0**
	V_2	49.177	2.854	134.633	**0.0**	2.854	**0.0**	6.459	**0.0**	12.248	3.319	18.939	**0.0**
	V_3	28.412	1.0	37.248	**0.0**	**1.881**	0.382	3.676	**0.0**	24.932	8.09	49.64	0.765
	V_4	94.240	14.049	142.226	**0.0**	9.471	**0.0**	27.64	**0.0**	22.531	4.680	27.321	1.640
F_4	S_1	0.501	0.358	0.508	**0.0**	0.133	0.098	0.472	0.001	0.163	0.054	0.213	**0.0**
	S_2	0.194	0.024	0.320	**0.0**	1.326	0.662	3.267	0.147	0.161	0.051	0.218	**0.0**
	S_3	0.415	0.114	0.545	**0.0**	2.730	1.484	4.046	0.001	0.869	0.503	0.973	0.001
	S_4	0.783	0.711	0.776	0.007	1.879	0.142	4.049	0.02	0.648	0.527	0.569	**0.0**
	V_1	0.078	0.027	0.105	**0.0**	**0.012**	**0.0**	**0.020**	**0.0**	0.037	0.033	0.042	**0.0**
	V_2	0.058	0.002	0.083	**0.0**	**0.012**	**0.0**	0.027	**0.0**	0.098	0.046	0.140	**0.0**
	V_3	0.138	0.085	0.165	**0.0**	0.015	**0.0**	0.049	**0.0**	0.081	0.027	0.103	**0.0**
	V_4	0.140	0.064	0.146	**0.0**	0.879	**0.0**	0.171	**0.0**	0.180	0.153	0.154	**0.0**

5 Conclusion

In this paper, a comparative study between binary evolutionary algorithms is performed and the effect of s and v-shaped transfer functions on the performance of single objective optimization problems is explored. The highest performing algorithm was BBA when compared with BGSA, BDA and other evolutionary algorithms in terms of avoiding local minima in both unimodal and multimodal functions. The results show the drastic improvement in performance with the introduction of v-shaped family of transfer functions for updating the position of agents. Our work shows the merit v-shaped functions and BBA have for use in binary algorithms. In the future, the current work aims to compare the performance of the transfer functions on other evolutionary and heuristic algorithms.

References

1. Larrañaga, P., Lozano, J.A.: Estimation of Distribution Algorithms: A New Tool for Evolutionary Computation, vol. 2. Springer, New York (2001)
2. Kennedy, J.: Particle swarm optimization. In: Encyclopedia of Machine Learning, pp. 760–766. Springer (2011)
3. Tsamardinos, I., Brown, L.E., Aliferis, C.F.: The max-min hill-climbing Bayesian network structure learning algorithm. Mach. Learn. 65(1), 31–78 (2006)
4. Ritthof, O., Klinkenberg, R., Fischer, S., Mierswa, I.: A hybrid approach to feature selection and generation using an evolutionary algorithm. In: UK Workshop on Computational Intelligence, pp. 147–154 (2002)
5. Raymer, M.L., Punch, W.F., Goodman, E.D., Kuhn, L.A., Jain, A.K.: Dimensionality reduction using genetic algorithms. IEEE Trans. Evol. Comput. 4(2), 164–171 (2000)
6. Rashedi, E., Nezamabadi-Pour, H., Saryazdi, S.: GSA: a gravitational search algorithm. Inf. Sci. 179(13), 2232–2248 (2009)
7. Rashedi, E., Nezamabadi-Pour, H., Saryazdi, S.: BGSA: binary gravitational search algorithm. Nat. Comput. 9(3), 727–745 (2010)
8. Yang, X.S.: A new metaheuristic bat-inspired algorithm. In: Nature Inspired Cooperative Strategies for Optimization (NICSO 2010), pp. 65–74. Springer (2010)
9. Mirjalili, S., Mirjalili, S.M., Yang, X.S.: Binary bat algorithm. Neural Comput. Appl. 25(3–4), 663–681 (2014)
10. Mirjalili, S.: Dragonfly algorithm: a new meta-heuristic optimization technique for solving single-objective, discrete, and multi-objective problems. Neural Comput. Appl. 27(4), 1053–1073 (2016)
11. Mirjalili, S., Lewis, A.: S-shaped versus v-shaped transfer functions for binary particle swarm optimization. Swarm Evol. Comput. 9, 1–14 (2013). http://www.sciencedirect.com/science/article/pii/S2210650212000648
12. Suganthan, P., Hansen, N., Liang, J., Deb, K., Chen, Y.p., Auger, A., Tiwari, S.: Problem definitions and evaluation criteria for the CEC 2005 special session on real-parameter optimization, pp. 341–357, January 2005

13. Liang, J.J., Qin, A.K., Suganthan, P.N., Baskar, S.: Comprehensive learning parti-
    cle swarm optimizer for global optimization of multimodal functions. IEEE Trans.
    Evol. Comput. **10**(3), 281–295 (2006)
14. Mendes, R., Kennedy, J., Neves, J.: The fully informed particle swarm: simpler,
    maybe better. IEEE Trans. Evol. Comput. **8**(3), 204–210 (2004)
15. Liang, J.J., Suganthan, P.N.: Dynamic multi-swarm particle swarm optimizer with
    local search. In: The 2005 IEEE Congress on Evolutionary Computation, vol. 1,
    pp. 522–528. IEEE (2005)

Ontology-Based Advertisement Recommendation in Social Networks

Francisco García-Sánchez[1(✉)] ⓘ, José Antonio García-Díaz[1] ⓘ,
Juan Miguel Gómez-Berbís[2] ⓘ, and Rafael Valencia-García[1] ⓘ

[1] Dpto. Informática y Sistemas, Facultad de Informática, Universidad de Murcia,
Murcia, Spain
{frgarcia, joseantonio.garcia8, valencia}@um.es
[2] Departamento de Informática,
Universidad Carlos III de Madrid, Leganés, Spain
juanmiguel.gomez@uc3m.es

Abstract. With the advent of the Web 2.0 era, a new source of a vast amount of data about users become available. Advertisement recommendation systems are among the applications that can benefit from these data since they can help gain a better understanding of the users' interests and preferences. However, new challenges emerge from the need to deal with heterogeneous data from disparate sources. Semantic technologies, in general, and ontologies, in particular, have proved effective for knowledge management and data integration. In this work, an ontology-based advertisement recommendation system that leverages the data produced by users in social networking sites is proposed.

Keywords: Ontology · Recommender system · Advertisement
Social network

1 Introduction

Recommender systems, as we know them, date back to the early 90's [1]. The goal of these systems is to serve the right items, be them news [1], movies [2], etc., to the right users, thus accomplishing a two-fold purpose: (i) success in the provision of items (e.g., increasing product sales) and (ii) user satisfaction. The emergence of social networks and the so called social Web has produced a tremendous impact on recommender systems becoming a new, vast source of data that can be leveraged to generate better suggestions [3]. Advertisement is one of the recommendation application domains that can benefit the most from the insights derived from that data. Social network advertising is a type of online promoting that focuses on social networking sites and takes advantage of the users' demographic information along with their interactions with others in the network [4].

The technologies associated to the Semantic Web [5] have proved effective for knowledge management in different contexts [6, 7]. The underlying ontological models, which are based on logical formalisms, enable computer systems to somehow interpret the information that is being managed. They also allow to carry out advanced reasoning and inferencing processes. The scientific community within these research fields has developed tools that make use of semantic technologies to improve

© Springer International Publishing AG, part of Springer Nature 2019
F. De La Prieta et al. (Eds.): DCAI 2018, AISC 800, pp. 36–44, 2019.
https://doi.org/10.1007/978-3-319-94649-8_5

recommender systems [8]. In this work, we propose an advertisement recommendation framework for social networking sites based on semantic technologies, which reduces the impact of the major issues hampering the performance of traditional recommendation engines. The main idea behind our approach is the use of a shared ontological model to represent both users' interests and ads. While advertisements' profiles are statically defined to accurately represent their content, users' profiles are dynamic, evolving on the basis of their behavior and activities in the social network.

The rest of the paper is organized as follows. In Sect. 2 background information on recommender systems and the application of semantic technologies is provided. The framework proposed in this work to serve recommendations about advertisements to users in social networks using ontologies is described in Sect. 3. In Sect. 4, a preliminary validation of our approach is shown. Finally, conclusions and future work are put forward in Sect. 5.

2 Related Work

A recommendation system is a tool that helps in finding matches between users and products of any kind that might be of their interest [9]. For this, historical data from which to infer the likes and dislikes of users becomes essential. Recommendation systems have traditionally been classified as content-based, collaborative, knowledge-based, demographic and hybrid [10]. Social recommendation systems incorporate both the social context and the contributions of the users in the social networking sites to improve the accuracy of recommendation [10]. The items suggested by these tools can be from products/content (e.g., movies [2]) to nodes and links within the social network (e.g., other users with similar interests [11]). Recommender systems that employ the social network structure to enhance the recommendation process help in alleviating major issues such as the cold-start problem and data sparsity through the extraction of implicit user preferences from the information already available on the site. However, new challenges arise with the need to exploit the ever evolving and augmenting users' data, namely, data variety, data volatility and data volume [3].

Ontologies and semantic technologies can help to overcome some of the challenges recommender systems face [2, 8, 12]. The authors in [2] describe a hybrid recommender system based on knowledge and social networks that is evaluated in the movies domain. In [8] the authors propose a new architecture for semantics-aware content-based recommender systems that includes a 'profile cleaner' component to update user profiles. Lastly, it is claimed that a recommender system to become 'intelligent' must show the following capabilities [12]: knowledge representation, learning capabilities, and reasoning mechanism. The framework in that work is an extension of knowledge-based recommender systems and consider the knowledge about users, items, domain, context and criticism to produce recommendations.

Our framework makes use of ontologies to model both the content of ads and the dynamic user profiles in a social networking site. Each user (and each ad) is represented as a vector in which each dimension corresponds to a separate ontological concept of the domain ontology. The recommendation is produced by calculating the similarity between the vectors representing users and those describing advertisements.

3 Ontology-Based Advertisement Recommendation in Social Networks

In this work, a semantically-enhanced recommender system that suggests ads to the users of social networking sites is proposed. Next, the architecture of the framework is presented and its main components are explained.

3.1 Proposed Framework

The architecture of the proposed system is shown in Fig. 1. It is composed of four main components, namely, the interests ontology, the ad profile generator, the user profile generator and the ads recommender. The input of the system consists of a collection of advertisements (their textual description) and a set of users registered on a social site with their textual posts. As output, the system produces user-ad recommendations aiming to find the suggestions that best suit each user. In a nutshell, the system works as follows. The textual description of advertisements is analyzed by means of a natural language processing (NLP) tool and the interests ontology elements with matches in the texts are highlighted. Then, the ad profile generator represents each ad as a vector in which each dimension corresponds to a separate ontological concept of the domain ontology and its value depends on the number of occurrences of the concept in the ad's descriptive text. Likewise, the user profile generator makes use of the interests ontology and creates a vector for each user with as many dimensions as the number of concepts in the domain ontology. In this case, a first version of the vector is produced by taking into account the registration information provided by the user, and the vector is updated as the user interacts with the social networking site. Finally, the ads recommender produces recommendations on the basis of the similarity found between the vectors representing users and those associated to ads.

3.2 Interests Ontology

Different ontologies and other less formal representation models to characterize users' profiles can be found currently in literature [13–15]. In [13], the interest ontology is automatically extracted by classifying user blog entries into some predefined service-domain ontologies. The algorithm described in that work assigns weights to each class and instance in the ontology using a keyword matching approach from the content of the entries. The authors in [14] claim that to achieve reliable recommendations it is first required to accurately detect the users' interests and they propose a technique to learn these preferences by analyzing the users social environment. The tagging behavior along with the tagged resources content are used as main data sources. Finally, a user modeling framework that maps the content of texts in social media to relevant categories in news media is described in [15]. User interest vectors represented by news categories are obtained by considering the similarities between user's messages and news categories based on Wikipedia.

The approach described in this manuscript is inspired by some of the referred previous works and aims at bringing together their benefits. A specific ontology model

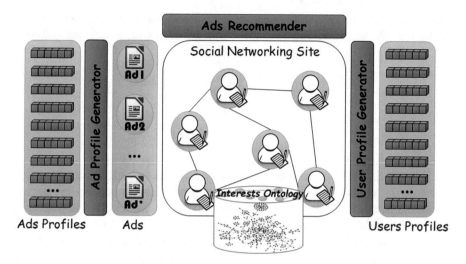

Fig. 1. Proposed framework functional architecture

has been built to describe interests in general terms and NLP tools are leveraged to gather relevant knowledge from textual documents. The Web Ontology Language[1] (OWL 2) was used to construct the interests ontology, which comprises the main concepts and relationships considered. The ontology encompasses 15 high class categories and is based on the Curlie Web directory[2] (previously known as the Open Directory Project and DMOZ). Only the first two levels of the hierarchy have been considered in our interests ontology, leaving the remaining subcategories as synonyms of the higher concepts by using the annotation property `rdfs:label`. In total, the ontology contains 620 classes organized as a fully-fledged taxonomy.

3.3   Ad Profile Generator

The ad profile generator focuses on the natural language description of the ads. When a new ad is added to the platform, the advertiser is asked to introduce a textual description of its content and main properties. Then, a set of NLP tools are used to semantically annotate the text in accordance with the domain ontology and WordNet. This process comprises two main stages, namely, the NLP phase and the semantic annotation phase. During the NLP phase, the morphosyntactic structure of each sentence in the text is extracted using the GATE framework[3]. The outcome of this first stage is a set of annotations representing the syntactic structure of the text. In the second phase, the ad description is annotated with the classes and instances of the ontology as follows. First, statistical approaches based on the syntactic structure of the

---

[1] https://www.w3.org/TR/owl2-overview/.

[2] https://curlie.org/.

[3] General Architecture for Text Engineering, https://gate.ac.uk/.

text are used to identify the most important linguistic expressions in the text. Then, the ontology individuals associated with each linguistic expression, if any, are gathered.

$$(v_1, v_2, \cdots, v_n) \text{ where } v_j = \sum_{i=1}^{n} \frac{(tf - idf)_{i,d}}{e^{dist(i,j)}} \tag{1}$$

$$(tf - idf)_{i,d} = \frac{n_{i,d}}{\sum_k n_{k,d}} * \log \frac{|D|}{N_i} \tag{2}$$

Once the generator module has obtained the knowledge annotations, it creates a vector for each ad with one component corresponding to each ontological concept in the ontology using Eq. 1, where $dist(i,j)$ is the semantic distance between the concept i and the concept j in the domain ontology calculated using the taxonomic (*subclass-of*) relationships of the concepts in the ontology, and $(tf - idf)_{i,d}$ is calculated using Eq. 2, where $n_{i,d}$ is the number of occurrences of the ontological entity i in the document d, $\sum_k n_{k,d}$ is the sum of the occurrences of all the ontological entities identified in the document d, $|D|$ is the set of all documents (textual description of the ads), and N_i is the number of all documents annotated with the ontological entity i.

3.4   User Profile Generator

The user profile generator is in charge of producing the ontology-based vector that represents each user's interests. Four main situations are considered: (i) the process in which the user signs up for the social site, (ii) when the user submits a post, (iii) each time the user interacts with an advertisement, and (iv) when the user updates his/her user profile in the social site. The vector is initialized when the user gets registered on the system and is fully restored if the user modifies his/her registration data. The knowledge that can be gathered at this stage depends on the information asked by the system. The user profile generator receives a text comprising the registration data as input. Then, it extracts the knowledge entities that are referred to in the text. Again, the GATE framework and the synonyms in Wordnet are used in this process. Once the relevant concepts in the domain ontology have been obtained, the user profile vector is generated as shown in Eq. 1 (in this case, the documents considered for the TF-IDF calculation are the registration details of all users).

Over time, users' interests tend to change. The recommendation engine keeps up with that variation by adjusting the user profile vector when either of the following actions occur: the user makes a comment on the site or the user clicks on an ad. Concerning the former, only textual posts are considered. Once more NLP tools are applied to gather the ontological elements that have been mentioned in the text. The vector associated to the user is renovated using Eq. 3, where v_i' is the previous value of the vector for the component i, and τ is a number between 0 and 1 denoting the mutation rate. The interpretation of the rest of the formula is the same as in Eq. 1, with the difference that in this case the documents considered for the TF-IDF calculation are all the previous textual contributions of the user.

$$v_i = \tau * v_i' + (1 - \tau) * \sum_{j=1}^{n} \frac{(tf - idf)_{j,d}}{e^{dist(i,j)}} , \text{ for } i = 1 \ldots n, \text{ with } 0 < \tau < 1 \qquad (3)$$

On the other hand, when the user clicks on an advertisement the system infers that the user is engaged by the theme of the advertisement and updates the vector accordingly. The revision process is described in Eq. 4, where v_i' represents the previous value of the vector for the component i, v_i^a is the value of the component i in the vector describing the advertisement a (the one that has attracted the attention of the user), and μ is a number between 0 and 1 denoting the mutation rate.

$$v_i = \mu * v_i' + (1 - \mu) * v_i^a, \text{ for } i = 1 \ldots n, \text{ with } 0 < \mu < 1 \qquad (4)$$

3.5 Ads Recommender

The objective of this module is to determine the ads that might be of interest for a given user. The process benefits from the semantic content and annotations previously generated. Specifically, the vectors representing users and advertisements, produced as described above, are used at this stage. These vectors are based on the same ontology, and thus they all have the same number of components, one for each separate concept of the domain ontology. With all, the ads recommender calculates the similarity between a user u and each advertisement a by using the cosine similarity as shown in Eq. 5, where a is the vector space model formed with Eq. 1 for a given advertisement, u is the vector that represents a user profile computed as described in Sect. 3.4, and θ is the angle that separates both vectors.

$$sim(u, a) = \cos \theta = \frac{u \cdot a}{|u| \times |a|} \qquad (5)$$

The advertisement that is ultimately suggested to the user is the one that (i) has a similarity score within the first quartile and (ii) has been recommended the less so far (the system records the number of times each ad is put forward for consideration). This way novelty, serendipity and diversity are fostered along with relevance.

4 Validation

We built a prototype of the framework and executed it on a simulated environment. An experiment was conducted in which 10 undergraduate students had to provide some basic registration data, published 20–30 comments (50–100 words long), and clicked on 5–10 advertisements (from a pool of 100 ads) in which they were interested. Then, our framework served a total of 20 ads to each student (from a different pool of other 100 ads), and the students indicated for each suggested ad whether it was relevant or not. An information systems-oriented evaluation based on the precision metric (see Eq. 6) was performed.

$$precision = (total\ relevant\ ads)/(total\ recommended\ ads) \qquad (6)$$

The results of the experiment are shown in Table 1. The precision of the framework increases as the number of users' comments and clicks gets larger, being the total average precision 78%.

Table 1. Results of the experiment.

User	No. comments	Total no. words	No. clicks	Relevant ads	Precision
1	27	1.948	7	15	75%
2	22	1.410	5	13	65%
3	29	2.407	8	18	90%
4	25	1.986	5	15	75%
5	23	1.495	7	14	70%
6	30	2.372	10	17	85%
7	26	1.648	6	15	75%
8	29	2.063	8	17	85%
9	25	1.681	6	15	75%
10	22	1.714	8	16	80%
Average					**78%**

5   Conclusions and Future Work

In the last few years, advertisement spending has shifted from high cost and medium-visibility media such as TV, radio and newspapers towards a low cost and higher visibility digital media such as the Web. One of the factors leading to this new trend is consumer-targeting capabilities, mainly fostered by the Web 2.0 movement Social sites such as Facebook are some of the leading vendors in the market. Ad recommender systems can learn a lot about users by inspecting their social activity. However, dealing with such a large and heterogeneous amount of data is a huge challenge. Semantic technologies have proved effective in similar situations, helping to handle information coming from disparate sources from a formal, knowledge perspective.

The framework proposed in this paper leverages ontologies to semantically model both users' interests and advertisements main features. As of today, the framework mainly manages textual content and make use of NLP tools to generate vectors representing user profiles as well as advertisement profiles in accordance with the domain ontology. Then, a similarity function is employed to rank the matching ads for each user, and diversity is promoted by suggesting the ad that has been less advocated up to this point. The proposed social recommender system is based on a hybrid approach that combines content-based, collaborative and knowledge-based recommendation strategies. With this approach, traditional problems hampering the performance of recommender systems such as the cold-start issue, sparsity and diversity, are avoided.

Our proposal does not yet fully exploit several information items inherently available in social sites that can help improve the accuracy and reliability of our framework. The interactions between users in the platform, along with their relationships and the multimedia content contributed by users are some of the elements planned to be integrated in the future. Also, a more robust validation, in a real environment and with large volumes of data, is required to check the scalability of the proposed approach.

Acknowledgements. This work has been supported by the Spanish National Research Agency (AEI) and the European Regional Development Fund (FEDER/ERDF) through project KBS4FIA (TIN2016-76323-R).

References

1. Resnick, P., Iacovou, N., Suchak, M., Bergstrom, P., Riedl, J.: GroupLens: an open architecture for collaborative filtering of netnews. In: Proceedings of the 1994 ACM Conference on Computer Supported Cooperative Work - CSCW 1994, pp. 175–186. ACM Press, New York (1994)
2. Carrer-Neto, W., Hernández-Alcaraz, M.L., Valencia-García, R., García-Sánchez, F.: Social knowledge-based recommender system. Application to the movies domain. Expert Syst. Appl. **39**, 10990–11000 (2012)
3. Eirinaki, M., Gao, J., Varlamis, I., Tserpes, K.: Recommender systems for large-scale social networks: a review of challenges and solutions. Futur. Gener. Comput. Syst. **78**, 413–418 (2018)
4. Jin, S.V.: "Celebrity 2.0 and beyond!" Effects of Facebook profile sources on social networking advertising. Comput. Hum. Behav. **79**, 154–168 (2018)
5. Shadbolt, N., Berners-Lee, T., Hall, W.: The semantic web revisited. IEEE Intell. Syst. **21**, 96–101 (2006)
6. Lagos-Ortiz, K., Medina-Moreira, J., Paredes-Valverde, M.A., Espinoza-Morán, W., Valencia-García, R.: An ontology-based decision support system for the diagnosis of plant diseases. J. Inf. Technol. Res. **10**, 42–55 (2017)
7. del Pilar Salas-Zárate, M., Valencia-García, R., Ruiz-Martínez, A., Colomo-Palacios, R.: Feature-based opinion mining in financial news: an ontology-driven approach. J. Inf. Sci. **43**, 458–479 (2017)
8. Boratto, L., Carta, S., Fenu, G., Saia, R.: Semantics-aware content-based recommender systems: design and architecture guidelines. Neurocomputing **254**, 79–85 (2017)
9. Kunaver, M., Požrl, T.: Diversity in recommender systems – a survey. Knowl.-Based Syst. **123**, 154–162 (2017)
10. Aggarwal, C.C.: An introduction to recommender systems. In: Recommender Systems, pp. 1–28. Springer International Publishing, Cham (2016)
11. Lalwani, D., Somayajulu, D.V.L.N., Krishna, P.R.: A community driven social recommendation system. In: 2015 IEEE International Conference on Big Data (Big Data), pp. 821–826. IEEE (2015)
12. Aguilar, J., Valdiviezo-Díaz, P., Riofrio, G.: A general framework for intelligent recommender systems. Appl. Comput. Inform. **13**, 147–160 (2017)

13. Nakatsuji, M., Yoshida, M., Ishida, T.: Detecting innovative topics based on user-interest ontology. Web Semant. Sci. Serv. Agents World Wide Web **7**, 107–120 (2009)
14. Mezghani, M., Péninou, A., Zayani, C.A., Amous, I., Sèdes, F.: Producing relevant interests from social networks by mining users' tagging behaviour: a first step towards adapting social information. Data Knowl. Eng. **108**, 15–29 (2017)
15. Kang, J., Lee, H.: Modeling user interest in social media using news media and Wikipedia. Inf. Syst. **65**, 52–64 (2017)

Hand Gesture Detection
with Convolutional Neural Networks

Samer Alashhab[1(✉)], Antonio-Javier Gallego[1], and Miguel Ángel Lozano[2]

[1] Department of Software and Computing Systems,
University of Alicante, Alicante, Spain
salashhab@ua.es, jgallego@dlsi.ua.es
[2] Department of Computer Science and AI,
University of Alicante, Alicante, Spain
malozano@ua.es

Abstract. In this paper, we present a method for locating and recognizing hand gestures from images, based on Deep Learning. Our goal is to provide an intuitive and accessible way to interact with Computer Vision-based mobile applications aimed to assist visually impaired people (e.g. pointing a finger at an object in a real scene to zoom in for a close-up of the pointed object). Initially, we have defined different hand gestures that can be assigned to different actions. After that, we have created a database containing images corresponding to these gestures. Lastly, this database has been used to train Neural Networks with different topologies (testing different input sizes, weight initialization, and data augmentation process). In our experiments, we have obtained high accuracies both in localization (96%–100%) and in recognition (99.45%) with Networks that are appropriate to be ported to mobile devices.

1 Introduction

Visual impairment determines to a large extent the life of a person, being considered the major sensory disability. Currently, there are approximately 285 million people with visual impairment [1]. According to the ONCE foundation, 80% of the information necessary for our daily life implies the organ of vision and, in Spain, 58,031 people maintain a quantifiable visual rest [2]. This exposes the importance of assisting visually impaired people in common tasks, such as obstacle detection [3] (on both ground level or at a certain height), guidance, finding objects, or reading signs or text, among others.

Applying computer vision on smartphones became a key element for assisting visually impaired people, we find works such as *Indoor Navigation System* [4], *Crosswalk Localization* [5] and *The vOICe* [6], an app that provides spoken OCR, color identifier, compass, face detector and GPS locater. We also mention some of our previous work, such as *Aerial Obstacle Detection* [7], which uses smartphones endowed with a 3D camera to detect the distance to an object and alert the user when an obstacle is found in the walking direction, and

© Springer International Publishing AG, part of Springer Nature 2019
F. De La Prieta et al. (Eds.): DCAI 2018, AISC 800, pp. 45–52, 2019.
https://doi.org/10.1007/978-3-319-94649-8_6

Super Vision for Cardboard,[1] which turns a smartphone and a Google Cardboard device into a low cost electronic glasses device, allowing to magnify or change the color of the image captured by the camera.

Lately, image classification methods have focused on algorithms known as Deep Learning [8]. Convolutional Neural Networks (CNNs) have excellent results in solving image recognition problems. Recent networks such as AlexNet [9], GoogleNet [10] and ResNet [11] have afforded a significant breakthrough in image classification. We find also some recent work on hand gestures recognition: in [12], with approximately 90% of accuracy, or [13] that obtained 77,50% of accuracy.

However, our goal is to develop a Computer Vision method that is able to locate and recognize gestures using smartphones, in order to use these gestures to interact with the application. We will focus our work on studying different CNNs applied to this problem. For this purpose, we created our own image dataset of hand gestures, that will be used to train and test different architectures of CNNs.

The remainder of this paper is organized as follows. Section 2 describes the proposed method, the different CNN that will be evaluated, and the technique used to locate gestures. Section 3 presents the dataset of hand gestures and the results obtained with the different CNNs. Finally, our conclusions and future work are described in Sect. 4.

2   Method

The scheme of the proposed method for gesture detection and localization is shown in Fig. 1. First, a CNN is trained for gesture classification. Then, the new images are forwarded through this CNN to the network prediction. Lastly, if a gesture is detected, its position is calculated using the technique known as *heat map*. Below we will see in detail each of these steps.

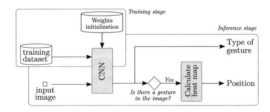

Fig. 1. Graphical scheme of the methodology considered in the present work.

In this work, we have evaluated seven state-of-the-art CNN topologies, two classic (VGG-16/19) [14], three recent network models: ResNet [11], Inception V3 [10] and Xception [15], that outperform the classic networks in object recognition tasks, and two lightweight architectures aimed at obtaining good results with a small number of parameters: MobileNet [16] and SqueezeNet [17]. Details

---

[1] http://supervisioncardboard.com.

regarding the implementation and the parameters used in these networks can be found in the corresponding references.

Another important issue for us was to use a network with few parameters, because the more parameters they have, the slower is their operation. Table 1 shows the number of parameters of the considered networks in descending order. It is important to have a trade off between the efficiency of the network and its accuracy, so in the experimentation we will study this aspect in detail.

Table 1. Number of parameters of each network in descending order.

Model	#Parameters	Model	#Parameters
ResNet	29,887,365	VGG 16	17,868,613
Inception v3	28,102,437	MobileNet	7,431,365
Xception	27,161,133	SqueezeNet	3,876,421
VGG 19	23,178,309		

Once a gesture is detected, its *heat map* is calculated in order to determine the activation zones and thereby identify its position. *Heat maps* are a class-discriminative localization technique helpful for producing visual explanations for CNNs. To calculate these heat maps we use the *Gradient-weighted Class Activation Mapping* (Grad-CAM) approach [18].

The training was done using Stochastic Gradient Descent [19] with the adaptive learning rate method proposed by Zeiler [20] and *categorical crossentropy.*

For training we considered two strategies: train the network starting from scratch (which we will call "uninitialized") and train the network starting from weights learned using the ILSVRC dataset (a 1,000-class subset from ImageNet [21]).

3   Experiments

3.1   Dataset

In order to validate the effectiveness of the proposed method, we manually collect a dataset of hand gestures using a simple mobile phone (see Fig. 2). We recorded

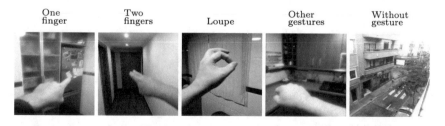

Fig. 2. Examples of images from our dataset.

five different states of a hand with different backgrounds, including: One finger, two fingers, loupe shape, other gestures (such as opened or closed hands), and no gesture (images without any hand).

The compiled dataset contains a total of 22,740 images, for which we attempted to balance the number of samples of each class. For training, a non-overlapped set with 70% of the images was used, leaving the remaining 30% for the test. The separation of training and validation data was made taking into account the image background. For validation we used images with backgrounds that had not been seen in training. Moreover, to evaluate the appropriate image size to be used, the images were scaled to different sizes commonly used by this type of networks: 128×128px, 160×160px, 192×192px, and 224×224px.

3.2   Evaluation Metrics

The performance of the proposed CNN models was evaluated using the weighted average of the F-measure (F_1) scores of each class. The F_1 is defined as:

$$\text{F-measure} = \frac{2 \cdot TP}{2 \cdot TP + FN + FP}$$

where TP (True Positives) denotes the number of correctly detected targets, FN (False Negatives) the number of non-detected or missed targets, and FP (False Positives or false alarms) the number of incorrectly detected targets.

Accuracy of the detected location was measured using Intersection over Union (IoU), by mapping each object proposal (op) to the ground truth (gt) bounding box with which it has a maximum IoU overlap. A detection is considered as TP if the area of overlap (a_o) ratio between the predicted bounding box (B_{op}) and the ground truth bounding box (B_{gt}) exceeds a certain threshold (λ) according to the following equation:

$$\left(a_o = \frac{area(B_{op} \cap B_{gt})}{area(B_{op} \cup B_{gt})} \right) > \lambda$$

where $area(B_{op} \cap B_{gt})$ depicts the intersection between the object proposal and the ground truth bounding box, and $area(B_{op} \cup B_{gt})$ depicts its union. By convention, we use a threshold of $\lambda = 0.5$ to set a TP candidate.

3.3   Results

This section presents the results obtained with the proposed methodology.

First we carried out an experiment varying the input size between 128×128px and 224×224px and the type of initialization. Table 2 shows the results in terms of F_1 (%). It was only possible to run ResNet with the largest input size, since it is a network with many pooling layers. In general, initialized networks with an input size of 224×224px obtained the best results: Xception, Inception v3, and MobileNet (above 99%); SqueezeNet also obtained competitive results (98.36%) with a smaller input (192×192px).

Data augmentation [9] was applied to the previous best results in order to artificially increase the size of the training set doing flips, zoom, and rotations

Table 2. F_1 (%) score comparison using the proposed CNNs with different input sizes and weights initialization.

Model	128 × 128		160 × 160		192 × 192		224 × 224	
	Uninit.	Init.	Uninit.	Init.	Uninit.	Init.	Uninit.	Init.
VGG 16	54.33	78.17	46.66	83.36	52.10	**85.47**	40.09	32.22
VGG 19	44.24	33.17	36.12	51.74	30.15	26.17	42.77	**64.04**
ResNet	–	–	–	–	–	–	67.91	**93.35**
Xception	68.44	98.90	70.93	99.41	74.29	99.21	75.55	**99.52**
Inception v3	72.67	95.28	70.65	97.03	78.91	98.11	76.26	**99.04**
MobileNet	54.63	97.32	54.82	98.70	59.96	98.98	64.11	**99.29**
SqueezeNet	28.05	92.31	31.07	93.73	34.26	**98.36**	36.13	97.93

Table 3. F_1 (%) score comparison using data augmentation techniques.

Model	Uninitialized			Initialized		
	Best input size	Without data aug.	With data aug.	Best input size	Without data aug.	With data aug.
VGG 16	**128 × 128**	54.33	68.57	**192 × 192**	85.47	76.79
VGG 19	**128 × 128**	44.24	54.52	**224 × 224**	64.04	**77.76**
ResNet	**224 × 224**	67.91	78.89	**224 × 224**	93.35	**93.59**
Xception	**224 × 224**	75.55	84.43	**224 × 224**	99.52	**99.77**
Inception v3	**192 × 192**	78.91	86.10	**224 × 224**	**99.04**	98.76
MobileNet	**224 × 224**	64.11	82.58	**224 × 224**	99.29	**99.45**
SqueezeNet	**224 × 224**	36.13	47.34	**192 × 192**	**98.36**	94.32

on the original images. Table 3 shows the results of this process. It can be seen that data augmentation improved the results slightly.

As stated before, we also intended to find the most efficient CNN to optimize its execution time on a mobile phone. From this point of view, it can be seen as a *Multi-objective Optimization Problem* (MOP) in which two functions have to be optimized: accuracy and efficiency. The F_1 allow us to compare the networks performance, however, to evaluate the efficiency we will use the number of parameters (see Table 1) weighted by the input size of the network. The common way of evaluating this kind of problems is by means of *non-dominance* concept. One solution is said to dominate another if, and only if, it is better or equal in each goal function and, at least, strictly better in one of them. Therefore, the best solutions (there might be more than one) are those that are non-dominated.

Figure 3 shows the MOP results, where the vertical axis represents the F_1 and the horizontal axis the number of parameters in percentage, considering as maximum the network with more parameters (ResNet) weighted by the largest image size (224 × 224px). In this graph, the training method is differentiated by shape and color, and the *non-dominated* elements are highlighted. Networks that optimize both objectives are SqueezeNet with input sizes of 128 × 128px and 192 × 192px, and MobileNet and Xception with an input size of 224 × 224px, and in all cases using data augmentation.

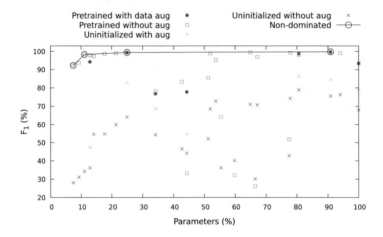

Fig. 3. Analysis of F_1 and efficiency as a *Multi-objective Optimization Problem* (MOP). Non-dominated elements are highlighted.

Evaluation of the Detected Position. As explained before (see Fig. 1), once a gesture is detected we calculate its position using the heat map obtained by the *Gradient-weighted Class Activation Mapping* (Grad-CAM) approach [18]. Figure 4 shows some heat maps samples for the MobileNet network. As can be seen, the red areas belong to the region of the image where the gesture appears, showing a good precision in the position detected.

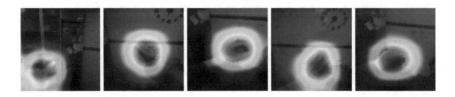

Fig. 4. Heat maps calculated using Grad-CAM [18], highlighting the localized class-discriminative regions. Localized classes from left to right are: One finger, Two fingers, Loupe, and two Other different gestures.

To evaluate the accuracy of the detected position we use the IoU metric described in Sect. 3.2. For this, we calculate the IoU between the bounding box that includes the zone of maximum activation of the heat map and the bounding box that contains the gesture of the corresponding image of the ground truth (manually labeled). Table 4 shows these results, where the Acc row depicts the percentage of correct detections (TP) using a threshold $\lambda = 0.5$ (common value). It can be seen that the proposed method obtains a good result for all the types of gestures, obtaining an average accuracy of 98%.

Table 4. Evaluation (in terms of IoU) of the obtained location.

Category	One finger	Two fingers	Loupe	Other gestures	Average
IoU	0.8234	0.8343	0.8254	0.8605	0.8359
Acc (%)	96	98	98	100	98

4   Conclusions

In this work, a new proposal for the detection and localization of hand gestures in color images obtained by a mobile phone camera is presented. To this end, different Deep Learning techniques have been analyzed, including seven state-of-the-art CNN topologies, the calculation of the heat maps using Grad-CAM, and the evaluation of different input sizes, weight initializations, and data augmentation techniques. In addition, a dataset with about 23,000 images including different types of hand gestures with different backgrounds has been compiled.

Best F_1 results were obtained by Xception (99.77%), MobileNet (99.45%) and Inception v3 (99.04%). When evaluating the accuracy of the detected position using the IoU metric, these networks obtain an average of 98% of correct detections.

These results were also analyzed considering both the F_1 and the number of parameters of the network. *Non-dominated* results (those that optimize both objectives) were obtained by SqueezeNet with input sizes 128×128px and 192×192px, and MobileNet and Xception with size 224×224px, in all cases using data augmentation.

As future work, we intend to improve the proposed algorithm by increasing the accuracy of the location process, and also combining this solution with other object detection algorithms in order to obtain information about the indicated objects and return that information to the end user.

Acknowledgements. This work was partially supported by the project TIN2015-69077-P of the Spanish Government.

References

1. Organization, W.H., et al.: Global Data on Visual Impairments 2010. World Health Organization Organization, Geneva (2012)
2. ONCE Foundation, Afiliados a la ONCE, junio 2017, June 2017. http://www.once.es/new/afiliacion/datos-estadisticos

3. Manduchi, R., Coughlan, J.: (Computer) Vision without sight. Commun. ACM **55**(1), 96–104 (2012)
4. Rituerto, A., Fusco, G., Coughlan, J.M.: Towards a sign-based indoor navigation system for people with visual impairments. In: Proceedings of the 18th International ACM SIGACCESS Conference on Computers and Accessibility, pp. 287–288. ACM (2016)
5. Ahmetovic, D., Manduchi, R., Coughlan, J.M., Mascetti, S.: Mind your crossings: mining GIS imagery for crosswalk localization. ACM Trans. Access. Comput. (TACCESS) **9**(4), 11 (2017)
6. The voice for android (2017). https://www.seeingwithsound.com/android.htm
7. Sáez, J.M., Escolano, F., Lozano, M.A.: Aerial obstacle detection with 3D mobile devices. IEEE J. Biomed. Health Inf. **19**(1), 74–80 (2015)
8. LeCun, Y., Bengio, Y., Hinton, G.: Deep learning. Nature **521**(7553), 436–444 (2015)
9. Krizhevsky, A., Sutskever, I., Hinton, G.E.: ImageNet classification with deep convolutional neural networks. In: Advances in Neural Information Processing Systems, pp. 1–9 (2012)
10. Szegedy, C., Vanhoucke, V., Ioffe, S., Shlens, J., Wojna, Z.: Rethinking the inception architecture for computer vision. CoRR, abs/1512.00567 (2015)
11. He, K., Zhang, X., Ren, S., Sun, J.: Deep residual learning for image recognition. In: Proceedings of the IEEE Conference on Computer Vision and Pattern Recognition, pp. 770–778 (2016)
12. Bheda, V., Radpour, D.: Using deep convolutional networks for gesture recognition in American sign language. In: CoRR, abs/1710.06836 (2017)
13. Molchanov, P., Gupta, S., Kim, K., Kautz, J.: Hand gesture recognition with 3D convolutional neural networks. In: 2015 IEEE Conference on Computer Vision and Pattern Recognition Workshops (CVPRW), pp. 1–7, June 2015
14. Simonyan, K., Zisserman, A.: Very deep convolutional networks for large-scale image recognition. In: ImageNet Challenge, pp. 1–10 (2014)
15. Chollet, F.: Xception: deep learning with depthwise separable convolutions. CoRR, abs/1610.02357 (2016)
16. Howard, A.G., Zhu, M., Chen, B., Kalenichenko, D., Wang, W., Weyand, T., Andreetto, M., Adam, H.: Mobilenets: efficient convolutional neural networks for mobile vision applications. arXiv preprint arXiv:1704.04861 (2017)
17. Iandola, F.N., Han, S., Moskewicz, M.W., Ashraf, K., Dally, W.J., Keutzer, K.: Squeezenet: alexnet-level accuracy with 50x fewer parameters and <0.5 mb model size. arXiv preprint arXiv:1602.07360 (2016)
18. Selvaraju, R.R., Das, A., Vedantam, R., Cogswell, M., Parikh, D., Batra, D.: Gradcam: why did you say that? Visual explanations from deep networks via gradient-based localization. arXiv preprint arXiv:1610.02391 (2016)
19. Bottou, L.: Large-scale machine learning with stochastic gradient descent. In: Proceedings of COMPSTAT 2010, pp. 177–186. Springer (2010)
20. Zeiler, M.D.: ADADELTA: an adaptive learning rate method. CoRR, abs/1212.5701 (2012)
21. Deng, J., Dong, W., Socher, R., Li, L.J., Li, K., Li, F.-F.: Imagenet: a large-scale hierarchical image database. In: 2009 IEEE Conference on Computer Vision and Pattern Recognition, pp. 248–255 (2009)

A Genetic Algorithm Model for Slot Allocation Optimization to Brazilian CTOP Approach

Natan Rodrigues, Leonardo Cruciol, and Li Weigang$^{(\boxtimes)}$

TransLab - Department of Computer Science, University of Brasília,
Brasília, DF 70910-900, Brazil
weigang@unb.br

Abstract. The Collaborative Trajectory Options Program (CTOP) makes each airline possible to share its route options to air traffic control center, and so achieve better business goals by reducing strategic operational costs. In Brazil, there are initial efforts to verify the benefits of CTOP implementation to improve the air traffic fluency and financial results. This paper presents a novel approach for Brazilian airspace using Genetic Algorithms to decrease the delay between available slots during CTOP. The slot optimization keeps improving in a safety-separating window of each aircraft en route. The case study presented an reducement about 70% of delay of a certain airline, when used this decision support system by air traffic control authority.

Keywords: Air Traffic Management
Collaborative Trajectory Options Program · Genetic Algorithm
Optimization · Slot allocation

1 Introduction

The Collaborative Trajectory Options Program (CTOP) is an Air Traffic Management (ATM) program, which aims to improve the air traffic flow and safety by considering the airlines' business goals. It makes possible each airline to improve its business goals by sharing route preferences with the air traffic control center and so, if possible, achieve better results. This program is used by the Federal Aviation Administration (FAA) in the USA and has increased the interaction to the airlines in the slot allocation decision process [1–3,8].

This program is an evolution of traditional programs of ATM and makes possible some new approaches to improve business results of the stakeholders. In Brazil, there are initial studies to verify how the CTOP could increase the airspace safety with a better financial operation index. However, the original CTOP proposal is not very compatible with Brazilian scenario, once there is few airways between main airports. Considering the Brazilian airspace and its main airports, it is possible to verify that CTOP could not be implemented as it works in the USA.

© Springer International Publishing AG, part of Springer Nature 2019
F. De La Prieta et al. (Eds.): DCAI 2018, AISC 800, pp. 53–60, 2019.
https://doi.org/10.1007/978-3-319-94649-8_7

Considering the difference between Brazilian and American airspace environment, it is proposed a new approach in CTOP by using a Genetic Algorithm (GA) model for slot allocation optimization and reducement of time window between each aircraft en route. When CTOP is initiated, a Flow Constrained Area (FCA) is defined, which will restrict geographically the current capacity, and so each slot to fly through the area is defined to allocate each aircraft. Considering this allocation approach, there are some solutions for CTOP in the literature, such as [8,9].

Once in Brazilian airspace there is no many airways between main airports to apply the original CTOP concept, it was considered an approach to reduce safely the time window between each aircraft en route, and so increase the capacity for every 15 min windows. Thus, the algorithm optimization is working in a slot definition time to reduce the infrastructure limitation.

The paper is divided as follows. Section 2 gives a description of the related works. In Sect. 3, the new CTOP approach is proposed using Genetic Algorithms for Brazilian scenario. Section 4 presents a case study and analysis of the results. Section 5 presents the conclusions and future works.

2   Related Works

2.1   Greedy Two-Step Optimization Approach

The FAA presented the CTOP with the main objective of improving the ATM for National Airspace System (NAS) members to achieve their business objectives considering the characteristics of each flight and the restrictions of airspace [4–7].

In the research presented by Kim and Clarke [8], a greedy model was developed to select the best Trajectory Options Set (TOS) for each flight captured in the CTOP, regarding to the existence of multiple FCAs and the negotiation process involving airlines for the best available slots.

The research focused on the optimization of flight allocation to certain airline for delay reduction. The model presented in Fig. 1 was developed in two steps using a greedy search algorithm, which achieved great results with a reduction up to 60% in operational costs.

2.2   Game Theory to Suggest Trajectories in CTOP

The proposal presented by [9] was developed using Game Theory to suggest trajectories to airlines for flights captured in CTOP. The allocation process was modelled as the Prisoners' Dilemma game with its solution.

The Fig. 2 presents a game as a matrix, where each cell contains the global delay for each airline based on the possible moves in the game. There are 3 possible moves for each player, resulting in 9 possible outcomes.

After generating and simulating more than 20,000 negotiations involving airline trajectories in the CTOP, the results presented better performance than the strategies currently used in the CTOP. The research obtained a reduction of 537 h of delays for a certain airline.

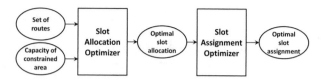

Fig. 1. Optimization procedures [8]

		Airline B		
		NOSLOT (1054)	NOSLOT (1054)	NOSLOT (1054)
		NOSLOT (510)	1 Trajectory + NOSLOT (513)	2 Trajectories + NOSLOT (471)
Airline A	1 Trajectory + NOSLOT (1012)	1 Trajectory + NOSLOT (1033)	1 Trajectory + NOSLOT (1102)	
		NOSLOT (510)	1 Trajectory + NOSLOT (477)	2 Trajectories + NOSLOT (489)
	2 Trajectories + NOSLOT (980)	2 Trajectories + NOSLOT (1002)	2 Trajectories + NOSLOT (1007)	
		NOSLOT (510)	1 Trajectory + NOSLOT (511)	2 Trajectories + NOSLOT (481)

Fig. 2. Game matrix model example [9]

2.3   Departure Management Using Genetic Algorithms

In ATM, Departure Management is a process to define the order of departures, especially during a scenario of delay and flights cancellation.

The research in [10] was set in Brazilian airspace, where the departure management was done empirically by the air traffic controllers, negotiating with the airlines and then generating an *First-Come First-Served (FCFS)* sequence.

Proposed by Holland in the 70's [11], Genetic algorithms are search methods that use mechanisms based on evolutionary and genetic theory. The approach of Genetic algorithms is an alternative method to search good solutions in a large and complex search space [12]. Genetic Algorithms codifies decision variables of search problems as finite strings formed by alphabets of certain cardinality. Strings that form possible solutions are called chromosomes, the used alphabets are called genes. [10] The Genetic Algorithm was used to work with a population of 1000 possible flight departure sequence solutions. In the approach, 100 generations were executed and the tests occurred simulating flight departures during normal and high flow schedules. The results achieved after the solution using Genetic Algorithms reduced up to 41% of the total possible delays at a high busy airport.

3 CTOP Approach Using GA to Brazilian Scenario

It was defined an approach for Brazilian scenario using Genetic Algorithms to verify the best group of slots to reduce the delay. The concepts of ATM and CTOP were configured in the definitions using GA to implement the algorithm and solution, as follow:

- **Gene**: In this solution, slot schedule creates the genes. A slot is the time interval that a particular aircraft can fly through an FCA, so the genes of a chromosome will be defined by the slot time, which is defined as: hour; minute; or second, as presented in Fig. 3.
- **Individual**: The chromosome or individual of the solution will be defined as each slot of the CTOP demand. Each chromosome (slot) will be composed of genes (schedules). Figure 3 presents the structure of the individual.
- **Population**: The set of all slots (individuals), including slots that have allocated and unallocated flights of the current CTOP demand.
- **Selection**: The elitist selection method was used to select individuals from the population. During the construction of the next generation R, it was selected the 60% of the individuals with the best fitness in the population P to compose the generation R.
- **Mutation**: The mutation genetic operator used in the proposed solution will perform a mutation with probability $P_M = 0.05$ to occur in each individual of the current generation. After several tests this value achieved satisfactory results, as described in the results.

  The mutation can add or subtract an interval of 30 min in the chromosome gene minutes. In the following example, a mutation occurred in the chromosome, adding 25 min in the minute gene of the chromosome: *before mutation: 20:22:00; after mutation: 20:47:00.*
- **Crossover**: The crossover used in the proposed solution is the uniform type, because an individual will inherit from the parent some range of time (hour, minute or second). The representation of the problem domain could be a difference between minutes, according to the following example: *parent: 12:50:23; children: 12:52:23.*

  The individual will inherit from only one parent of genes with a probability $P_C = 0.66$.

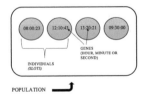

Fig. 3. Gene, individual and population structures of the proposed Genetic Algorithm.

– **Fitness function**: The fitness function is applied after executing the flight allocation in the set of slots using the CTOP slot allocation algorithm.

$$A_{slot} = t_{flight} - t_{slot} \tag{1}$$

$$A_{slot} = \begin{cases} 1, & t_{flight} - t_{slot} = 0 \\ t_{flight} - t_{slot}, & t_{flight} - t_{slot} \neq 0 \end{cases} \tag{2}$$

$$f = \frac{1}{A_{slot}} \tag{3}$$

Equation 1 is defined to obtain the delay of each slot after a flight allocation. In Equation, t_{flight} is the time that flight enters in corresponding FCA and t_{slot} is the slot time. The difference between the schedules is the delay that a flight will receive. As described in the equation, there is no difference if the delay is equal to 0 or 1 min.

The Eq. 3 defines the fitness function applied to the individual before to create a new generation. The fitness value may be between 0 and 1, which one is the best fitness.

4 Case Study

4.1 Scenarios Environment

A case study was developed to validate the GA model, regarding to Brazilian airspace with two FCAs: FCA 1 and FCA 2; two airlines, Airline A and Airline B, and 99 simulated flights from 9 Brazilian airports.

As shown in Fig. 4, flights depart from *Fortaleza International Airport - Pinto Martins* (FOR), *Recife International Airport - Gilberto Freyre* (REC), *Teresina Airport - Senator Petrônio Portella* (THE), *Salvador International Airport - Luiz Eduardo Magalhães* (SSA), *Brasilia International Airport - President Juscelino Kubitschek* (BSB), *Belo Horizonte-Confins International Airport - Tancredo Neves* (CNF), *Goiânia International Airport - Santa Genoveva* (GYN) and arrive at *São Paulo/Congonhas Airport - Deputy Freitas Nobre* (CGH) and *Guarulhos International Airport - Governor André Franco Montoro* (GRU) airports.

Considering the amount of CTOP captured flights, it was distributed in three scenarios: (1) 50% of flights from Airline A and 50% from Airline B; (2) 63% of flights from Airline A and 37% from Airline B; and (3) 75% of flights from Airline A and 25% from Airline B.

The CTOP demand was defined as an interval with 6 hours from 18:00 to 00:00, which the FCA's capacity was two aircraft every 15 min. Thus, it was created 48 slots for each FCA throughout the execution of CTOP. The amount of generations required for the GA to reach a convergence and an optimization of the general allocation algorithm was verified closer of 75 generations, as observed in the results.

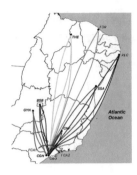

Fig. 4. Multiple routes between nine Brazilian airports.

4.2   Analysis of Results

It was considered three scenarios:

- **Scenario: 50% Airline A - 50% Airline B.** The general airspace delay decreases from the first generation and achieves a value smaller than the initial value, as presented in Fig. 5(a). After the execution of the algorithm, the minimum general delay, 499 min, was achieved in generation 76. For the specific delay for Airline A the delay decreases since the first generation achieves the lowest value, 205 min, in generation 74. For Airline B, the specific delay also decreases from the first generation, achieving the lowest value, 216 min, in generation 66.
- **Scenarios: 63% Airline A - 37% Airline B and 75% Airline A - 25% Airline B.** The actions were similar for both scenarios. The general airspace delay decreases from the first generation and achieved lower values in the final generation, obtaining delays smaller than the initial values, as presented in Fig. 5(b) and (c), In the specific delays by airline, the delay for airlines with more flights and with less distribution also follow the behavior of general delays for both cases.
- **All Scenarios.** Considering the general airspace delay and specific for each airline, the delay decrease from the first generation reaching lower delays in the final generation. It is important to note that there was a decrease of the delays in the GA approach compared to the general slot allocation approach. The developed approach can obtain an optimization of the slot allocation algorithm.

The delay behavior of the GA are represented in Table 1. After execution of the GA, the average of decrease was up to 67 % for the general airspace delays. For airline A and B delays, decreased up to 72%.

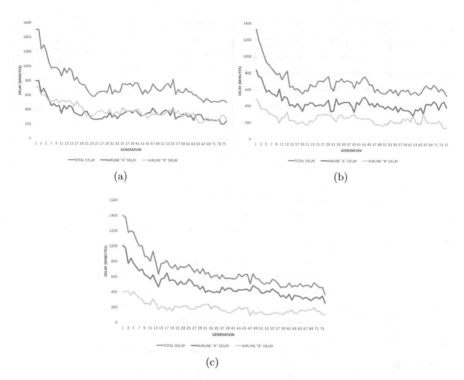

Fig. 5. Delays in scenarios during GA execution: (a) delays in scenario 50% Airline A - 50% Airline B ; (b) delays in scenario 63% Airline A - 37% Airline B; (c) delays in scenario 75% Airline A - 25% Airline B.

Table 1. Delays (minutes) in initial generation (general airspace approach) and the GA minimum delay.

Scenario	Initial delay (minutes)			GA minimum delay (minutes)		
	Airline A	Airline B	Total	Airline A	Airline B	Total
50% A - 50% B	792	708	1500	205	216	499
63% A - 37% B	834	487	1321	282	129	524
75% A - 25% B	1001	395	1396	259	94	371

5 Conclusions

The CTOP initiative has achieved notable results in the USA as an opportunity to make airlines possible to share their route preferences to FAA and create a chance to improve their business goals regarding to each flight and its environment.

The initial studies to verify how CTOP could increase the safety and fluency in Brazilian air traffic management has been developed last years. The Brazilian airspace could be improved by an adaptive CTOP approach, which must consider some particular aspects as the difficult to increase the number of airways.

This paper presented a GA model slot allocation algorithm to optimize the Brazilian approach to CTOP slot allocation. The GA approach achieved an optimal set of slots, which decreased about 67% of the general airspace delay and 72% of a particular airline compared to the CTOP slot allocation algorithm used in the USA.

The proposal of future work for this research is the execution of the GA model with more than two airlines and considering only high busy airports in Brazil including international flights. It is important to evaluate a new approach for Brazilian airspace with more airways and a better flow management, once there are an infrastructure limitation.

References

1. Infraero: História da Infraero (2017). http://www4.infraero.gov.br/sobre-a-infraero/historia/. Accessed October 2017
2. Infraero: Estatísticas de movimentos operacionais de aeron- aves nos aeroportos brasileiros (2017). http://www.infraero.gov.br/index.php/br/estatisticas/estatisticas.html. Accessed Oct 2017
3. Golibersuch, M.: CTOP assignment algorithm and substitution processing. FAA PMO Industry Forum 2012 AJM-222 (2012)
4. FAA: NextGen Implementation Plan. Federal Aviation Administration, Washington, DC (2012)
5. FAA: This is CTOP - An introduction to the Collaborative Trajectory Options Program (CTOP), and the benefits to the users who participate. Federal Aviation Administration (2012)
6. FAA: Collaborative Trajectory Options Program (CTOP). Federal Aviation Administration (2012)
7. Novak, M., Somersall, P., Wolford, D.: CTOP Industry Day. A Seminar on the Collaborative Trajectory Options Program. Technical report, US Department of Transportation Federal Aviation Administration (2010)
8. Kim, B., Clarke, J.-P.: Optimal airline actions during collaborative trajectory options programs (2014)
9. Cruciol, L., Clarke, J.-P., Weigang, L.: Trajectory option set planning optimization under uncertainty in CTOP. In: IEEE 18th International Conference on Intelligent Transportation Systems, pp. 2084–2089. IEEE (2015)
10. Ferreira, D.M., Rosa, L.P., Ribeiro, V.F., de Barros Vidal, F., Weigang, L.: Genetic algorithms and game theory for airport departure decision making: GeDMAN and CoDMAN. In: International Conference on Knowledge Management in Organizations, pp. 3–14 (2014)
11. Holland, J.H.: Genetic algorithms. Sci. Am. JSTOR **267**(1), 66–73 (1992)
12. Srinivas, M., Patnaik, L.M.: Genetic algorithms: a survey. Comput. IEEE **27**(6), 17–26 (1994)

AllergyLESS. An Intelligent Recommender System to Reduce Exposition Time to Allergens in Smart-Cities

José Antonio García-Díaz[1](✉) (iD), José Ángel Noguera-Arnaldos[2],
María Luisa Hernández-Alcaraz[1], Isabel María Robles-Marín[2],
Francisco García-Sánchez[1] (iD), and Rafael Valencia-García[1] (iD)

[1] Department of Informatics and Systems, Universidad de Murcia, Murcia, Spain
{joseantonio.garcia8,mlhernandez,frgarcia,
valencia}@um.es
[2] Proyectos y soluciones tecnológicos avanzadas, SLP (Proasistech),
Avda. Primero de Mayo, nº2, Torres Azules. Torre A, 4ª Planta,
Drcha, Murcia, Spain
{jnoguera,isabel.robles}@proasistech.com

Abstract. Allergic rhinitis affects between 10% and 30% of the worldwide population. It reduces the quality of life of the individuals and causes losses in the local economy due to absenteeism. AllergyLESS is a recommendation system to solve mobility issues of the citizens by informing them which walking routes minimizes the exposure time to allergens. The system collects air-quality and allergens metrics from wireless pollution stations and open-data sources. In the cases when the pollution stations do not cover the whole area of interest, an ontology for the healthcare domain along with a set of data mining processes are used to forecast the presence of allergens. The system was validated by carrying out a number of controlled simulations of real situations.

Keywords: Recommender systems · Ontologies · Internet of Things
Data mining · Healthcare · Allergies

1 Introduction

When investments in human and social capital are built upon communication infrastructure, it results in a sustainable economic growth and a high quality of life, with a wise management of natural resources, through participatory governance [1]. In the last years, many cities have undertaken projects to become smart cities focusing on several domains such as safety, education, social services, mobility, etc. Applying the smart cities concept to the healthcare sector can reduce related cost and improve the overall efficiency [2]. Internet of Things (IoT) is the core of Smart Cities [3]. IoT consist in the process to add Internet connectivity to ordinary objects providing them with new features and characteristics.

Air pollution is a risk factor that can be extremely harmful to the environment, in general, and to humans, in particular. Given this, it is necessary to build models representing the air pollutant dispersion to support environmental decision making [4]. Research suggests a causative relationship between air pollution and the increased incidence of allergic disorders due to the exposure to traffic-related air pollution [5]. Allergic rhinitis (a.k.a. hay fever) affects between 10% and 30% of the worldwide population. This percentage has increased over the last 50 years and sensitization rates to one or more common allergens among school children are currently approaching 40%–50% [6]. The manifestations of allergy in students reduces the quality of life and academic performance. In addition, in some cases there is a risk of severe reactions and, exceptionally, the cause of death in the school due to environmental issues. Finally, it is worth to mention that in industry, allergic rhinitis causes sickness absences. In [7], the authors performed a study to measure the cost of absenteeism associated with allergic rhinitis. The results estimated a cost around $601 million and productivity losses between $2.4 and $4.6 billion. In order to tackle this problem, some strategic actions can be taken by the different political agents of the society such as developing a coordinated national strategy to facilitate allergy training by establishing and standardizing education programs [8].

The objective of the AllergyLESS project is to create a recommender system to help users to minimize the exposure time to allergens during their daily walking routes. The system gathers air quality and pollution metrics from (1) low-cost air quality monitoring stations strategically located within the city and (2) open-data sources related to meteorology. This information is used to populate an ontological model, which represents geographic areas with high concentrations of pollution and/or allergens. The geographical position of the user, together with his/her daily mobility patterns, is contrasted against the model to suggest alternative routes. The suggested routes should reduce, as much as possible, the exposition to allergens. The healthcare ontology constitutes the backbone of our approach. An ontology is a knowledge representation with aims to share, reuse and apply automatic reasoning. It provides a set of concepts to represent the domain, the relationships and constraints between them [9]. Ontologies have been already successfully applied to the medical and biomedical domains [10, 11] and to the meteorological forecasting domain [12].

The rest of the paper is organized as follows. In Sect. 2, the system AllergyLESS is presented, including a description of how its different modules interact to solve the objectives of the system. A summary of the evaluation of the system is put forward in Sect. 3. Finally, conclusions and future work are discussed in Sect. 4.

2   System Architecture

2.1   Overall Design

The system architecture is depicted in Fig. 1. The system is composed of three main modules: (1) the information gathering module, (2) the processing module, and (3) the user interface module.

The information gathering module consists of a set of pollution stations installed on streetlights or other places throughout the city. Each station has different monitoring sensors measuring a unique parameter related with air quality (e.g., humidity, ozone,

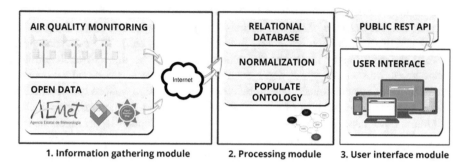

Fig. 1. Overall AllergyLESS architecture

carbon monoxide, etc.) or with the presence of an allergen (e.g., pollen). This module extends this information from open-data repositories containing meteorological information. The processing module, on the other hand, is responsible for the treatment of the information. It applies different techniques such as data mining, rule-based systems and ontology reasoning to forecast the presence of allergens in the city. The generated models can be accessed through a public REST API. Finally, the user interface module can be seen as a multi-device web interface that makes the information accessible to the user. The user has to create a profile with a list of his/her allergic diseases. Once the profile is created, the user only needs to indicate the final destination and the interface exploits the geographical information gathered from the device to suggest the best route. Next, these modules are explained in detail.

2.2   Information Gathering Module

The information gathering module takes over the acquisition of air quality metrics and pollution rates from (1) a network of pollution stations installed throughout the city and (2) open-data sources. On the one hand, we designed and deployed low-cost air quality monitoring stations to measure air quality metrics and allergens concentrations. Each station consists of a PC-embed node with a set of plugged sensors, which retrieve metrics related to air quality and concentration of allergens. The architecture of the air quality monitoring system is composed of (1) a matrix of sensors related to air quality and allergens concentrations, (2) an equipment for signal adaptation and processing, (3) a coordinator node, and (4) the cloud service.

Each sensor of an air-quality monitoring station can measure one parameter related to air quality. These measures are linked to the ontology, so the relationship between air quality parameters and risk factors concerning each allergic disease can be traced. A list of the parameters gathered from the matrix of sensors is shown in Table 1. These metrics were selected according to the factors identified in Air Pollution Ontology [13]. Each sensor transmits the measures taken to the coordinator node. Then, the equipment for signal adaptation converts the different analogic signals from the sensors to I2C or ModBus communication protocols. The coordinator node, which is an embedded PC with an ARM microprocessor, stores the data collected by the sensors into an internal database. The use of this database helps keep the data in case of temporal network failures. In addition, the coordinator node is aware of the status of each sensor and, when a failure is detected, it sends descriptive logging messages to report the problem and its nature (e.g., calibrating issues, functional failures, etc.). The communication between the coordinator node and the cloud server can be done through wireless technologies, thus allowing the deployment of the stations in places with no wired connection. The system collects data in real time. The pollution stations can be installed within the city in street lights or monuments following Smart City and IoT approaches.

On the other hand, as the resources needed to buy or deploy the pollution stations can be limited due to economic and geographical factors, the system obtains meteorological information from other open-data sources. For this, we first performed an exhaustive search for open-data repositories with climatological information for Spain. Three sources were found, namely, AEMET[1], OpenWeather[2], and APIXU[3]. The information in these open-data sources was used to extend and validate the metrics obtained by the air-quality monitoring stations. AEMET, the Spanish Meteorological Agency, has 26 stations placed in the Region of Murcia. The system collects wind speed and wind direction metrics in real time to build a model for forecasting the presence of allergens in geographical areas outside the range of the installed pollution stations. In addition, OpenWeather provides the system with information about ultra-violet radiation that cannot be measured with the any of the available sensors.

2.3 Processing Module

This module uses the information previously gathered from the network of pollution stations and open-data sources to create a data structure ready-to be-used by the user. This data structure is then made available to other software components through a public REST API. The process flow of this module can be divided into the following phases: (1) normalization, (2) ontology population, and (3) data mining.

In the normalization stage, different measurement units and different time intervals are integrated from the information gathered. The ontology population step is concerned with the instantiation of an ontology in the allergy domain taken the normalized data as input. This healthcare ontology links allergic diseases with risk factors and air-quality parameters. To create this ontology, an exhaustive search in the literature was performed to obtain base ontologies related to allergic issues within the healthcare domain. We identified (1) the Asthma Ontology[4], which collects risk factors, diseases and symptoms related to asthma, (2) the Air Pollution Ontology [13], which contains contaminants included in ISO 37120[5], and (3) the Weather Ontology [14], which models meteorological information in different time domains. As far as we are concerned, at present there are no ontologies for representing the air quality monitoring stations domain. So, an ontology was created comprising the previously identified

---

[1] http://www.aemet.es.

[2] https://openweathermap.org/.

[3] https://www.apixu.com/.

[4] http://aber-owl.net/ontology/AO/#/.

[5] https://www.iso.org/standard/62436.html.

Table 1. Air quality and allergic parameters measured by sensors

Sensor type	Measure	Unit	Range	Type
Ozone	O3	ppm	0–5	Air quality
Carbon monoxide	CO	ppm	20–2000	Air quality
Nitrogen dioxide	NO2	ppm	0–5	Air quality
Carbon dioxide	CO2	ppm	10–1000	Air quality
Sulfur dioxide	SO2	ppm	20–2000	Air quality
Fine particles	PM2.5	V/(0.1 mg/m3)	0–1,5	Air quality
Large particles	PM10	pcs/cm3	12–35	Air quality
Pollen	Pol	Grains/m3	50–200	Allergic
Humidity	Hum	%	0–100	Air quality
Temperature	T	°C	0–100	Air quality

concepts (i.e., diagnosis, diseases, risk factors, etc.), and entities to represent the monitoring stations, the metrics that they can measure, their location within a geographical area, and their ratio were included. The result of this process was an ontology that contains 286 concepts, 8 object properties and 10 data properties. An excerpt of the ontology is shown in Fig. 2.

Finally, data mining techniques, such as support vector machines (SVM) and Bayes Network (BayesNet), were applied to detect and forecast high concentrations of allergens. This information is sent to the user through the user interface module via alerts and notifications allowing the use of different channels such as emails, SMS or push notifications. An example of alert is: "Tomorrow, the concentration of pollen in the air may exceed the permissible health limits at the University of Murcia". This kind of alerts help the users of the application to prevent the exposure to the allergens.

2.4   User Interface

The user interface of the application relies on web technologies allowing a multi-device access through web browsers. The system requires users to create an account and provide personal information about their allergic diseases. This information is stored into the ontological model by creating user profiles. The communication between the user interface and the server is done through the public REST API. The security and confidentiality of user data are ensured by encrypting the data with JSON Web Tokens and transmitting the information under TLS.

The user interface is divided into two sections, namely, the user profile area, and the map area. In the user profile area, users register their allergic diseases. In the map area, users can view a heatmap of their nearby area with the current concentrations of allergens and air-quality. In this section of the interface, users can also specify where they want to go by manually entering the location. Once the users press the "Start Route" button, they are shown the best route, in terms of minimum exposure time to allergens, as generated by the processing module.

The user interface is also in charge of notifying users when risk factors, according to the calculations in the processing module, are present.

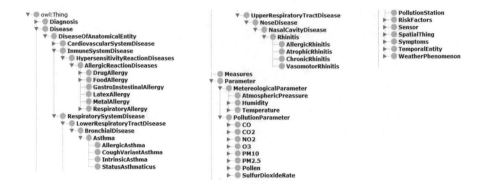

Fig. 2. Schema of the main classes of the ontology

3   Evaluation

To evaluate the prototype, a real testbed was designed in the Campus of Espinardo of the University of Murcia (Spain). For this experiment, two air-quality monitoring stations were placed within the campus to collect real data. The locations of these stations, as well as the data collected from the Information Gathering Module, can be seen in Fig. 3, where the blue areas represent the location with the highest concentration of allergens. Then, students at the university were asked to perform a daily routine that included walking around 1.3 km each day for a period of 5 days. The origin of the route was the Faculty of Computer Science and destination was the Faculty of Economics. Before each walk, the students should manually select in the prototype the Faculty of Economics as the journey's end. As they were walking towards the destination, the application showed a hotspot map including the air pollution metrics, and suggested the route through which the subject is less time exposed to allergens.

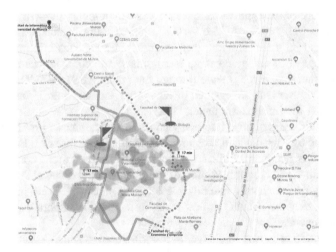

Fig. 3. A map representation of the validation experiment

To validate the model, we extracted the air-quality and pollution metrics from each station and manually checked if the route was chosen correctly. On the other hand, to evaluate the user interface, we conducted a survey to each student to obtain feedback. In the survey the students should rate a number of features from 1 to 5. The issues considered in the survey were related to the usefulness of the app, its usability and performance. Despite the fact that the results obtained seemed promising, we are already working on new, more complex real scenarios encompassing high traffic concentration.

4   Conclusions and Future Work

In this work, a functional prototype of the AllergyLESS system is presented. The main aim of this system is to suggest users with walking routes in which pollution levels are as low as possible. This prototype is fully functional with all components finished and working. When the validation period ends, we will have a final prototype ready to be offered to the market.

Even though the results of the experiments carried out so far are promising, it is necessary to design new, more complex real scenarios encompassing high traffic concentration. The goal would be to reproduce real life situations. It is also left for future work to evaluate the suitability of the application for users of different age ranges. So far only university students have been surveyed. A more representative sample of the population should be considered. Finally, it may be beneficial to publish the information gathered in an anonymous way, allowing public authorities to take strategical decisions. In this way, the system will be contributing to social corporate responsibility.

Acknowlegements. This work has been supported by the Comunidad Autónoma de la Región de Murcia (CARM) and the European Regional Development Fund (FEDER/ERDF) through project 2I16SAE00025 under the RIS3MUR program.

References

1. Caragliu, A., Del Bo, C., Nijkamp, P.: Smart cities in Europe. J. Urban Technol. **18**(2), 65–82 (2011)
2. Solanas, A., Patsakis, C., Conti, M., Vlachos, I.S., Ramos, V., Falcone, F., Martinez-Balleste, A.: Smart health: a context-aware health paradigm within smart cities. IEEE Commun. Magaz. **52**(8), 74–81 (2014)
3. Zanella, A., Bui, N., Castellani, A., Vangelista, L., Zorzi, M.: Internet of things for smart cities. IEEE Internet Things J. **1**(1), 22–32 (2014)
4. Metral, C., Falquet, G., Karatzas, K.: Ontologies for the integration of air quality models and 3D city models. arXiv preprint arXiv:1201.6511 (2012)
5. Brandt, E.B., Myers, J.M.B., Ryan, P.H., Hershey, G.K.K.: Air pollution and allergic diseases. Curr. Opin. Pediatr. **27**(6), 724 (2015)

6. World Health Organization: White Book on Allergy 2011-2012 Executive Summary. By Prof. Ruby Pawankar, MD, PhD, Prof. Giorgio Walkter Canonica, MD, Prof. Stephen T. Holgate, BSc, MD, DSc, FMed Sci and Prof. Richard F. Lockey, MD (2012)
7. Crystal-Peters, J., Crown, W.H., Goetzel, R.Z., Schutt, D.C.: The cost of productivity losses associated with allergic rhinitis. Am. J. Manag. Care **6**(3), 373–378 (2000)
8. Muraro, A., Clark, A., Beyer, K., Borrego, L.M., Borres, M., Carlsen, K.L., Wickman, M.: La atención al niño alérgico en la escuela: Grupo de Trabajo EAACI/GA2LEN sobre el niño alérgico en la escuela (2010)
9. Chandrasekaran, B., Josephson, J.R., Benjamins, V.R.: What are ontologies, and why do we need them? IEEE Intell. Syst. Appl. **14**(1), 20–26 (1999)
10. Ruiz-Martínez, J.M., Valencia-García, R., Martínez-Béjar, R., Hoffmann, A.: BioOntoVerb: a top level ontology based framework to populate biomedical ontologies from texts. Knowl. Based Syst. **36**, 68–80 (2012)
11. Rodríguez-González, A., Alor-Hernández, G.: An approach for solving multi-level diagnosis in high sensitivity medical diagnosis systems through the application of semantic technologies. Comput. Biol. Med. **43**(1), 51–62 (2013)
12. Bally, J., Boneh, T., Nicholson, A.E., Korb, K.B.: Developing an ontology for the meteorological forecasting process. In: Decision Support in an Uncertain and Complex World: The IFIP TC8/WG8, vol. 3 (2004)
13. Oprea, M.M.: AIR_POLLUTION_Onto: an ontology for air pollution analysis and control. In: IFIP International Conference on Artificial Intelligence Applications and Innovations, pp. 135–143. Springer, Boston (2009)
14. Kofler, M.J., Reinisch, C., Kastner, W.: An ontological weather representation for improving energy-efficiency in interconnected smart home systems. In: ASM 2012. ACTA Press, Naples (2012)

Guided Evolutionary Search for Boolean Networks in the Density Classification Problem

Thiago de Mattos[1,2($\boxtimes$)] and Pedro P. B. de Oliveira[2]

[1] PPGEEC, Universidade Presbiteriana Mackenzie, São Paulo, SP, Brazil
`thiagode.mattos@mackenzista.com.br`
[2] FCI, Universidade Presbiteriana Mackenzie, São Paulo, SP, Brazil
`pedrob@mackenzie.br`

Abstract. Boolean networks consist of nodes that represent binary variables, which are computed as a function of the values represented by their adjacent nodes. This local processing entails global behaviors, such as the convergence to fixed points, a behavior found in the context of the density classification problem, where the aim is the network's convergence to a fixed point of the prevailing node value in the initial global configuration of the network; in other words, a global decision is targeted, but according to a constrained, non-global action. Here, we rely on evolutionary searches in order to find rules and network topologies with good performance in the task. All nodes' neighborhoods are assumed to be defined by non-regular and bidirectional links, and the Boolean function of the network initialized by the local majority rule. Two evolutionary searches are carried out: first, in the space of network topologies, guided by a parameter (ω) related to the 'small-worldness' of the networks, and then, in the space of Boolean functions, but constraining the network topologies to the best family identified in the previous experiment. The results clearly make it evident the key and successful role of the ω parameter in looking for solutions to the task at issue.

Keywords: Boolean networks · Cellular automata
Density classification · Evolutionary computation

1 Introduction

It is common in nature to observe systems that exhibit behaviors or characteristics that emerge globally out of small interactions among their constituent parts. These are systems where no central entities are responsible for controlling and transmitting information. Boolean networks are a suitable model for those systems [9], and for the present work they are used to address the well-studied density classification task, where, from an initial configuration, the node values are reevaluated by a predetermined number of time steps, until a fixed point is achieved in the network, that should reflect the predominance of node values

F. De La Prieta et al. (Eds.): DCAI 2018, AISC 800, pp. 69–77, 2019.
https://doi.org/10.1007/978-3-319-94649-8_9

in the initial network configuration. Clearly, this represents the solution, so to speak, of a global problem, constrained by the non-global action taken by the network's nodes.

Our first experiment is an evolutionary search towards network topologies that would perform well in the density classification. Initially, small-world networks are obtained through connection rewirings [9], where some small extent connections are replaced by larger ones. In [4], it is possible to observe how the information flow is affected by the presence of rewired edges in small-world networks and how the topology is related to individual and collective behaviors.

Differently from the experiments in [4], where a homogeneous neighborhood size was adopted to make the Boolean functions evolve easier, here the evolutionary searches start from regular networks with ring topology, which are transformed into small-world networks with irregular neighborhood sizes, similarly to [8]. Also, the network Boolean function is kept fixed, which in this case is represented by the local majority rule. Unlike the experiments in [1], that used evolutionary searches from regular rings and edge additions with probability of 10% for each node, our approach relied on an evolutionary search for network topologies belonging to a specific value range of the ω parameter, which is related to the 'small-worldness' of the networks. All networks contain the same number of edges of a regular graph of degree $k = 8$, avoiding forced perturbations such as additions/removal of edges, and allowing only edges rewirings conditioned to a certain probability.

From the previous results, that indicated the best family of network topologies, an evolutionary search for Boolean functions (local rules) was then performed, aiming at the performance improvement in the task. It is relevant to notice that in order to decrease the search effort, all nodes in the network were initialized by a stochastic version of the local majority rule, where, in the case of a tie, one of the binary values is chosen with the same probability.

In the next section, the fundamental concepts of small-world networks and their particularities are discussed. Following, the density classification task and the evolutionary mechanisms are briefly presented, outlining how those mechanisms can be used in large search spaces, where the use of traditional algorithms is unpractical. While Sect. 3 details the evolutionary mechanism adopted in the experiments, Sect. 4 details the experiments performed, as well as their results. We conclude in Sect. 5, discussing our findings and raising perspectives for possible follow-ups.

2   Background Knowledge

2.1   Small-World Networks

Small-world networks are distinct from other network topologies mainly because of two properties: low average minimum path length $L(p)$, just like in random networks, and high average clustering coefficient $C(p)$, a property found in regular networks [9]. The average minimum path is a property calculated by averaging the size of all possible minimum paths between two nodes, and measures the efficiency of information spread on networks. The network's average cluster-ing coefficient is a property calculated by averaging the clustering coefficients from each node $(C(v))$, which is a local property that measures the cohesion degree of node v with the nodes it relates to. As an example, a social network with some individuals belonging to small friendship circles (clusters), permits the existence of friendship bonds between individuals from different circles, ensuring that information transits quickly through the entire network.

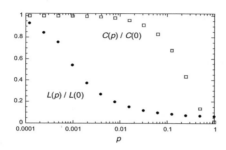

Fig. 1. The average minimum path length $L(p)$ and the average clustering coefficient $C(p)$ as a function of the probability variation p used in the edge rewiring operation, adopted from [9]. All plotted points have been normalized by the values L(0) and C(0), from a regular lattice.

A model to generate small-world networks is proposed in [9], where edges from a regular graph can be rewired based on a certain probability $0 \leq p \leq 1$. Those resulting networks can be situated at some point between the structural spectrum of regular and random networks (Fig. 1).

A difficulty arises in ranking small-world networks not obtained by means of the approach in [9], since it is not possible to quantify the network 'small-worldness' in the absence of the probability p. To circumvent this situation, net-works with the desired characteristics were artificially obtained, having their L and C values compared to the ones related to the networks subject of study. Among those benchmark measurements, the omega parameter (ω) has been favored [7]. It is calculated as a function of the average minimum path value of a random network with the same node degree distribution (L_{random}), and the average clustering coefficient of a ring type network (C_{ring}), with the desir-able characteristics for comparison. The ω value is defined as $\frac{L_{random}}{L} - \frac{C}{C_{ring}}$ and is situated in the range from -1 to 1. Networks with the most remarkable small-world features ($L \approx 0$ and $C \approx 1$) have $\omega \approx 0$. Positive ω values denote networks with more random characteristics ($L \approx L_{random}$ e $C \ll C_{ring}$), and negative ω values refer to networks with closer characteristics to ring networks ($L \gg L_{random}$ and $C \approx C_{ring}$).

2.2   Density Classification Task

The Density Classification Task (*DCT*) is a problem that measures the computational capacity of a Boolean network to evaluate the state predominance of their nodes. Basically, given any Boolean network, it is asked what it is the predominant state in the whole network, or, in other terms, the state with the highest density. This question might be trivially answered, by simply counting all nodes states, provided this possibility would be allowed; however, the problem specifies that such an answer must be obtained by evaluating each node individually, using its transition rule by a predetermined number of time steps, so that the whole network should converge to a fixed point representing the prevailing state of the initial configuration. In other words, given a Boolean network with any binary initial combination (nodes in state 0 or 1), processed by a predetermined number of time steps, it will converge all nodes to state 0 if the initial configuration has a predominance of zeros, or to state 1, otherwise.

It is known that the *DCT* cannot be perfectly solved with two state cellular automata (a Boolean network specialization) with neighborhood of any size [6]. However, by relaxing some of the restrictions, it is possible to find ways that provide almost perfect solutions; An extensive survey on the various possible reformulations of the problem is presented in [3].

2.3   Evolutionary Search

For certain classes of problems where the search space is very large, good solutions cannot be found in acceptable time by traditional algorithms. According to [5], evolutionary mechanisms present in acceptable time, solutions close to the optimal, if not the optimal solution itself. Each mechanism iteration represents a generation, where the evaluation, selection, reproduction and mutation of individuals is carried out, with the reproduction and mutation operators conditioned to a probability. As the generations advance, the best individuals are preserved, making the search converge to a near-ideal solution. The mechanism automatically discards unsuitable individuals, just as natural selection deprives less adapted living beings in nature.

3   Methodology

Based on the evolutionary mechanisms described in [1,2,8], two experiments are carried out herein: evolutionary searches for network topologies (Sect. 4.1) and rules (Sect. 4.2), where the objective in both is to find networks with good performance in the density classification task.

In the topology evolutionary search, an initial population containing n individuals is created, each one representing a regular Boolean network with degree k, containing 149 nodes and 1000 different initial state configurations. Then, for each of the individuals, the edge rewiring operation described in [9] is performed and each individual fitness is evaluated, this value being represented by the average of correct answers for the density classification task. An elitism operation is then carried out, keeping the individuals with the best fitness for the next generation; then, a standard *roulette wheel* selection [5] is performed in the individuals out of the elite. Next, the mutation operation is performed, where, for each network node, one of its neighbors is randomly changed, conditioning such operation to a probability p_r. All the previous process is repeated for z generations. Similarly, the same strategy is adopted in the rule evolutionary search, where individuals are rules, instead of topologies.

For both searches, a population with 100 individuals is processed by 200 generations in the topology evolutionary search and by 100 generations in the rule evolutionary search. Regarding statistical distribution of densities in the initial configurations, two types are employed: binomial and nearly-balanced [3], the latter meaning that the number of 0s and 1s differs only by one unit.

Another point to highlight about the topology evolutionary search is that only the network topology undergoes evolution, while the node transition rule (Boolean function) remains fixed, and it is represented by the majority rule over a node's neighborhood, which leads to the majority state among those in the adjacent nodes (in the case of a tie, 0 or 1 is chosen, with 50% probability for each one).

In the rule evolutionary search, the rule itself undergoes evolution and the network topology is kept fixed, namely, the ones found in the previous experiment. At the starting point, the Boolean function is represented by the neighborhood majority state, using the lexicographical representation of cellular automata rules described in [10].

In both experiments, we have chosen the neighborhood majority rule as the starting point, instead of random rules, as a way to reduce the computational effort for the searches.

4   Experiments and Results

4.1   Topology Evolutionary Search in Specific ω Value Ranges

In this experiment, the evolutionary mechanism is executed with 10% of elitism and mutation probability of 1%, using networks belonging to specific ω value ranges. This experiment tries to identify ω value ranges where networks with good performance in the density classification task exist. The C_{ring} value used in ω calculation (Sect. 2.1) is obtained from a regular network of degree $k = 8$, with ring topology, that will act as a benchmark individual for all other topologies found by the evolutionary mechanism.

According to the initial configurations, the binomial distribution type yields results of up to 88.2% for DCT (Table 1), and the results get higher as $\omega \to 1$, that is, ω values referring to networks with characteristics closer to the random ones. In all executions, we have identified small performance decreases across generations. This is naturally due to the renewal of the initial configurations at each generation, which do not compromise fitness convergence.

As expected, the nearly-balanced distribution yields lower values for DCT compared to the binomial distribution. That happens because the nearly-balanced distribution constitutes the most difficult configuration for DCT [3].

Table 1. Results for the topology evolutionary search. Each execution starts around the lower range of ω, and evolution proceeds respecting the ω range. At the final generation, all executions reach an ω value close to the upper bound.

Execution	p	ω range	Binomial		Nearly Bal.	
			ω	DCT (%)	ω	DCT (%)
1	0.00000	[-0.75, -0.50]	-0.503	0.2	-0.731	0
2	0.00267	[-0.50, -0.25]	-0.250	0.3	-0.497	0
3	0.00969	[-0.25, 0.00]	-0.001	0.7	-0.243	0
4	0.02710	[0.00, 0.25]	0.248	3.6	0.249	0.2
5	0.05320	[0.25, 0.50]	0.500	28.4	0.497	2.2
6	0.08965	[0.50, 0.75]	0.750	79.6	0.750	38.5
7	0.14500	[0.75, 1.00]	0.936	88.2	0.883	52.6

4.2   Rule Evolutionary Search

According to the previous results, the topology evolutionary search using exclusively the majority rule undergoes eventually into some sort of limitation. In order to overcome this, a rule evolutionary search is carried out, starting from the neighborhood majority rule and using the network topologies found in the first experiment (Sect. 4.1). The rules are now represented by bit sequences, accordingly to the lexicographical representation defined in [10]. Since the search is dealing with irregular neighborhoods, each neighborhood size will have an associated bit sequence, allowing the rule evolution for each specific neighborhood size.

Differently from the first experiment, this time the mechanism performs reproduction, that yields all possible pairs among individuals. For each pair, a single random point in both bit sequences is used to perform the crossover [5], creating in this way two new individuals. All individuals will undergo mutation, with each bit being flipped with probability of 0.1%.

As displayed in Table 2, the rule evolutionary search was effective in improving the performance for the DCT. For the binomial distribution type, the most substantial performance improvement can be found in execution 5 (Fig. 2), that used a network with $\omega = 0.5$. It can be noticed a performance transition from 28.4% in the first experiment, to 85.8% in the second. The same network showed for the nearly-balanced distribution type a really substantial performance improvement, going from 2.2% in the first experiment to 54.1% in the second.

Table 2. Results for the rule evolutionary search.

Execution	Binomial		Nearly Bal.		Execution	Binomial		Nearly Bal.	
	ω	DCT (%)	ω	DCT (%)		ω	DCT (%)	ω	DCT (%)
1	-0.503	52.9	-0.731	54.8	5	0.500	85.8	0.497	54.1
2	-0.250	53.8	-0.497	53.4	6	0.750	88.8	0.750	61.0
3	-0.001	53.1	-0.243	54.9	7	0.936	89.5	0.883	62.0
4	0.248	53.6	0.249	54.0					

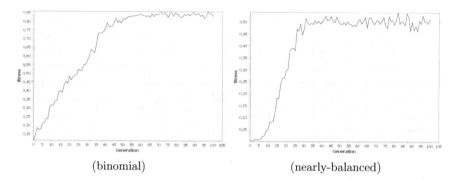

(binomial)   (nearly-balanced)

Fig. 2. Fitness evolution for execution 5 in the rule evolutionary search. For the binomial distribution, the fitness starts around 0 and not at 0.284 as expected (see Table 1). This happens because now the Boolean function is encoded as a bit sequence, instead of the simple state counting adopted in the first experiment, which yields different results even if the encoded bit sequence initially represents the local majority rule. Only a few generations are needed to adjust the bit setting in order to yield values around 0.284 again.

4.3   *DCT* Stress Test for the Best Individuals

A *DCT* stress test is performed on the best individuals obtained in the previous experiments. This is carried out in order to guarantee consistency of the results.

Table 3. *DCT* stress test for the best individuals from both experiments. Each individual was tested over $1,000,000$ (one million) initial configurations with binomial and nearly-balanced distributions.

Search for Network Topologies				Search for Rules			
ω	Binomial (%)	ω	Nearly Bal. (%)	ω	Binomial (%)	ω	Nearly Bal. (%)
-0.503	0.0095	-0.731	0	-0.503	50.0435	-0.731	50.0017
-0.250	0.0267	-0.497	0	-0.250	50.0018	-0.497	49.9969
-0.001	0.1065	-0.243	0	-0.001	50.0354	-0.243	49.9693
0.248	1.8039	0.249	0.0015	0.248	50.0869	0.249	50.0217
0.500	24.2732	0.497	1.0218	0.500	82.0799	0.497	50.0681
0.750	75.6350	0.750	33.5844	0.750	84.1779	0.750	55.6484
0.936	84.4725	0.883	47.0992	0.936	85.3244	0.883	55.7184

In all executions, we noticed that the DCT evaluation remains as expected, with small oscillations in the values, which guarantees the consistency of the previously obtained results (Table 3).

5   Conclusions

We have identified the effectiveness of using the omega parameter (ω) as a guide to evolutionary searches for network topologies in the context of the density classification task (DCT). The results show that the adopted approach explores the vast space of possible topologies more assertively than just observing and comparing the values of the average minimum path (L) and the average clustering coefficient (C). The best topologies found are close to a random network, without actually being one. In the DCT context, the information spreading on the network is a key success factor, and random networks do that efficiently, but the problem with them is that they lack robustness, in the sense that a failure at some edge or node of a random network $(\omega = 1)$ can affect badly the final DCT result, since the nodes are not well clustered like in the topologies with $\omega < 1$.

In all experiments, we have identified the need for a small number of time steps to obtain a correct value in the density classification task. On average, it took 20 time steps, much less than the 298 time steps limit that was used in [1,8]. The value of 298 time steps is reached only in cases when the DCT fails, and the reason for this is that at certain moments the density information transmitted over the network is compromised, due to some particularity of the topology and/or the evaluation rule (the Boolean function).

The evolutionary search for rules initialized with the local majority rule led to a substantial improvement in the DCT performance, demonstrating that sequencing the evolutionary search for topologies with the latter is a definitely effective search strategy. As for the fitness values found in the search for topologies, the evolutionary mechanism combined with the omega parameter led to a score of 88.2%, thus exceeding the 82.3% found in [1].

Acknowledgements. We express our gratitude to CAPES (Coordenação de Aperfeiçoamento de Pessoal de Nível Superior - Brazil), for providing a student grant to the first author.

References

1. Chira, C., Andreica, A.: Network topologies for cellular automata computation. Emerg. Complex. Comput. **8**, 271–281 (2014)
2. Darabos, C., Giacobini, M., Tomassini, M.: Performance and robustness of cellular automata computation on irregular networks. Adv. Complex Syst. **10**(supp01), 85–110 (2007)
3. De Oliveira, P.P.B.: On density determination with cellular automata: results, constructions and directions. J. Cell. Autom. **9**(5–6), 357–385 (2014)
4. Godoy, A., Tabacof, P., Von Zuben, F.J.: The role of the interaction network in the emergence of diversity of behavior. PLoS ONE **12**, e0172073 (2017)

5. Jong, K.A.D.: Evolutionary Computation - A Unified Approach. MIT Press, Cambridge (2006)
6. Land, M., Belew, R.: No two-state CA for density classification exists. Phys. Rev. Lett. **74**, 5148–5151 (1995)
7. Telesford, Q.K., Joyce, K.E., Hayasaka, S., Burdette, J.H., Laurienti, P.J.: The ubiquity of small-world networks. Brain Connect. **1**, 367–375 (2011)
8. Tomassini, M., Giacobini, M., Darabos, C.: Evolution and dynamics of small-world cellular automata. Complex Syst. **15**(4), 261–284 (2005)
9. Watts, D.J., Strogatz, S.H.: Collective dynamics of 'small-world' networks. Nature **393**, 440–442 (1998)
10. Wolfram, S.: Universality and complexity in cellular automata. Phys. D **10**(1–2), 1–35 (1984)

Peculiarity Classification of Flat Finishing Motion Based on Tool Trajectory by Using Self-organizing Maps

Masaru Teranishi$^{(\boxtimes)}$, Shimpei Matsumoto, and Hidetoshi Takeno

Hiroshima Institute of Technology, 2-1-1, Miyake, Saeki-ku, Hiroshima, Japan
teranisi@cc.it-hiroshima.ac.jp

Abstract. The paper proposes an unsupervised classification method for peculiarities of flat finishing motion with an iron file, measured by a 3D stylus. The proposed method extract personal peculiarities based on trajectory of an iron file. The classified peculiarities are used to correct learner's finishing motions effectively for skill training. In the case of such skill training, the number of classes of peculiarity is unknown. A torus type Self-Organizing Maps is effectively used to classify such unknown number of classes of peculiarity patterns.

Experimental results of the classification with measured data of an expert and sixteen learners show effectiveness of the proposed method.

Keywords: Motion analysis · Feature extraction
Self-organizing maps · Clustering

1 Introduction

In the technical education of junior high schools in Japan, new educational tools and materials are in development, for the purpose to transfer kinds of crafting technology. When a learner studies technical skills by using the educational tools, two practices are considered to be important: (1) to imitate motions of experts and (2) to notice their own "Peculiarity", and correct it with appropriate aids.

However, present educational materials are not yet effective to assist the practices because most materials consist still or motion pictures of tool motions of experts. Even though the learners could read out rough outlines of the correct motion from these materials, it is difficult to imitate detailed motion due to less information these materials have. Especially, it is most difficult to imitate fine motions and postures of the tool by only looking the expert motions from a fixed viewpoint. Furthermore, the learners could not recognize their own "peculiarity" as a difference between the expert's motion and the learner's motion from the materials. To solve the problem, a new assistant system for a brush coating skill has developed [1] as the related work of this study. The system presents a learner corrective suggestion by play-backing the learner's motion by using animated 3D Graphics.

© Springer International Publishing AG, part of Springer Nature 2019
F. De La Prieta et al. (Eds.): DCAI 2018, AISC 800, pp. 78–85, 2019.
https://doi.org/10.1007/978-3-319-94649-8_10

The paper describes feature extraction and classification functional part of the development of a new technical educational assistant system [2] that let learners acquire a flat finishing skill with an iron file. The system measures a flat finishing motion of a learner by a 3-D stylus device, and classifies the learner's "peculiarity". The system assists the learners how to correct bad peculiarities based on classified "peculiarity". The paper mainly describes the learner's peculiarities classification method which will be implemented in the proposed system. The authors have already proposed a peculiarity classification method based on velocity information of an iron file motion [3]. The paper focuses on trajectory of an iron file in flat finishing motion as another learners' peculiarity information. A torus type Self-Organizing Maps(SOM) [4,5] is proposed to classify learners' "peculiarity" effectively.

2   Motion Measuring System for Flat Finishing Skill Training

Figure 1(a) shows an outlook of the motion measuring devices of the system for flat finishing skill training. The system simulates a flat finishing task that flatten a top surface of an pillar object by an iron file. The system measures a 4D(time series of 3D)motion data of the file by using a 3D stylus. We use the PHANTOM Omni (SensAble Technologies) haptics device as the 3D stylus component of the system. The file motion is measured by attaching the grip of the file to the encoder stylus part of the haptics device. Assuming that the system will have a force feed back teaching function, we use a light weight mock file made of an acrylic plate which imitates a real 200 mm length iron file [2].

In the measurement task, a learner operates the mock file in order to flatten the top surface of a dummy work whose area has 25 mm width and 25 mm depth, at 80 mm height. The system measures the motion of the mock file. The measured motion is recorded as a time series of the position of the file with X, Y, and Z coordinate values, and the posture with the tilt angles along the three axes with Tx, Ty, Tz as the radians. The spatial axes of the operational space and tilt angles of the file are assigned as shown in Fig. 1(b).

(a)

(b)

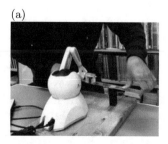
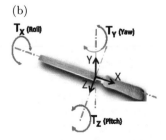

Fig. 1. Flat finishing skill measuring system: (a) outlook, (b) coordinates of measuring.

Figure 2 shows an example of an expert's motion measured by the system. In these plots, the expert operates the file with reciprocating motion 4 times within 20 s. The main direction of the reciprocating motion is along X axis. The expert pushes the file to the direction that X reduces, and pull the file to the opposite direction. Since the file works only in the pushing motion, we focused the pushing motion in the classification task in the rest part of the paper. The rest axes Y, Z are: Y and Z axes correspond vertical and horizontal directions of file along main motion direction of X axis.

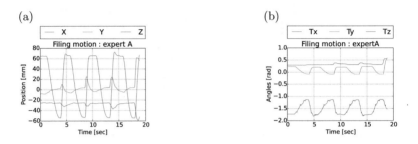

Fig. 2. Filing motion of an expert: (a) tool position, (b) posture of the tool.

To classify every file-pushing motion as the same dimensional vector by the SOM, we have to process measured data by two steps: (1) clipping out each time series potion of file-pushing motions, and (2) re-sampling the time series portions. **(1) Clipping:** The clipping is done according to the file-pushing motion range which is defined in X coordinates, beginning with X_{begin}, and ending with X_{end}. **(2) Re-sampling:** The resulted clipped time series of the pushing motions have different time lengths. So we could not use the series directly to the SOM because the dimensions of each series differ. Therefore, we should arrange dimensions of all series to the same number. Every series are re-sampled in order to have the same number of the sampling time points, by using the linear interpolation, as shown in Fig. 3(a). Since the stylus samples the motion enough fast, we can use the linear interpolation in the re-sampling with less lose of location precision. Figure 3(b) shows the re-sampled result of the expert motion, Fig. 3(c) is that of a learner. In each plot, all four motions are imposed.

There is relevant differences between the expert's motions and the learner's motions. Every motion of the expert draws almost the same shaped curve, but the learner's don't. Each curves of learner's motion differs each time, in spite of identical person's actions. Among the three coordinate components of a motion, we have extracted peculiarity for X coordinate motion based on velocity information in our former work [3], because the X coordinate motion are main direction of file motion. The rest Y and Z coordinates motions are also useful as peculiarity: it might offset from the correct motion, or some meaningful motion along main motion direction. In this paper, we use Y and Z coordinates values of the measured data for the classification. The aim of the classification study is to form the SOM to organize such different motion curves in the map. After that, taking and using code-books of the learner's motion as the "bad peculiarities".

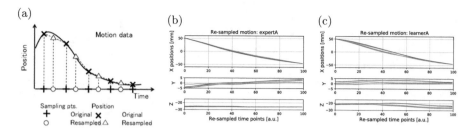

Fig. 3. (a) Re-sampling of a motion pattern, and re-sampled motion: (b) an expert, (c) a learner.

3 New Feature Extraction Based on Y-Z Trajectory of Tool

Although the filing motion data in Y, Z direction have useful information about peculiarities, it is difficult to tell the learner their peculiarities exactly by only using the motion curve. Instead, to display the peculiarities on the basis of trajectory is an effective way. Each position data (y, z) at a time point denotes the grip position in a Y-Z intersection at the time. Since Y and Z motions are smaller than X motion, we obtain the Y-Z trajectory as the time series of (y, z) positions. The resulted Y-Z trajectory is also projection of a 3D trajectory onto Y-Z plane. The Y-Z trajectories of an expert and a learner are shown in Fig. 4. In Fig. 4, the beginning positions of Y-Z trajectories are aligned to $(0,0)$, by subtracting the beginning (y, z) coordinate value from the rest part of (y, z) values for each Y-Z trajectory. As shown in Fig. 4, by plotting Y-Z trajectory in a Y-Z plane, leaners could well recognize Y and Z motion rather than plotting Y and Z time series respectively as shown in Fig. 3.

In our former work [3], the peculiarity of a learner is represented by taking difference between the learner and experts' ideal motion. In this paper, we use Y-Z trajectory directory to peculiarity classification and visualization. Because the expert's Y-Z motion of Fig. 4 tells that the ideal motion consists of a certain move in vertical direction(Y) and no motion in horizontal direction(Z). We think that in case of the Y-Z motion, taking difference between the leaners Y-Z motion and that of expert may let the learner get difficult to recognize their peculiarity. So we use the Y-Z trajectory data directly, without taking difference from that of the expert.

4 Peculiarity Classification by Torus Type SOM

We classify the file motion data by using the Self-Organizing Maps (SOM). Technical issues about the classification of the file motion are considered in three major points: (1)peculiarities are implicitly existing among the motion data, (2) the number of peculiarity variations, i.e., the number of classes is unknown and (3) every motions are not exactly same even though in the same learner, there are some fluctuation in each motion.

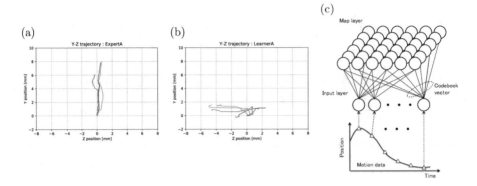

Fig. 4. Y-Z trajectories of (a) an expert, (b) a learner, and (c) structure of SOM.

4.1   Structure of SOM

SOM is one of effective classification tools for patterns whose number of classes is unknown and whose classification features are implicit. Figure 4(c) shows the structure of the SOM. The SOM is a kind of neural networks which have two layers: one is the input layer, the other is the map layer. The input layer has n neuron units, where n is the dimension of input vector $\boldsymbol{x} = (x_1, x_2, \cdots, x_n)^T$. In this paper, the input vector $\boldsymbol{x}$ consists of a Y-Z trajectory with l re-sampled points, $\boldsymbol{x} = (y_1, y_2, \cdots, y_l, z_1, z_2, \cdots, z_l)^T$, where $n = 2l$. The map layer consists of neuron units, which is arranged in 2D shape. Every unit of the map layer has full connection to all units of the input layer. The ith map unit u_i^{map} has full connection vector $\boldsymbol{m}_i$ which is called "code-book vector". The SOM classifies the input pattern by choosing the "firing code-book" $\boldsymbol{m}_c$ which is the nearest to the input vector $\boldsymbol{x}$ in the meaning of the distance defined as: $\|\boldsymbol{x} - \boldsymbol{m}_c\| = \min_i\{\|\boldsymbol{x} - \boldsymbol{m}_i\|\}$.

The SOM classifies the high dimensional input pattern vector according to the similarity to the code-book vectors. The map units also arranged in two dimensional grid like shape, and neighbor units have similar code-book vectors. Therefore, the SOM is able to "map" and visualize the distribution of high dimensional input patterns into a simple two dimensional map easily. Since the SOM also could form a classification of patterns automatically by using "Self-Organizing" process described in the later section. Therefore, we need not to consider the number of classes.

The SOM organizes a map by executing the following three steps onto every input pattern: (1) first, the SOM input a pattern, (2) then it finds a "firing unit" by applying Eq. (4.1) to every code-book vector $\boldsymbol{m}_i$, (3) and it modifies code-book vectors of the "firing unit" and its neighbors. In the step (3), code-book vectors are modified toward the input pattern vector. The amount of modification is computed by the following equations, according to a "neighbor function" h_{ci} which is defined based on a distance between each unit and the firing unit. $\boldsymbol{m}_i(t+1) = \boldsymbol{m}_i(t) + h_{ci}(t)\{\boldsymbol{x}(t) - \boldsymbol{m}_i(t)\}$ where t is the current and $t+1$ is the next count of the modification iterations. The neighbor function h_{ci} is a function to limit modifications of code-book vectors to local map units which are neighbor the firing unit. The proposed method uses "Gaussian" type neighbor function. The Gaussian type modifies code-book vectors with varying amounts that decays like Gaussian function, proportional to the distance from the firing unit [4]. The reason why we use the Gaussian function is based on the assumption that the Y-Z trajectories distribute continuously in the feature space.

4.2   Torus Type SOM

When we use the SOM straightforward, a "Map edge distortion" problem often occurs. The problem is observed as over-gathering of input data to code-books which are located edges of the map. Therefore, the classification performance become worth due to the each of the edge code-books represents more than one appropriate class of the input data. We have confirmed the problem in the early development of the system [2]. The problem is caused based on the fact that there is no code-book outside the edge, therefore the edge code-book can't be modified to appropriate direction in the feature space. We solve the problem by introducing the torus type SOM [4,5]. In the torus SOM, each code-book has cyclic neighbor relation in the feature map. In Fig. 4(c), every code-book at the edges of the map adjoins ones at the opposite edges. Since the torus SOM has no map edge, the map is free from edge distortion problem.

5   Clustering Torus SOM Result

The distribution of code-books of the torus type SOM help the skill trainers to look out the peculiarities distribution of learners. We think that a grouping facility of the code-books gives the trainers more efficiency in case of many learners they have. Therefore, the resulted code-books of the torus type SOM are divided into some countable clusters. The proposed system could be more helpful for the trainers by providing such clustering information as types of peculiarities. The trainer could teach according to each peculiarity type effectively in short time. Additionally, such clustering result resolves ambiguous boundary of feature map, which one of disadvantage the torus type SOM has. For this purpose, we introduce automatic clustering method of code-books of the torus type SOM [6].

The clustering method divides code-books into clusters automatically according to densities of the code-books in the feature space. The densities of code-book is named "cluster map". A cluster map value $d(i, j)$ of the code-book $\boldsymbol{m}_{i,j}$ is calculated by

$$d(i,j) = \frac{1}{|D(i,j)|} \sum_{(\mu,\nu)} \in D(i,j)(\boldsymbol{m}_{i,j} - \boldsymbol{m}_{i-\mu,j-\nu})^T \times (\boldsymbol{m}_{i,j} - \boldsymbol{m}_{i-\mu,j-\nu}) \quad (1)$$

where $D(i,j)$ is the first order neighbor region of the map location (i, j), which includes six code-books in the case of hexagonal map topology. The amount of a cluster map value is in inverse proportion to the density of the code-book.

The clustering is done by labeling code-book location (i, j) based on the cluster map by the following algorithm:

Step 1: Sort all $d(i, j)$ and index them with numbers $q = 1, 2, \cdots$ in ascending order. Let $s(i, j) = q$ and define $(i, j) = s^{-1}(q)$.
Step 2: Set $L = 0$.
Step 3: Iterate Step 3-1 for $q = 1, 2, \cdots$.
  Step 3-1: if $d(s^{-1}(q))$ is the smallest value among its first order neighbor **then** $L = L + 1$, and assign label $\gamma(s^{-1}(q)) = L$.
  else assign $\gamma(s^{-1}(q))$ with the same label of neighbor which has the smallest d.

6   Classification Experiment

To evaluate of the effectiveness of the proposed method, we carried out experiments of the filing motion peculiarities classification. The motion data were measured with an expert and sixteen learners. Every person operated three of four filing motions. Totally we used 66 motion data for classification. We obtain Y-Z trajectories from each re-sampled motion data with $n = 100$ sampling points. The torus SOM consists the input layer with 200 units and the map layer with 49 code-book vectors, which are arranged in 2D formation with 7 by 7, hexatopology. The self-organizing process started with initial relaxation coefficient $\alpha = 0.01$, initial extent of the neighbor function $\sigma = 7$, and iterated 10,000 times with the Gaussian type neighbor function.

The Y-Z trajectories of an expert and learners are classified as shown in Fig. 5. By using the cluster map value, all data are classified into three clusters as shown in Fig. 5. The resulted three clusters display typical learners peculiarities: move

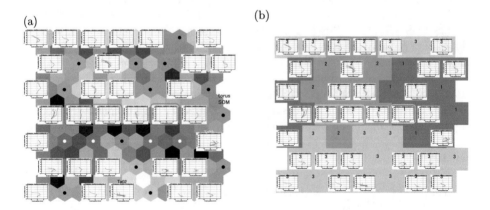

Fig. 5. (a) Resulted feature map of torus SOM. The color of hexagons display distances between two code-book vectors: dark = long, light = short. (b) Clustering result based on cluster map.

right in the former part and then return to center (Cluster 1), move left in the later part (Cluster 2), and have less move in vertical direction and move left in the later part (Cluster 3). From these results, the torus SOM works effectively to classify and visualize motion peculiarities of the learners.

7 Conclusion

The paper proposed a new motion classification method of individual learner's "peculiarities" as a part of the development of the new technical educational tool of flat finishing skill based on Y-Z trajectory of file motions. A torus Self-Organizing Maps (SOM) has proposed in order to classify the filing motion by using Y-Z trajectories into an expert's motions and "peculiarities" of learners. The automatic clustering method by using cluster map value also classify the all codebook vectors of the SOM into appropriate typical three peculiarities clusters.

To evaluate the usefullness of peculiarity feature, more other experts' motion measurements are required, and to evaluate further effectiveness of torus SOM, classification evaluation is required with more number of learner person data, and application of the method to other data components (Tx, Ty, Tz) are remained as future works.

We thank anonymous reviewers for their very useful comments and suggestions. The work was supported by JSPS KAKENHI Grant Number 17K04827.

References

1. Matsumoto, S., Fujimoto, N., Teranishi, M., Takeno, H., Tokuyasu, T.: A brush coating skill training system for manufacturing education at Japanese elementary and junior high schools. Artif. Life Robot. **21**, 69–78 (2016)
2. Teranishi, M., Matsumoto, S., Takeno, H.: Classification of personal variation in tool motion for flat finishing skill training by using self-organizing maps. In: The 16th SICE System Integration Division Annual Conference, vol. 3L2-1, pp. 2655–2669 (2015)
3. Teranishi, M., Matsumoto, S., Fujimoto, N., Takeno, H.: Personal peculiarity classification of flat finishing skill training by using torus type self-organizing maps. In: Proceedings of 14th international conference DCAI 2017, Porto, Portugal, June 2017, pp. 231–238 (2017)
4. Kohonen, T.: Self-organizing Maps. Springer, Heidelberg (2001). https://doi.org/10.1007/978-3-642-56927-2
5. Ito, M., Miyoshi, T. and Masuyama, H.: The characteristics of the torus self-organizing map. In: 6th International Conference on Soft Computing (IIZUKA 2000), pp. 239–244 (2000)
6. Tanaka, M., Furukawa, Y., Tanino, T.: Clustering by using self organizing map. J. IEICE **J79-D-II**(2), 301–304 (1996)

Classification of Human Body Smell by Learning Vector Quantization

Sigeru Omatu$^{(\boxtimes)}$

Osaka Institute of Technology, Osaka 530-8568, Japan
omtsgr@gmail.com
https://www.oit.ac.jp

Abstract. In this paper we consider classification of human body smell using learning vector quantization (LVQ). Smells of human body are classified as sweaty lockerroom smell, middle-aged smell, and age-of-smell. The first one is mainly detected for persons from teenagers to twenties, the second one is for persons from thirties to fifties, and the third one is for persons over fifties. The aim of this paper is to classify smells into three smalles stated above. The sweaty smell is a smell similar to ammonia and isovaleric acid, middle-aged smell is similar to diacetyl, and the age-of-smell is similar to nonenaar. Using a special sampling box, we train the smell sensing data such that each of those smells could be classified into true smell using LVQ. After that, we develop a hardware (Kunkun body) to classify various smell data into each smell.

Keywords: Smell measurement · Smell classification
Learning vector classification

1 Introduction

The sense of smells is one of five senses of human. Compared with other senses, its study has not been paid attention due to chemical reaction although it has been recognized for various applications in human life and industrial sectors. Based on the progress of medical research on the system of olphactory organs, artificial electronic nose (E-nose) systems have been developed. From technical and commercial viewpoints, various E-nose systems have been developed in the field of quality control of food industry [1], public safety [2], and space applications [3].

Historically, Milke [4] proved that two kinds of metal-oxide semiconductor gas sensor (MOGS) could have the ability to classify several sources of fire more precisely compared with a conventional smoke detector. However, his results achieved only 85% of correct classification by using a conventional statistical pattern classification.

An E-nose has been developed for smell classification of various sources of fire such as household burning materials, cooking smells, the leakage from the liquid petroleum gas (LPG) in [5,6]. by using neural networks of layered type.

© Springer International Publishing AG, part of Springer Nature 2019
F. De La Prieta et al. (Eds.): DCAI 2018, AISC 800, pp. 86–93, 2019.
https://doi.org/10.1007/978-3-319-94649-8_11

The purpose of this paper focuses on the classification of human body smell by using neural network of learning vector quantization (LVQ). Human body smells are the main three types such as sweaty smell, middle-aged smell, and age-of-smell according to our age. We adopt the LVQ to make easy to extend the number of neurons to achieve the pre-assigned accuracy. The research purpose is to develop new hardware to classify these three smells and their visualization using smart-phone.

2   Smell Sensors and Sensing System

Gas sensors using a tin oxide were produced in 1968 [5,6]. There are tin, iron oxide, and tungsten oxide as typical gas sensing substances which are used in metal oxide semiconductor gas sensors. The principle of MOGS is based on oxidation and reduction of metals. We will explain the principle in case of carbon oxide and tin. If we heat the oxidedtin about 250 °C, the following oxidation will occur

$$\frac{1}{2}O_2 + (SnO_{2-x})^* \rightarrow O^- ad(SnO_{2-x})$$

Here, $x = 1$ or zero, $(\cdot)^*$ denotes active state and Aad(B) means A and B are adsorbed. In this case, electron barrier is very high and large resistance appears as shown in Fig. 1. If CO gas will come, then the following reduction will occur and electron barrier is very low, which means the resistance of MOGS becomes low resistance as shown in Fig. 2.

$$CO + O^- ad(SnO_{2-x}) \rightarrow CO_2 + (SnO_{2-x})^*.$$

Therefore, using the above oxidation and reduction, resistances of MOGS become variable, that is, the MOGS works to detect whether a specific gas will be present or not. Figure 3 show the framework of measurement circuit.

In this paper, MOG sensors are used to measure the single smell. Table 1 shows the five sensors in the experiment.

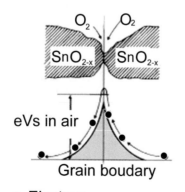

Fig. 1. Oxidation of sensor.

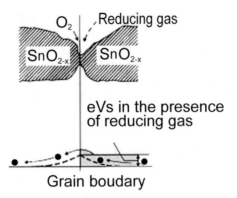

Fig. 2. Oxidation of sensor.

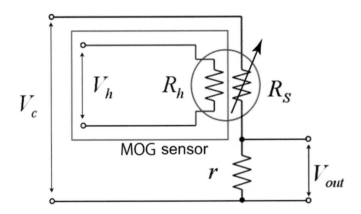

Fig. 3. Oxidation of sensor.

Table 1. Gas sensors used in the experiment.

Sensor number (Type)	Gas detection types	Applications
Sensor 1 (SB-AQ8)	Volatile, organic compound	Air quality
Sensor 2 (SB-15)	Propane, butane	Flammable gas
Sensor 3 (SB-42A)	Freon	Refrigerant gas
Sensor 4 (SB-31)	Alcol, organic solvent	Solvent

3   Neuron Model and Neural Network

We show the neuron structure in Fig. 4. The left hand side denotes inputs $(x_1, x_2, x_3, \ldots, x_n)$ and the right hand side is the output y where $w_k, k = 0, 1, \ldots, n$ show weighting coefficients. $f(\cdot)$ is an output function of the neuron. Note that $x_0 = -1$ and $w_0 = \theta$ where θ shows a threshold of the neuron. Although there are several types of the output function $f(\cdot)$, we adopt the sigmoid function.

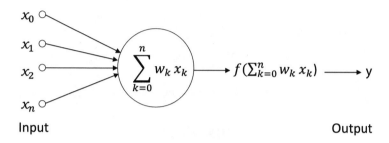

Input   Output

Fig. 4. Neuron model.

4   Measurement of Smell Data

We have measured four types of smells as shown in Table 2. The sampling frequency is 500[ms], that is, the sampling frequency is 0.5[s]. The temperatures of smell gases were 18–$24[°C]$ and the humidities of gases are 20–$30[\%]$. Note that those factors in Table 2 are main smells included in our body. Bad smell of our body will make the neighboring persons unpleasant although the person himself might not notice it.

Table 2. Training data.

Label	Substance	Samples
A	Ammonia	100
B	Diacetyl	100
C	Nonenaar	100

5   Learning Algorithm of LVQ

Let us denote the weighting vector by $\mathbf{w_j} = (w_{j1}, w_{j2}, \ldots, w_{jn})$, $j = 1, 2, \ldots, k$ where w_{ji} denotes a weighting coefficient from input i to output j. The learning algorithm of LVQ is to change the weighting vector in the following way.

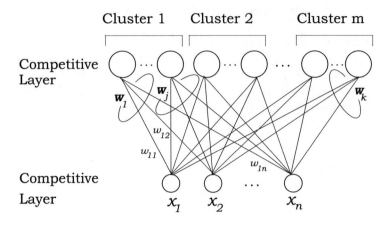

Fig. 5. Layered neural network.

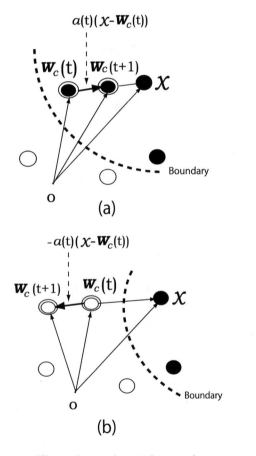

Fig. 6. Layered neural network.

Step 1. Find the nearest weight vector $\mathbf{w_c}$ which is called winner vector such that

$$\mathbf{w_c}, \quad c = \min_j d_j$$

where

$$d_j = ||\mathbf{x} - \mathbf{w_j}||.$$

Step 2. We denote the winner vector $\mathbf{w_c}$ at iteration time t like $\mathbf{w_c(t)}$.
Step 3. If $|\mathbf{x}$ and $\mathbf{w_c}$ belong to the same class as in Fig. 6(a), then set

$$\mathbf{w_c(t + 1)} = \mathbf{w_c(t)} + \alpha(\mathbf{t})(|\mathbf{x} - \mathbf{w_c})$$

where T denotes the total iteration

$$\alpha(t) = \alpha_0(1 - \frac{t}{T}).$$

If $\mathbf{x}$ and $\mathbf{w_c}$ belong to a different class as in Fig. 6(b), then set

$$\mathbf{w_c(t + 1)} = \mathbf{w_c(t)} - \alpha(\mathbf{t})(|\mathbf{x} - \mathbf{w_c}).$$

6   Smell Classification of Body

We must extract the features of the sample paths. Those sample paths of Figs. 7 and 8 show that at t = 0, dry airs blow and at 100 s later gasses are injected. We can see that after about 60 s since dry airs blew in and after about 60 seconds later since smell gasses were injected at t = 100 s later all sensors would become almost steady. Therefore, we decided the changes from t = 60 to t = 160 for each sensor reflect the features of the gasses and take the difference values at t = 60 s and t = 160 s for each sensor as features.

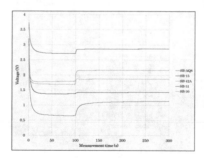

Fig. 7. Sample path of acet-aldehyde.

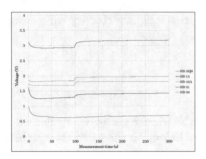

Fig. 8. Sample path of Ethylene.

7   Training for Classification of Smells

In order to classify the feature vector, we allocate the desired output for the input feature vector where it is four dimensional vector as shown in Table 2. When the converse is not so well, we add the number of output neurons such that perfect classification could be obtained. In our case, the training has been performed until the total error becomes less than or equal to $.5 \times 10^{-2}$ where $\eta = .2$. Note that the training data set is the first sample data among three repeated data of A, B, and C. After that, the second sample data are selected as the training data set and the third samples data are selected as the training data set. The test data sets are selected as the remaining data sets except for the training data set.

8   Smell Classification Results

Using a layered neural network with five inputs and four output, we trained the neural network for one training data until the total error becomes 0.001.

Fig. 9. Hardware system of Kunkun body.

After training, we checked the remaining two data set. For two test data, correct classification rates became 100% and using the leave-one-out cross validation check, correct classification rates were 100%.

9   Hardware System

We have made the hardware of Kunkun body, which could be classified the body smell with high accuracy as shown in Fig. 7. HANA is the prototype of Kunkun body for training.

10   Conclusions

We have proposed a classification of human body smells by using neural network of LVQ. Furthermore, we have developed its hardware of small size in order to take it easy to bring with us all the time.

Acknowledgment. This research has been partially supported by JKA Foundation (2017M-144).

References

1. Norman, A., Stam, F., Morrissey, A., Hirschfelder, M., Enderlein, D.: Packaging effects of a novel explosion-proof gas sensor. Sens. Actuator B. **114**, 287–290 (2003)
2. Baric, N., Bucking, M., Rapp, M.: A novel electronic nose based on minimized saw sensor arrays coupled with same enhanced headspace analysis and its use for rapid determination of volatile organic compounds in food quality monitoring. Sens. Actuator B. **114**, 482–488 (2006)
3. Young, R., Buttner, W., Linnel, B., Ramesham, R.: Electronic nose for space program applications. Sens. Actuator B. **93**, 7–16 (2003)
4. Milke, J.: Application of neural networks for discriminating fire detectors. In: 1995 International Conference on Automatic Fire Detection, AUBE 1995, Duisburg, Germany (1995)
5. Charumpom, B., Yoshioka, M., Fujinaka, T., Omatu, S.: An e-nose system using artificial neural networks with an effective initial training data set. IEE J. Trans. EIS **123**, 1638–1644 (2003)
6. Fujinaka, T., Oshioka, M., Omatu, S., Kosaka, T.: Intelligent electronic nose systems for fire detection systems based on neural networks. Int. J. Adv. Intell. Syst. **2**, 268–277 (2009)

A Multi-objective Evolutionary Proposal for Matching Students to Supervisors

Victor Sanchez-Anguix[1]([✉]), Rithin Chalumuri[1], and Vicente Julian[2]

[1] Coventry University, Coventry CV1 2JH, UK
ac0872@coventry.ac.uk, chalumuv@uni.coventry.ac.uk
[2] Departamento de Sistemas Informáticos y Computación,
Universitat Politècnica de València,
Camí de Vera s/n, 46022 Valencia, Spain
vinglada@dsic.upv.es

Abstract. In the last few years there has been a growing interest in the use of artificial intelligence to improve different areas of education such as student team formation, learning analytics, intelligent tutoring systems, or the recommendation of learning resources. This paper presents a genetic algorithm that aims to improve the allocation of students to supervisors while taking both the students' and supervisors' preferences with regards to research topics, and by providing a balanced allocation for supervisors' workload. A Pareto optimal genetic algorithm has been designed and tested for the resolution of this problem.

Keywords: Evolutionary computing · Genetic algorithms · Education

1 Introduction

The education sector is rapidly changing in response to external and internal factors such as globalization, mass education, and a stronger competitors' base [3,5]. The later has forced higher education institutions to adopt strategies and plans to offer a higher quality education. For instance, institutions have adopted student-centered strategies to teaching and learning like cooperative learning, flipped learning, gamification, etc. [2,8,10].

While there have been extensive efforts to improve education by using artificial intelligence (AI) [1,6,9], in comparison, there has been little attention on a key and important element in both undergraduate and postgraduate degrees: dissertations. Dissertations offer a unique opportunity to carry out research and projects in the students' desired field of study, and employ a variety of skills acquired during their degrees. This piece of individual work is carried out under the supervision of a tutor that offers guidance, technical expertise, and academic advice on the development of the project. The question that arises is how higher education institutions can improve students' experience with their dissertations' process. It seems logical that choosing an appropriate supervisor for a student is key to the success of the dissertation. However, both students and supervisors

© Springer International Publishing AG, part of Springer Nature 2019
F. De La Prieta et al. (Eds.): DCAI 2018, AISC 800, pp. 94–102, 2019.
https://doi.org/10.1007/978-3-319-94649-8_12

have imperfect knowledge and they are not aware about the expertise available in the faculty or the students' needs.

In this paper we propose a tool that helps faculties to match students to supervisors according to the students' interests, the available expertise, the supervisors' interests, and the maximum workload for supervisors. The tool gathers the aforementioned information and produces an allocation of students to supervisors by means of a multi-objective genetic algorithm. The remainder of this paper is structured as follows. First we discuss how this work compares with related work in the state-of-the-art. Next, we provide a formal description of the problem that we aim to solve in Sect. 3, and describe the proposed genetic algorithm in Sect. 4. Then, we describe how empirical experiments have been designed and carried out to test the empirical performance of the proposed genetic algorithm under a variety of configurations. Finally, we conclude by discussing the conclusions of this paper and future work.

2 Related Work

As mentioned, there have been multiple efforts to improve education by using artificial intelligence techniques. For instance, Alberola et al. [1] propose an optimization method based on Bayesian learning and linear programming to divide students into high performing teams according to emerging Belbin roles. The AI community has also provided solutions to other problems like for instance that of creating intelligent tutoring systems adapted to the students' levels and needs [6]. In the last few years, we have also seen an increase in the number of works proposing recommender systems to suggest learning resources to students based on several characteristics like their skills and learning needs [9].

A related problem to the one presented in this paper is the allocation of students to projects [7]. Harper et al. propose a genetic algorithm to match students to summer projects with competitive results compared to integer programming methods. The problem is that of matching two sets where only one of the sets has preferences over the other. In our setting, we not only consider the students' preferences but also the supervisors' preferences with regards to areas of research and their maximum workload.

3 Problem Formalization

The problem that is tackled by our proposal is that of matching students to supervisors in the context of a school or faculty. Although the allocation is implemented for just a centre, it can be easily extended to support multiple centres or faculties. For allocation purposes, there are two main actors that should be considered:

- **Students:** For the sake of simplicity, students require a main supervisor that will guide them through the dissertation process. Ideally, the supervisor should have the expertise in the areas that are of interest for the student.

– **Supervisors:** A tutor can act as a supervisor for multiple students according to the needs of the faculty or school. Ideally, supervisors are more inclined to supervise projects that are closer to their expertise and research interests. In addition to this, any allocation of students to supervisors should adhere to the teaching workload of supervisors. Depending on the number of courses taught, some supervisors can supervise more students than others.

Hence, the problem can be viewed as a one-to-many matching problem with multiple objectives: satisfying the students' and supervisors' preferences, and adhering to supervisors' workload. In the next subsection we define how students and supervisors' preferences are elicited, and the optimization problems.

3.1   Eliciting Preferences

Most academic disciplines can be divided in a wide variety of topics and subtopics. In practice, a discipline may contain thousands of different subcategories. In this work, we assume that neither students or supervisors can provide explicit preferences about all of those topics and subtopics. Moreover, we assume that the topics in a discipline can be organized in a tree-like structure similar to that proposed by the ACM classification.[1] For the purpose of this application, we assume that an agent a (either a student or a supervisor) submits an explicit rank/order $>_a = \{kw_1, \ldots, kw_n\}$ over n different (sub)categories in the tree, where kw_j represents the j-th most preferred (sub)category in the tree for a.

3.2   Interest Matching Quality

An allocation of students to supervisors defines a many-to-one matching problem, as a student has to be assigned to a supervisor, but a supervisor can be assigned multiple students. In order to measure the quality of this allocation, the quality of the matching between a student and a supervisor should be quantified from both the students' and the supervisors' perspective. We define the quality of matching a_j interests to a_i interests from a_i perspective as:

$$f(a_i, a_j) = \sum_{kw_k \in >_{st_i}} w_k \times sim_{rnk}(k, l) \times sim(kw_k, kw_l')$$ (1)

where w_k represents the importance given to the matching of the k-th subcategory in the student ranking $>_{a_i}$, kw_l' is the most keyword in rank $>_{a_j}$ to kw_k, sim is a function that calculates the similarity between two subcategories in the tree-like taxonomy, and sim_{rnk} represents the similarity between the position occupied by both categories in the rank of both the student and the supervisor.

On the one hand, the similarity between two (sub)categories in the tree-like taxonomy, sim is calculated as follows:

$$sim(kw_k, kw_l) = \frac{|path(kw_k) \cap path(kw_l)|}{|path(kw_k)|}$$ (2)

---

[1] https://www.acm.org/publications/class-2012.

where $path(.)$ returns the path of a subcategory in the tree-like taxonomy. On the other hand, the similarity between the rank of kw_k and kw_l is measured as:

$$sim_{rnk}(k,l) = \frac{1}{1+|k-l|} \tag{3}$$

The quality of the allocation of a supervisor to a student from the student perspective, f_{st}, is equivalent to Eq. 1. Considering the fact that multiple students can be assigned to a single supervisor, we define the quality of matching students to a supervisor, f_s, as the average quality defined by Eq. 1 for the pairs defined by the supervisor and the students assigned to that supervisor.

3.3 Supervisors' Workload

As mentioned above, some supervisors may be capable of supervising more students than others due to varying teaching commitments. This constraint is represented as a maximum number that can be supervised by a tutor. Given a supervisor, s_i, we define c_i as the capacity or maximum number of students that can be supervised by the supervisor. Therefore, the workload level of a supervisor, wl_i, is defined as the number of students supervised in the allocation divided by his/her capacity, c_i. Moreover, $\mathcal{WL}_\mathcal{A} = \{wl_1, \ldots, wl_{|\mathcal{S}|}\}$ is the vector of workload levels for supervisors in an allocation.

3.4 Allocation Quality

Let us define $\mathcal{A}$ as an allocation of students $\mathcal{ST}$ to supervisors $\mathcal{S}$. For the sake of simplicity, let us assume that $\mathcal{ST}_{\mathcal{A},s_i}$ defines the set of students allocated to supervisor s_i in allocation $\mathcal{A}$. Similarly, we assume that $\mathcal{S}_{\mathcal{A},st_j}$ denotes the supervisor allocated to student st_j in allocation $\mathcal{A}$. Then, the overall quality of the allocation $\mathcal{A}$ from the students' perspective is defined as:

$$\mathcal{F}_{st}(\mathcal{A}) = \frac{\sum\limits_{st_j \in \mathcal{ST}} f_{st}(st_j, \mathcal{S}_{\mathcal{A},st_j})}{|\mathcal{ST}|} \tag{4}$$

Similarly, we can also define the overall quality of an allocation $\mathcal{A}$ from the supervisors' perspective:

$$\mathcal{F}_s(\mathcal{A}) = \frac{\sum\limits_{s_i \in \mathcal{S}} f_s(st_I, \mathcal{ST}_{\mathcal{A},s_i})}{|\mathcal{S}|} \times \frac{1}{1+std(\mathcal{WL}_\mathcal{A})^2} \tag{5}$$

where the function takes the average interest match of supervisors and it is weighted by the squared of the standard deviation of the supervisors' workload level. By applying this weight, we attempt to avoid solutions where supervisors' interests are matched at the cost of some supervisors taking many students while some other supervisors take just a few.

4   Genetic Algorithm Design

In this section, we describe how a genetic algorithm has been designed to solve the aforementioned problem. Considering the fact that there are two different objectives to be maximized (i.e., $\mathcal{F}_s$, $\mathcal{F}_{st}$), we chose a NSGA-II scheme [4] for finding an approximate Pareto optimal frontier defined by both objectives. Next, we define how the genetic algorithm has been designed.

- **Solution representation:** In order to represent an allocation, an array is employed where each array position represents a student, and the content of such array positions represents the supervisor allocated to that student.
- **Population initialization:** We generate $|pop|$ random valid solutions, where $|pop|$ defines the maximum population size.
- **Mutation:** A special mutation operation was designed for this problem. This operator is applied over selected parents from the current population. More specifically, the operator mutates each gene in a solution with a probability of p_{mut}. The mutation is carried out by either swapping the supervisors between two students, or transferring the student from the current supervisor to another one that has enough capacity (i.e., changing the value of the array position). The latter operation is carried out with a probability of p_{tra} if it is possible. Otherwise, the gene selected to be mutated undergoes a swap operation with a probability of p_{swa}. The mutation operation never generates invalid solutions.
- **Crossover:** The crossover operator is a modification of the classic k-point crossover [11] that takes two parents and generates one child. This operator is applied over selected parents from the current population. After generating a child, it is likely that the new solution will be invalid due to violation of one or several supervisors' capacity. Therefore, a random mechanism that takes into consideration the supervisors that can take students, and those that are over their capacity is applied to fix solutions.
- **Selection:** The selection mechanism employed to determine individuals that take part in genetic operations is tournament selection. It follows the non-domination rank and the crowding distance as defined by [4].
- **Stopping criteria:** The genetic algorithm runs for a fixed number of iterations defined by gen_{max}. After that, the algorithm stops.
- **Population:** After every iteration, the population is limited at $|pop|$ individuals. This is done by (i) taking the current population and children arising from crossover and mutation operators; (ii) calculating the Pareto optimal solutions in the set; (iii) and then adding as many Pareto optimal solutions as possible to the new generation population. Steps (ii) and (iii) are repeated while the new generation population has not reached the maximum size.

5 Experiments

5.1 Experiment Design

First we describe how the experiments were designed to test the practical performance of the genetic algorithm. A synthetic dataset was created taking the ACM classification as a reference.

We focused on a matching scenario where 200 students are matched to 60 supervisors. These are numbers that resemble the scenario faced by the School of Computing, Electronics, and Mathematics at Coventry University every year when undergraduate students from computing courses have to be matched to supervisors in the school. In order to simulate the students' and supervisors' preferences, we randomly selected 5 keywords for each supervisor and for each student from the first three levels of the ACM classification. Therefore, there are 2071 possible keywords from which supervisors' and students' preferences can be chosen. With respect to the workload capacity of supervisors, the total capacity adding all supervisors was set 240 students, and the individual capacity of each supervisor followed $c_i \sim \mathcal{U}(4, 10)$ where $\mathcal{U}$ represents a uniform distribution. A total of 5 random scenarios were generated using this setting.

In addition to that, a total of $|pop| = 128$ solutions were randomly generated as the initial population for each scenario. This way, the effect of the other parameters of the genetic algorithm can be compared in equal conditions. Similarly, each variation of the genetic algorithm run for a fixed number of $gen_{max} = 200$ generations to give similar conditions to the different configurations. The rest of the parameters of the genetic algorithm were varied as follows:

- k: The number of crossover points for the crossover operator, which was chosen from a set of values of [2,4,6,8].
- p_{mut}: The probability of mutating a gene in a solution, which took values from [0.05, 0.10, 0.15].
- p_{swa}: The probability of swapping students from a supervisor to another in mutation actions. This parameter took values from [0.25, 0.50, 0.75].

A grid search was carried out over the aforementioned configuration values. The metric recorded to test the performance of the different configurations was the maximum joint fitness[2] found in the Pareto optimal frontier.

5.2 Results

The experiments were carried out according to the previous design. Table 1 gathers the results for the grid search carried out in the previous experimental design. The best experimental configuration was found for a value of $k = 8$, $p_{mut} = 0.05$, and $p_{swa} = 0.75$. This implies that the genetic algorithm found better solutions by mixing parents to a greater extent (i.e., more crossover points) and, therefore, taking smaller inherited segments from parents. In addition, in that setting, the genetic algorithm tends to obtain better results when little disruption

---

[2] Defined as the product of the fitness functions.

Table 1. Grid search carried out over the genetic algorithm's parameters

k	p_{mut}	p_{swa}	Best joint	k	p_{mut}	p_{swa}	Best joint	k	p_{mut}	p_{swa}	Best joint
2	0.05	0.25	0.130	4	0.05	0.25	0.147	8	0.05	0.25	0.157
2	0.05	0.5	0.132	4	0.05	0.5	0.143	8	0.05	0.5	0.160
2	0.05	0.75	0.130	4	0.05	0.75	0.156	8	0.05	0.75	0.164
2	0.1	0.25	0.122	4	0.1	0.25	0.132	8	0.1	0.25	0.157
2	0.1	0.5	0.127	4	0.1	0.5	0.147	8	0.1	0.5	0.162
2	0.1	0.75	0.109	4	0.1	0.75	0.137	8	0.1	0.75	0.154
2	0.15	0.25	0.118	4	0.15	0.25	0.142	8	0.15	0.25	0.157
2	0.15	0.5	0.114	4	0.15	0.5	0.130	8	0.15	0.5	0.160
2	0.15	0.75	0.110	4	0.15	0.75	0.135	8	0.15	0.75	0.163

is introduced in the genetic material of parents' solutions (i.e., low mutation rate). A more detailed representation of these observations can be appreciated in Fig. 1. The graphs represent the evolution of the average best joint utility as the number of crossover points increase (left), the average best joint utility as the probability of swapping students from one supervisor to another in mutation operations increases (right), and the best joint utility as the probability of mutating a gene in a solution to be mutated increases (bottom). In these graphs, it can be observed that, overall, (i) the genetic algorithm benefits from taking more crossover points; (ii) it benefits from a small mutation rate that introduces small perturbations in the solution; and (iii) it usually requires a balance between swapping students (exploring the same solution structure) and transferring students (exploring new solutions' structures).

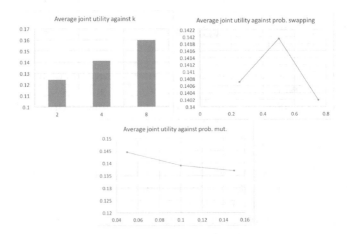

Fig. 1. Average best joint utility (vertical axis) found in the Pareto optimal set against the number of crossover points k (left), p_{swa} (right), and p_{mut} (bottom).

6 Conclusions

In this paper we have presented a genetic algorithm to match students to supervisors in an educational setting. The algorithm takes into consideration the preferences of students with regards to project topics, the preferences of supervisors with regards to research areas, and the workload of supervisors and their maximum supervision capacity. We have employed a NSGA-II scheme using classic crossover and selection operators and we have proposed a specific mutation operator that can swap students between supervisors and transfer students from one supervisor to another without generating invalid solutions. The experimental analysis that we have carried out with simulated data shows that the problem benefits from highly mixing gene material from parents, and introducing low perturbations in parents via mutation operators. Our future work includes analyzing the performance of this algorithm with real data that we are collecting from students and supervisors, creating new crossover operators that are more apt for this problem, and exploring more sophisticated ways of representing students' and supervisors' preferences [12].

Acknowledgements. This work is partially supported by funds of the Faculty of Engineering and Computing at Coventry University, and funds from EU ICT-20-2015 Project SlideWiki granted by the European Commission.

References

1. Alberola, J.M., Del Val, E., Sanchez-Anguix, V., Palomares, A., Teruel, M.D.: An artificial intelligence tool for heterogeneous team formation in the classroom. Knowl. Based Syst. **101**, 1–14 (2016)
2. Bergmann, J., Sams, A.: Flipped learning: gateway to student engagement. International Society for Technology in Education (2014)
3. Biggs, J., Tang, C.: Teaching for quality learning in higher education. The Society for Research into Higher Education (2011)
4. Deb, K., Pratap, A., Agarwal, S., Meyarivan, T.: A fast and elitist multiobjective genetic algorithm: NSGA-II. IEEE Trans. Evol. Comput. **6**(2), 182–197 (2002)
5. Fry, H., Ketteridge, S., Marshall, S.: A Handbook for Teaching and Learning in Higher Education: Enhancing Academic Practice, Routledge (2015)
6. Graesser, A.C., Chipman, P., Haynes, B.C., Olney, A.: AutoTutor: an intelligent tutoring system with mixed-initiative dialogue. IEEE Trans. Educ. **48**(4), 612–618 (2005)
7. Harper, P.R., de Senna, V., Vieira, I.T., Shahani, A.K.: A genetic algorithm for the project assignment problem. Comput. Oper. Res. **32**(5), 1255–1265 (2005)
8. Kapp, K.M.: The Gamification of Learning and Instruction: Game-Based Methods and Strategies for Training and Education. Wiley, San Francisco (2012)
9. Manouselis, N., Drachsler, H., Vuorikari, R., Hummel, H., Koper, R.: Recommender systems in technology enhanced learning. In: Recommender Systems Handbook, pp. 387–415. Springer (2011)
10. Slavin, R.E.: Synthesis of research of cooperative learning. Educ. Leadersh. **48**(5), 71–82 (1991)

11. Spears, W.M., De Jong, K.A.: An analysis of multi-point crossover. In: Foundations of Genetic Algorithms, vol. 1, pp. 301–315. Elsevier (1991)
12. Visser, W., Aydogan, R., Hindriks, K.V., Jonker, C.M.: A framework for qualitative multi-criteria preferences. In: International Conference on Agents and Artificial Intelligence, pp. 243–248 (2012)

Real-Time Conditional Commitment Logic and Duration Communication Interpreted Systems

Bożena Woźna-Szcześniak[1]([✉]) and Ireneusz Szcześniak[2]

[1] IMCS, Jan Długosz University in Częstochowa,
Al. Armii Krajowej 13/15, 42-200 Częstochowa, Poland
b.wozna@ajd.czest.pl
[2] ICIS, Częstochowa University of Technology,
ul. Dąbrowskiego 69, 42-201 Częstochowa, Poland
iszczesniak@icis.pcz.pl

Abstract. We propose a real-time conditional and unconditional commitment logic (RTCTLC) with semantics defined over the *duration communication interpreted system* – a system with arbitrary integer durations on transitions. The transitions with durations allow us to model different levels of temporal deadlines and to reduce extra verification work resulting from the use of unit measure steps. The whole framework allows us to formally model the behaviour of agents using (conditional, unconditional, and group) commitments and real-time constraints in order to permit reasoning about qualitative and quantitative requirements.

Keywords: Real-time conditional commitment logic
Duration communication interpreted system · The Escrow protocol

1 Introduction

Agents have the capabilities of responding to external changes within their own environment, controlling their own behaviour, and interacting with other agents (social ability). Thereby they are intelligent entities. Following [7] we define *multi-agent systems* (MASs) as *"systems in which several interacting, autonomous agents pursue some set of goals or perform some set of tasks"*. Thus, the key property of MASs is the ability of agents to interact (communicate, negotiate, coordinate) with one another as well as their environments.

The formalism of *interpreted systems* (IS) [5] provides a useful framework to model MASs and to verify various classes of temporal and epistemic properties only. The formalism of *communication interpreted systems* (CIS) [3] is an extension of ISs, which makes possible reasoning about agents' social abilities as well. In the paper we propose a new formalism of *duration communication interpreted systems* (DCIS), which compared to ISs and CISs has two properties

© Springer International Publishing AG, part of Springer Nature 2019
F. De La Prieta et al. (Eds.): DCAI 2018, AISC 800, pp. 103–111, 2019.
https://doi.org/10.1007/978-3-319-94649-8_13

that provide increased expressive power: *long* and *instantaneous* steps. The long
steps property means that DCISs allow transitions to take a long time, e.g. 1000
time units. Such transitions can be obviously encoded in IS's by inserting 999
intermediate states. This, however, greatly increases the global state space of a
given MAS. The instantaneous steps property means that DCISs allow transitions to have zero duration. Transitions with zero-duration are a very convenient
way of counting specific actions only.

In the literature of agent communication, the social commitment has two
forms: unconditional and conditional [2]. The basic idea of unconditional commitment is that the debtor agent makes a contractual obligation and he directs it
towards the creditor agent, to bring about a certain fact. For example, the seller
unconditionally commits to the buyer to ship the requested goods. The main
idea of conditional commitments is that the debtor agent can merely commit
towards the creditor agent to bring about consequences when specific conditions
are met. For example, the seller commits to the buyer to ship the requested
goods if the buyer sends the agreed payment.

The main objective of this paper is to provide a new agent communication
language, called RTCTLC, with semantics defined on DCISs. The RTCTLC language extends the RTCTL^{CC} [6] language with group conditional and unconditional social commitments modalities and their fulfillment modalities. The proposed semantics allows to consider arbitrary duration in our model's transitions
and thereby to have different levels of temporal deadlines for unconditional and
conditional commitments and their fulfillments.

The rest of the paper is organised as follows. In Sect. 2 we present the DCIS
formalism together with its Kripke model, and we illustrate it by means of the
Escrow protocol [1]. In Sect. 3 we define the syntax and semantics of RTCTLC
and we illustrate it, among others, by means of the properties of the Escrow
protocol. We conclude and identify future research directions in Sect. 4.

2   Duration Communication Interpreted System

A MAS is a non-empty and finite set of agents ($\mathbb{A} = \{1, \ldots, n\}$) together with
the environment $\mathcal{O}$ (a special agent), in which the agents operate. *Duration
communication interpreted system* (DCIS) extends CIS by taking into account
the duration that occurs during the execution of MAS. Formally, let $\mathcal{PV} = \bigcup_{c \in \mathbb{A}} \mathcal{PV}_c \cup \mathcal{PV}_\mathcal{O}$ be a set of propositional variables such that $\mathcal{PV}_{c_1} \cap \mathcal{PV}_{c_2} = \emptyset$
for all $c_1, c_2 \in \mathbb{A} \cup \{\mathcal{O}\}$, and each agent $c \in \mathbb{A}$ and the environment $\mathcal{O}$ be
defined by:

- Var_c ($Var_\mathcal{O}$) - a finite set of at most n non-negative local *integer variables*
  that represent communication channels through which messages are sent and
  received, and are used to define the *social accessibility* relation.
- L_c ($L_\mathcal{O}$) - a non-empty and finite set of *local states* modelling the instantaneous configuration of the agent c (the environment $\mathcal{O}$). Each local state is
  associated with different values obtained from different assignments to vari-
  ables in Var_c ($Var_\mathcal{O}$). It is assumed that the local states of $\mathcal{O}$ are "public".

- $Act_{\mathbf{c}}$ ($Act_{\mathcal{O}}$) - a non-empty and finite set of *actions* such that the special *null* action $\epsilon_{\mathbf{c}}$ ($\epsilon_{\mathcal{O}}$) belongs to $Act_{\mathbf{c}}$ ($Act_{\mathcal{O}}$); it is assumed that actions are "public", and each element of $Act = Act_1 \times \ldots \times Act_n \times Act_{\mathcal{O}}$ is called the *joint action*.
- $D_{\mathbf{c}} : Act_{\mathbf{c}} \mapsto \mathbb{N}$ ($D_{\mathcal{O}} : Act_{\mathcal{O}} \mapsto \mathbb{N}$) - a *duration* function that assigns to every action a natural number, called the *duration* of the action.
- $P_{\mathbf{c}} : L_{\mathbf{c}} \mapsto 2^{Act_{\mathbf{c}}}$ ($P_{\mathcal{O}} : L_{\mathcal{O}} \mapsto 2^{Act_{\mathcal{O}}}$) - a *protocol* function that assigns to every local state a set of actions that can be fired at that state.
- $t_{\mathbf{c}} : L_{\mathbf{c}} \times L_{\mathcal{O}} \times Act \mapsto L_{\mathbf{c}}$ ($t_{\mathcal{O}} : L_{\mathcal{O}} \times Act \mapsto L_{\mathcal{O}}$) - a (partial) local *evolution* function. We assume that: (1) if $\epsilon_{\mathbf{c}} \in P_{\mathbf{c}}(\ell_{\mathbf{c}})$, then $t_{\mathbf{c}}(\ell_{\mathbf{c}}, \ell_{\mathcal{O}}, (a_1, \ldots, a_n, a_{\mathcal{O}})) = \ell_{\mathbf{c}}$ for $a_{\mathbf{c}} = \epsilon_{\mathbf{c}}$; (2) if $\epsilon_{\mathcal{O}} \in P_{\mathcal{O}}(\ell_{\mathcal{O}})$, then $t_{\mathcal{O}}(\ell_{\mathcal{O}}, (a_1, \ldots, a_n, \epsilon_{\mathcal{O}})) = \ell_{\mathcal{O}}$.
- $V_{\mathbf{c}} : L_{\mathbf{c}} \mapsto 2^{\mathcal{PV}_{\mathbf{c}}}$ ($V_{\mathcal{O}} : L_{\mathcal{O}} \mapsto 2^{\mathcal{PV}_{\mathcal{O}}}$) - a *valuation function* assigning to every local state a set of propositional variables that are true at that state.

A set of all *global states* is defined as $S = L_1 \times \ldots \times L_n \times L_{\mathcal{O}}$ [5], and each element $s \in S$ represents some instantaneous configuration of the given MAS. For a given set of agents $\mathbb{A}$, the environment $\mathcal{O}$, and a set of initial global states $\iota \subseteq S$, a *duration communication interpreted system* is a tuple $\mathcal{D} = (\{L_{\mathbf{c}}, Act_{\mathbf{c}}, Var_{\mathbf{c}}, P_{\mathbf{c}}, D_{\mathbf{c}}, t_{\mathbf{c}}, V_{\mathbf{c}}\}_{\mathbf{c} \in \mathbb{A} \cup \{\mathcal{O}\}}, \iota)$.

Let $s = (\ell_1, \ldots, \ell_n, \ell_{\mathcal{O}})$ be a global state and $l_{\mathbf{c}}(s)$ denote the local state of $\mathbf{c} \in \mathbb{A} \cup \{\mathcal{O}\}$ in the state s. A *global evolution function* $t : S \times Act \mapsto S$ is defined as usually by $t(s, a) = s'$ iff $t_{\mathbf{c}}(l_{\mathbf{c}}(s), l_{\mathcal{O}}(s), a) = l_{\mathbf{c}}(s')$ for all $\mathbf{c} \in \mathbb{A}$ and $t_{\mathcal{O}}(l_{\mathcal{O}}(s), a) = l_{\mathcal{O}}(s')$. In brief we write the above as $s \xrightarrow{a} s'$. Furthermore, following [3], we denote the value of a variable $x \in Var_{\mathbf{c}}$ at local state $l_{\mathbf{c}}(s)$ by $l_{\mathbf{c}}^x(s)$, and we assume that if $l_{\mathbf{c}}(s) = l_{\mathbf{c}}(s')$, then $l_{\mathbf{c}}^x(s) = l_{\mathbf{c}}^x(s')$ for all $x \in Var_{\mathbf{c}}$. Next, for $\mathbf{c}_1, \mathbf{c}_2 \in \mathbb{A} \cup \{\mathcal{O}\}$, $\sim_{\mathbf{c}_1 \to \mathbf{c}_2} \subseteq S \times S$ is a serial *social accessibility* relation defined as: $s \sim_{\mathbf{c}_1 \to \mathbf{c}_2} s'$ iff the following conditions are true: (1) $l_{\mathbf{c}_1}(s) = l_{\mathbf{c}_1}(s')$, and (2) $s \xrightarrow{a} s'$ for some $a \in Act$, and (3) $Var_{\mathbf{c}_1} \cap Var_{\mathbf{c}_2} \neq \emptyset$ and $\forall x \in Var_{\mathbf{c}_1} \cap Var_{\mathbf{c}_2}$ we have $l_{\mathbf{c}_1}^x(s) = l_{\mathbf{c}_2}^x(s')$, and (4) $\forall y \in Var_{\mathbf{c}_2} - Var_{\mathbf{c}_1}$ we have $l_{\mathbf{c}_2}^y(s) = l_{\mathbf{c}_2}^y(s')$. The intuition behind the social accessibility relation $\sim_{\mathbf{c}_1 \to \mathbf{c}_2}$ from a global state s to another global state s' is the following. Since $\mathbf{c}_1$ initiates the communication and it does not learn any new information, the states s and s' are indistinguishable for $\mathbf{c}_1$ ($l_{\mathbf{c}_1}(s) = l_{\mathbf{c}_1}(s')$). The global state s' is reachable from state s via some joint action. There is a communication channel between $\mathbf{c}_1$ and $\mathbf{c}_2$ ($Var_{\mathbf{c}_1} \cap Var_{\mathbf{c}_2} \neq \emptyset$). The channel is filled in by $\mathbf{c}_1$ in state s, and in state s' $\mathbf{c}_2$ receives the information, which makes the value of the shared variable the same for $\mathbf{c}_1$ and $\mathbf{c}_2$ ($l_{\mathbf{c}_1}^x(s) = l_{\mathbf{c}_2}^x(s')$). The states s and s' are indistinguishable for $\mathbf{c}_2$ with regard to the variables that have not been communicated by $\mathbf{c}_1$, i.e., unshared variables (($\forall y \in Var_{\mathbf{c}_2} - Var_{\mathbf{c}_1}$) $l_{\mathbf{c}_2}^y(s) = l_{\mathbf{c}_2}^y(s')$).

With each $\mathcal{D}$ we associate the following Kripke model $M = (Act, S, \iota, T, D, V, \{\sim_{\mathbf{c}_1 \to \mathbf{c}_2}\}_{\mathbf{c}_1, \mathbf{c}_2 \in \mathbb{A} \cup \{\mathcal{O}\}})$, where

- Act is the set of joint actions, S is a set of global states, and $\iota \subseteq S$.
- $T \subseteq S \times Act \times S$ is a transition relation on S defined by: $(s, a, s') \in T$ iff $s \xrightarrow{a} s'$. We assume that the relation T is total, i.e., for any $s \in S$ there exists $s' \in S$ and an action $a \in Act \setminus \{\bar{\epsilon}\}$ such that $s \xrightarrow{a} s'$ and $\bar{\epsilon} = (\epsilon_1, \ldots, \epsilon_n, \epsilon_{\mathcal{O}})$.
- $D : Act \to \mathbb{N}$ is the *global duration* function defined as $D((a_1, \ldots, a_n, a_{\mathcal{O}})) = max$, where max is the maximum value of the set $\{D_1(a_1), \ldots, D_n(a_n)\}$.

- $V : S \to 2^{\mathcal{PV}}$ is the *global valuation* function defined as $V(s) = \bigcup_{\mathbf{c} \in \mathbb{A} \cup \{\mathcal{O}\}} V_{\mathbf{c}}(l_{\mathbf{c}}(s))$.
- $\sim_{\mathbf{c}_1 \to \mathbf{c}_2} \subseteq S \times S$ is the social accessibility relation for $\mathbf{c}_1 \in \mathbb{A}$ and $\mathbf{c}_2 \in \mathbb{A}$.

Observe that each transition in our quantitative temporal model M may take an arbitrary time unit for execution from one state to another state. For example, during a transition $s \xrightarrow{a} s'$ with $D(a) = 8$ the system moves from "s at some time t" to "s' at $t + 8$". Between t and $t + 8$ there is no state (or time) where the system is in. The underlying real-time model M is discrete, has a tree-like structure, and can be unfolded into a set of execution *paths* in which each *path* $\pi = s_0 \xrightarrow{a_1} s_1 \xrightarrow{a_2} s_2 \xrightarrow{a_3} \dots$ is an infinite sequence of linked transitions. For such a path and for $m \in \mathbb{N}$, we denote by $\pi(m)$ the m-th state s_m. Finally, for $j \le m \in \mathbb{N}$, we denote by $\pi[j..m]$ the finite sequence $s_j \xrightarrow{a_{j+1}} s_{j+1} \xrightarrow{a_{j+2}} \dots s_m$ with $m - j$ transitions and $m - j + 1$ states. The (cumulative) duration $D\pi[j..m]$ of the finite sequence $\pi[j..m]$ is $D(a_{j+1}) + \dots + D(a_m)$ (hence 0 when $j = m$). We write $\Pi(s)$ for the set of all the paths that start at $s \in S$.

Escrow Protocol. The escrow protocol [1] is a three-party protocol involving a buyer, a seller and a trusted third-party escrow, which works as follows:

1. Either the Buyer or the Seller begins a transaction. Buyer and Seller agree to the terms of the transaction via Escrow.com.
2. The Buyer deposits the payment to the secure Escrow account.
3. Escrow.com verifies the payment and notifies the Seller that funds have been secured in Escrow.
4. The Seller ships the goods to the Buyer - Upon payment verification, the Seller is authorised to send the goods and submit tracking information. Escrow.com verifies that the Buyer receives the goods.
5. The Buyer has a set number of days to inspect the goods and the option to accept or reject it. If the Buyer is satisfied, he authorises the Escrow to pay the Seller. Escrow.com then pays the Seller. However, if the Buyer is not satisfied, in addition to notifying the escrow, he returns the goods to the Seller. When the Seller notifies the escrow about the goods being received back, the Escrow refunds the deposit to the Buyer with a set number of days.

Modelling the Escrow Protocol. In the DCIS formalism, it is convenient to see the Buyer ($\mathcal{B}$) and the Seller ($\mathcal{S}$) as agents (i.e., $\mathbb{A} = \{\mathcal{B}, \mathcal{S}\}$), and the Escrow as the environment $\mathcal{E}$. The associated duration communication interpreted system is the following: $\mathcal{D} = (\{L_{\mathbf{c}}, Act_{\mathbf{c}}, Var_{\mathbf{c}}, P_{\mathbf{c}}, D_{\mathbf{c}}, t_{\mathbf{c}}, V_{\mathbf{c}}\}_{\mathbf{c} \in \{\mathcal{B}, \mathcal{S}, \mathcal{E}\}}, \iota)$, where

- $L_{\mathcal{B}} = \{b_0, b_1, b_2, b_3, b_4, b_5, b_6\}$, $L_{\mathcal{S}} = \{m_0, m_1, m_2, m_3, m_4, m_5\}$, and $L_{\mathcal{E}} = \{e_0, e_1, e_2, e_3, e_4, e_5, e_6, e_7\}$. Thus the global set of states $S = L_{\mathcal{B}} \times L_{\mathcal{S}} \times L_{\mathcal{E}}$. The meaning of local states is the following: b_0 - initiate the transaction (contract) by submitting the payment to $\mathcal{E}$; b_1 - wait for the goods; b_2 - inspect the goods and make a decision on their acceptance or rejection; b_3 - authorize $\mathcal{E}$ to pay $\mathcal{S}$. The contract is fulfilled successfully; b_4 - return the goods to the $\mathcal{S}$; b_5 - wait for the refund from $\mathcal{E}$; b_6 - the contract is violated by the $\mathcal{B}$. m_0 - wait for the payment to $\mathcal{E}$ from $\mathcal{B}$; m_1 - shipping the goods to $\mathcal{B}$; m_2 - wait for the decision of $\mathcal{B}$; m_3 - wait for the payment from $\mathcal{E}$; m_4 - wait for the goods from $\mathcal{B}$ and notify $\mathcal{E}$ about the goods being received; m_5 - the contract is violated by $\mathcal{B}$. e_0 - wait for $\mathcal{B}$ to make the payment; e_1 - notify $\mathcal{S}$ that the payment was made; e_2 - wait for $\mathcal{S}$ to ship the goods and $\mathcal{B}$ to make a decision; e_3 - pay $\mathcal{S}$; e_4 - wait for $\mathcal{B}$ to return goods; e_5 - the contract is fulfilled successfully; e_6 - refund the deposit to $\mathcal{B}$; e_7 - the contract is violated by $\mathcal{B}$.

- the sets of natural (non-negative) variables available to the agents are: $Var_{\mathcal{B}} = \{x_1, x_2\}$, $Var_{\mathcal{S}} = \{x_1, x_3\}$ and $Var_{\mathcal{E}} = \{x_2, x_3\}$. The variable x_1 represents the communication channel (CC) between $\mathcal{B}$ and $\mathcal{S}$. The variable x_2 represents CC between $\mathcal{B}$ and $\mathcal{E}$. The variable x_3 represents CC between $\mathcal{S}$ and $\mathcal{E}$.

- $Act = Act_{\mathcal{B}} \times Act_{\mathcal{S}} \times Act_{\mathcal{E}}$ with the following sets of local actions: $Act_{\mathcal{B}} = \{deposit, returnGoods, goodsOk, goodsNotOk, end_B, \epsilon_B\}$, $Act_{\mathcal{S}} = \{sendGoods, release, end_S, \epsilon_S\}$, $Act_{\mathcal{E}} = \{notify, refund, pay, end_{\mathcal{E}}, \epsilon_{\mathcal{E}}\}$, where ϵ_B, ϵ_S, and $\epsilon_{\mathcal{E}}$ stand for the null actions.

- the local protocols of the agents and the environment are: $P_{\mathcal{B}}(b_0) = \{deposit\}$; $P_{\mathcal{B}}(b_1) = \{\epsilon_B\}$; $P_{\mathcal{B}}(b_2) = \{goodsOk, goodsNotOk\}$; $P_{\mathcal{B}}(b_3) = \{end_B, \epsilon_B\}$; $P_{\mathcal{B}}(b_4) = \{returnGoods\}$; $P_{\mathcal{B}}(b_5) = \{\epsilon_B\}$; $P_{\mathcal{B}}(b_6) = \{end_B\}$. $P_{\mathcal{S}}(m_0) = \{\epsilon_S\}$; $P_{\mathcal{S}}(m_1) = \{sendGoods\}$; $P_{\mathcal{S}}(m_2) = \{\epsilon_S\}$; $P_{\mathcal{S}}(m_3) = \{end_S\}$; $P_{\mathcal{S}}(m_4) = \{release\}$; $P_{\mathcal{S}}(m_5) = \{end_S\}$; $P_{\mathcal{E}}(e_0) = \{\epsilon_{\mathcal{E}}\}$; $P_{\mathcal{E}}(e_1) = \{notify\}$; $P_{\mathcal{E}}(e_2) = \{\epsilon_{\mathcal{E}}\}$; $P_{\mathcal{E}}(e_3) = \{pay\}$; $P_{\mathcal{E}}(e_4) = \{\epsilon_{\mathcal{E}}\}$; $P_{\mathcal{E}}(e_5) = \{end_{\mathcal{E}}\}$; $P_{\mathcal{E}}(e_6) = \{refund\}$; $P_{\mathcal{E}}(e_7) = \{end_{\mathcal{E}}\}$;

- the local durations of the agents and the environment are: $D_{\mathcal{B}}(goodsOk) = D_{\mathcal{B}}(goodsNotOk) = 14$, $D_{\mathcal{B}}(deposit) = D_{\mathcal{B}}(return Goods) = D_{\mathcal{B}}(end_B) = 0$. This means that $\mathcal{B}$ has 14 days to make a decision. $D_{\mathcal{S}}(sendGoods) = 3$, $D_{\mathcal{S}}(release) = 2$, $D_{\mathcal{S}}(end_S) = 0$. This means that $\mathcal{S}$ has 2 days to release the deposit and 3 days to ship goods. $D_{\mathcal{E}}(notify) = D_{\mathcal{E}}(refund) = D_{\mathcal{E}}(pay) = D_{\mathcal{E}}(end_{\mathcal{E}}) = 0$.

- Let $\bar{\epsilon}$ be the joint null action, ℓ a local state of an agent $\mathbf{c} \in \{\mathcal{B}, \mathcal{S}\}$, $\ell_{\mathcal{E}}$ a local state of the Environment, $act_{\mathcal{B}}(a)$, $act_{\mathcal{S}}(a)$, and $act_{\mathcal{E}}(a)$ denote, respectively, a local action of $\mathcal{B}$, $\mathcal{S}$, and $\mathcal{E}$. We assume the following local evolution functions. The Buyer: $t_{\mathcal{B}}(\ell, \ell_{\mathcal{E}}, a) = \ell$ if $a \neq \bar{\epsilon}$ and $act_{\mathcal{B}}(a) = \epsilon_{\mathcal{B}}$. $t_{\mathcal{B}}(b_0, \ell_{\mathcal{E}}, a) = b_1$ if $act_{\mathcal{B}}(a) = deposit$. $t_{\mathcal{B}}(b_1, \ell_{\mathcal{E}}, a) = b_2$ if $act_{\mathcal{S}}(a) = sendGoods$. $t_{\mathcal{B}}(b_2, \ell_{\mathcal{E}}, a) = b_3$ if $act_{\mathcal{B}}(a) = goodsOk$. $t_{\mathcal{B}}(b_2, \ell_{\mathcal{E}}, a) = b_4$ if $act_{\mathcal{B}}(a) = goodsNotOk$. $t_{\mathcal{B}}(b_4, \ell_{\mathcal{E}}, a) = b_5$ if $act_{\mathcal{B}}(a) = returnGoods$. $t_{\mathcal{B}}(b_5, \ell_{\mathcal{E}}, a) = b_6$ if $act_{\mathcal{E}}(a) = refund$. $t_{\mathcal{B}}(b_6, \ell_{\mathcal{E}}, a) = b_0$ if $act_{\mathcal{B}}(a) = end_B$ and $act_{\mathcal{E}}(a) = end_{\mathcal{E}}$. $t_{\mathcal{B}}(b_3, \ell_{\mathcal{E}}, a) = b_0$ if $act_{\mathcal{B}}(a) = end_B$ and $act_{\mathcal{E}}(a) = end_{\mathcal{E}}$. The Seller: $t_{\mathcal{S}}(\ell, \ell_{\mathcal{E}}, a) = \ell$ if $a \neq \bar{\epsilon}$ and $act_{\mathcal{S}}(a) = \epsilon_{\mathcal{S}}$. $t_{\mathcal{B}}(m_0, \ell_{\mathcal{E}}, a) = m_1$ if $act_{\mathcal{E}}(a) = notify$. $t_{\mathcal{B}}(m_1, \ell_{\mathcal{E}}, a) = m_2$ if $act_{\mathcal{S}}(a) = sendGoods$. $t_{\mathcal{B}}(m_2, \ell_{\mathcal{E}}, a) = m_3$ if $act_{\mathcal{E}}(a) = pay$. $t_{\mathcal{B}}(m_2, \ell_{\mathcal{E}}, a) = m_4$ if $act_{\mathcal{B}}(a) = returnGoods$. $t_{\mathcal{B}}(m_4, \ell_{\mathcal{E}}, a) = m_5$ if $act_{\mathcal{S}}(a) = release$. $t_{\mathcal{B}}(m_5, \ell_{\mathcal{E}}, a) = m_0$ if $act_{\mathcal{S}}(a) = end_S$ and $act_{\mathcal{E}}(a) = end_{\mathcal{E}}$. $t_{\mathcal{B}}(m_3, \ell_{\mathcal{E}}, a) = m_0$ if $act_{\mathcal{S}}(a) = end_S$ and $act_{\mathcal{E}}(a) = end_{\mathcal{E}}$ and $act_{\mathcal{B}}(a) = end_B$. The Escrow: $t_{\mathcal{E}}(\ell_{\mathcal{E}}, a) = \ell_{\mathcal{E}}$ if $a \neq \bar{\epsilon}$ and $act_{\mathcal{E}}(a) = \epsilon_{\mathcal{E}}$. $t_{\mathcal{E}}(e_0, a) = e_1$ if $act_{\mathcal{B}}(a) = deposit$. $t_{\mathcal{E}}(e_1, a) = e_2$ if $act_{\mathcal{E}}(a) = notify$. $t_{\mathcal{E}}(e_2, a) = e_3$ if $act_{\mathcal{B}}(a) = goodsOk$.

$t_\mathcal{E}(e_2, a) = e_4$ if $act_\mathcal{B}(a) = goodsNotOk$. $t_\mathcal{E}(e_3, a) = e_5$ if $act_\mathcal{E}(a) = pay$. $t_\mathcal{E}(e_4, a) = e_6$ if $act_\mathcal{S}(a) = release$. $t_\mathcal{E}(e_6, a) = e_7$ if $act_\mathcal{E}(a) = refund$. $t_\mathcal{E}(e_7, a) = e_0$ if $act_\mathcal{B}(a) = end_B$ and $act_\mathcal{E}(a) = end_\mathcal{E}$. $t_\mathcal{E}(e_5, a) = e_0$ if $act_\mathcal{B}(a) = end_B$ and $act_\mathcal{E}(a) = end_\mathcal{E}$.

The set of possible global states S for the Escrow protocol is defined as the product $L_\mathcal{B} \times L_\mathcal{S} \times L_\mathcal{E}$, and we consider the following set of initial states $\iota = \{(b_0, m_0, e_0)\}$. Furthermore, in the Kripke model of the Escrow protocol, we assume the following set of proposition variables: $\mathcal{PV} = \{deposit, delivered, goodsOk, goodsNotOk, returnGoods, refund\}$ with the following interpretation:

- $(M, s) \models deposit$ if $l_\mathcal{B}(s) = b_1$, - $(M, s) \models refund$ if $l_\mathcal{B}(s) = b_6$,
- $(M, s) \models delivered$ if $l_\mathcal{B}(s) = b_2$ and $l_\mathcal{S} = m_2$ and $l_\mathcal{E} = e_2$,
- $(M, s) \models goodsNotOk$ if $l_\mathcal{B}(s) = b_4$,
- $(M, s) \models goodsOk$ if $l_\mathcal{B}(s) = b_3$ and $l_\mathcal{S} = m_2$,
- $(M, s) \models returnGoods$ if $l_\mathcal{B}(s) = b_5$ and $l_\mathcal{S} = m_4$.

3   The Real-Time Conditional Commitment Logic

The Real-Time Conditional Commitment Logic (RTCTLC) is a combination of RTCTL [4], a branching temporal logic with timing constraints, with the commitment modality and its fulfillment [3], the group commitment modality [8] and its fulfillment, the conditional commitment modality and its fulfillment [6], the conditional group commitment modality and its fulfillment.

Syntax of RTCTLC. Let $p \in \mathcal{PV}$ be a propositional variable, $\mathbf{c}, \mathbf{c}_1, \mathbf{c}_2, \Gamma \subseteq \mathbb{A} \cup \{\mathcal{O}\}$, and I is an interval in $\mathbb{N}$ of the form: $[a, b)$ and $[a, \infty)$, for $a, b \in \mathbb{N}$ and $a \neq b$. We define RTCTLC formulae inductively as follows:
$$\varphi ::= \mathbf{true} \mid \mathbf{false} \mid p \mid \neg\varphi \mid \varphi \wedge \varphi \mid \mathrm{EX}\varphi \mid \mathrm{E}(\varphi \mathrm{U}_I \varphi) \mid \mathrm{A}(\varphi \mathrm{U}_I \varphi) \mid CC \mid \mathrm{Fu}(CC)$$
$$CC ::= \mathrm{C}_{\mathbf{c}_1 \to \mathbf{c}_2}(\varphi) \mid \mathrm{CC}_{\mathbf{c}_1 \to \mathbf{c}_2}(\varphi, \varphi) \mid \mathrm{C}_{\mathbf{c} \to \Gamma}(\varphi) \mid \mathrm{CC}_{\mathbf{c} \to \Gamma}(\varphi, \varphi), \text{ where}$$
$\neg$ and $\wedge$ are the standard Boolean connectives for negation and conjunction, respectively. The other propositional connectives are defined as abbreviations, i.e., $\vee$ for disjunction, $\Rightarrow$ for implication, and $\Leftrightarrow$ for logical equivalence. E and A are the *existential* and *universal* quantifiers on paths. X and U_I are RTCTL path modalities standing for the "neXt state" and the "bounded Until". Other RTCTL path modalities are defined in a standard way.

The operators $\mathrm{C}_{\mathbf{c}_1 \to \mathbf{c}_2}$, $\mathrm{CC}_{\mathbf{c}_1 \to \mathbf{c}_2}$, $\mathrm{C}_{\mathbf{c} \to \Gamma}$, and $\mathrm{CC}_{\mathbf{c} \to \Gamma}$ stand for *commitment, conditional commitment, group commitment* and *group conditional commitment*. The formulae $\mathrm{C}_{\mathbf{c}_1 \to \mathbf{c}_2}(\varphi)$, $\mathrm{CC}_{\mathbf{c}_1 \to \mathbf{c}_2}(\psi, \varphi)$, $\mathrm{C}_{\mathbf{c} \to \Gamma}(\varphi)$, and $\mathrm{CC}_{\mathbf{c} \to \Gamma}(\psi, \varphi)$ are read, respectively, as "agent $\mathbf{c}_1$ commits towards agent $\mathbf{c}_2$ that φ", "agent $\mathbf{c}_1$ commits towards agent $\mathbf{c}_2$ to consequently satisfy φ once the antecedent ψ holds", "agent $\mathbf{c}_1$ commits towards agents in Γ that φ", and "agent $\mathbf{c}_1$ commits towards agents in Γ to consequently satisfy φ once the antecedent ψ holds".

The operators $\mathrm{Fu}(\mathrm{C}_{\mathbf{c}_1 \to \mathbf{c}_2})$, $\mathrm{Fu}(\mathrm{CC}_{\mathbf{c}_1 \to \mathbf{c}_2})$, $\mathrm{Fu}(\mathrm{C}_{\mathbf{c} \to \Gamma})$, and $\mathrm{Fu}(\mathrm{CC}_{\mathbf{c} \to \Gamma})$ stand for *fulfillments* of commitment, conditional commitment, group commitment, and *group conditional commitment*, respectively. The formula $\mathrm{Fu}(\mathrm{C}_{\mathbf{c}_1 \to \mathbf{c}_2}(\varphi))$ is read as "the commitment $C_{c_1 \to c_2}(\varphi)$ is fulfilled". The other formulae are read in the analogous manner.

Semantics of RTCTLC. The RTCTLC formulae are interpreted over models generated by DCIS. Let $M = (Act, S, \iota, T, D, V, \{\sim_{c_1 \to c_2}\}_{c_1, c_2 \in \mathbb{A} \cup \{0\}})$ be such a model, $s \in S$, φ an RTCTLC formula, and $s \sim_{c_1 \to \Gamma} s'$ if $s \sim_{c_1 \to c_2} s'$ for all $c_2 \in \Gamma$. $M, s \models \varphi$ denotes that φ is true at the state s in M. The relation $\models$ is defined inductively as follows:

- $M, s \models$ **true**, $- M, s \not\models$ **false**, $- M, s \models p$ iff $p \in V(s)$,
- $M, s \models \neg\varphi$ iff $M, s \not\models \varphi$, $- M, s \models \varphi \wedge \psi$ iff $M, s \models \varphi$ and $M, s \models \psi$,
- $M, s \models EX\varphi$ iff $(\exists \pi \in \Pi(s))M, \pi(1) \models \varphi$,
- $M, s \models E(\varphi U_I \psi)$ iff $(\exists \pi \in \Pi(s))(\exists m \geq 0)(D\pi[0..m] \in I$ and $M, \pi(m) \models \psi$ and $(\forall j < m)M, \pi(j) \models \varphi)$,
- $M, s \models A(\varphi U_I \psi)$ iff $(\forall \pi \in \Pi(s))(\exists m \geq 0)(D\pi[0..m] \in I$ and $M, \pi(m) \models \psi$ and $(\forall j < m)M, \pi(j) \models \varphi)$,
- $M, s \models C_{c_1 \to x}\varphi$ iff $(\forall s' \in S)(s \sim_{c_1 \to x} s'$ implies $M, s' \models \varphi)$, where $x \in \{c_2, \Gamma\}$,
- $M, s \models CC_{c_1 \to x}(\psi, \varphi)$ iff $(\forall s' \in S)((s \sim_{c_1 \to x} s'$ and $M, s' \models \psi)$ implies $M, s' \models \varphi)$, where $x \in \{c_2, \Gamma\}$,
- $M, s \models Fu(C_{c_1 \to x}\varphi)$ iff $(\exists s' \in S)(s' \sim_{c_1 \to x} s$ and $M, s' \models C_{c_1 \to x}\varphi$ and $M, s \models \varphi \wedge \neg C_{c_1 \to x}\varphi)$, where $x \in \{c_2, \Gamma\}$,
- $M, s \models Fu(CC_{c_1 \to x}(\psi, \varphi))$ iff $(\exists s' \in S)(s' \sim_{c_1 \to x} s$ and $M, s' \models CC_{c_1 \to x}(\psi, \varphi)$ and $M, s \models \varphi \wedge \neg CC_{c_1 \to x}(\psi, \varphi))$, where $x \in \{c_2, \Gamma\}$.

An RTCTLC formula φ is *valid* in M, denoted by $M \models \varphi$, iff for each $s \in \iota$, $M, s \models \varphi$, i.e., φ holds at every initial state of M.

For the propositions, Boolean connectives and temporal modalities, the relation $\models$ is defined in the standard manner. The state formulae $C_{c_1 \to c_2}\varphi$ and $C_{c_1 \to \Gamma}\varphi$ are satisfied in the model M at state s iff the formula φ holds in every accessible state obtained by the accessibility relations $\sim_{c_1 \to c_2}$ and $\sim_{c_1 \to \Gamma}$, respectively. The state formulae $CC_{c_1 \to c_2}(\psi, \varphi)$ and $CC_{c_1 \to \Gamma}(\psi, \varphi)$ are satisfied in the model M at state s iff the formula φ holds in every state that satisfies formula ψ and which is accessible by the accessibility relations $\sim_{c_1 \to c_2}$ and $\sim_{c_1 \to \Gamma}$, respectively. Observe that unconditional commitments are a special case of conditional commitments when the antecedents are always true.

The state formulae $Fu(C_{c_1 \to c_2}\varphi)$ and $Fu(C_{c_1 \to \Gamma}\varphi)$ are satisfied in the model M at state s iff s satisfies φ and the negation of the commitments $C_{c_1 \to c_2}\varphi$ and $C_{c_1 \to \Gamma}\varphi$, respectively, and there exists a state s' satisfying the commitment from which the state s is reachable via the accessibility relations $\sim_{c_1 \to c_2}$ and $\sim_{c_1 \to \Gamma}$, respectively. The state formulae $Fu(CC_{c_1 \to c_2}(\psi, \varphi))$ and $Fu(CC_{c_1 \to \Gamma}(\psi, \varphi))$ are satisfied in the model M at state s iff s satisfies φ and the negation of the commitments $CC_{c_1 \to c_2}(\psi, \varphi)$ and $CC_{c_1 \to \Gamma}(\psi, \varphi)$, respectively, and there exists a state s' satisfying the commitment from which the state s is reachable via the accessibility relations $\sim_{c_1 \to c_2}$ and $\sim_{c_1 \to \Gamma}$, respectively. The idea behind this semantics is to say that a commitment is fulfilled when we reach an accessible state from the commitment state in which the formula φ holds and the commitment becomes no longer active.

We conclude this section by illustrating how the basic RTCTLC modalities could be used to express important correctness properties of commitment protocols that must place an explicit bound on the time between events. First, observe that $AF_I p$ specifies the bounded inevitability of p, i.e., p must hold along all paths whose cumulative duration satisfies I. Thus, the RTCTLC formula:

1. $AG(deposit \rightarrow AF_{[0,4)} delivered)$ specifies that whenever the Buyer pays for the goods, then the Seller will deliver the goods within cumulative duration of 3 days. Observe that $AF_I \varphi := A(true U_I \varphi)$, $AG_I \varphi := \neg EF_I(\neg\varphi)$.
2. $AG(delivered \Rightarrow (EF_{[0,15)} goodsOk \vee EF_{[0,15)} goodsNotOk))$ specifies that whenever the Seller ships the goods, then the Buyer accepts or rejects the goods within cumulative duration of 14 days. Note that $EF_I \varphi := E(true U_I \varphi)$.
3. $AG_{[0,4)}(CC_{S \rightarrow B}(deposit, AF_{[0,4)} delivered))$ specifies that along all paths within cumulative duration of 3 days the Seller S commits to the Buyer B to ship the requested goods within cumulative duration of 3 days if B has sent the agreed payment.
4. $AG_{[0,4)}(Fu(CC_{S \rightarrow B}(deposit, AF_{[0,4)} delivered)))$ specifies the fulfillment of the above Seller's commitment.
5. $AG_{[0,4)}(CC_{S \rightarrow \{E,B\}}(deposit, AF_{[0,4)} delivered))$ specifies that along all paths within cumulative duration of 3 days the Seller S commits to both to the Escrow E and to the Buyer B to ship the requested goods within cumulative duration of 3 days if B has sent the agreed payment.
6. $AG_{[0,4)}(Fu(CC_{S \rightarrow \{E,B\}}(deposit, AF_{[0,4)} delivered)))$ specifies the fulfillment of the above Seller's commitment.
7. $AG(goodsNotOk \Rightarrow CC_{E \rightarrow B}(returnGoods, AF_{[0,3)} refund))$ specifies that along all paths if the Buyer is not satisfied, then the Escrow commits to the Buyer to refund the deposit within cumulative duration of 2 days under condition that the Buyer has returned the goods to the Seller.

4   Conclusion

We shown how to extend the logic $RTCTL^{CC}$ [6] to the group quantitative conditional commitment logic RTCTLC, which is suitable not only for reasoning about conditional commitments and their fulfillments of a single real-time agent but also for reasoning about conditional commitments and their fulfillments by groups of agents. We proposed a new semantics of RTCTLC that is defined over the DCIS that allows for arbitrary durations over the global transitions. Thereby, we are able to reason about different levels of temporal deadlines and to reduce extra verification work resulting from the use of unit measure steps. We illustrated the proposed formalism by means of the Escrow protocol.

In future, we plan to develop a new model checking algorithm for RTCTLC and a new bounded model checking algorithm for its existential version.

References

1. What is escrow? How does escrow work? https://www.escrow.com/
2. Bentahar, J., Moulin, B., Meyer, J., Lespérance, Y.: A new logical semantics for agent communication. In: Proceedings of the CLIMA VII. LNAI, vol. 4371, pp. 151–170. Springer-Verlag (2007)
3. El-Menshawy, M., Bentahar, J., Kholy, W.E., Dssouli, R.: Reducing model checking commitments for agent communication to model checking ARCTL and GCTL*. Auton. Agent. Multi-agent Syst. **27**(3), 375–418 (2013)
4. Emerson, E.A., Mok, A., Sistla, A.P., Srinivasan, J.: Quantitative temporal reasoning. Real-Time Syst. **4**(4), 331–352 (1992)
5. Fagin, R., Halpern, J.Y., Moses, Y., Vardi, M.Y.: Reasoning about Knowledge. MIT Press, Cambridge (1995)
6. Kholy, W.E., Menshawy, M.E., Laarej, A., Bentahar, J., Al-Saqqar, F., Dssouli, R.: Real-time conditional commitment logic. In: Proceedings of the PRIMA 2015. LNCS, vol. 9387, pp. 547–556. Springer (2015)
7. Weiss, G.: Multi-agent Systems: A Modern Approach to Distributed Artificial Intelligence. MIT Press, Cambridge (1999)
8. Woźna-Szcześniak, B.: Trends in contemporary computer science. In: Formal Methods and Data Mining. On the SAT-based Verification of Communicative Commitments, pp. 175–186. Białystok University of Technology Publishing Office (2014)

Moodsically. Personal Music Management Tool with Automatic Classification of Emotions

Jorge García Vicente, Ana B. Gil$^{(\boxtimes)}$, Ana de Luis Reboredo,
Diego Sánchez-Moreno, and María N. Moreno-García

Dpto. de Informática y Automática – Facultad de Ciencias,
University of Salamanca, Plaza de la Merced s/n, 37008 Salamanca, Spain
{ull6625,abg,adeluis,mmg}@usal.es,
sanchezhh@gmail.com

Abstract. It is a fact that music is directly linked to emotions. Various researches study the link between musical characteristics and the feeling produced or even induced. This work shows a web tool that allows the automatically extraction of musical characteristics of songs including the emotional classification and it uses this metadata to manage the user playlist in streaming. The objective of this work has been to contribute to improve a streaming music tool with perceptual characteristics associated with emotions and musical descriptors elements. The tool provides profile management, such as search engine, customizable playlist generation and song's recommender alignment with emotional elements associated with music characteristics.

Keywords: Sentiment analysis · Music streaming
User profiling and personalization

1 Introduction

Music is fundamental for the human being as a natural way of expression and communication. It allows the construction of a personal identity and collective identities. It is a fact that music directly links to emotions. The music stimulates the cerebral centers that move the emotions and allow manifesting our musical feeling. Although the field of emotions is very subjective, several investigations have tried to analyze which musical characteristics influence the emotional reactions of people, reaching in certain cases to concrete conclusions. From these studies, various algorithms have been developed and purpose the automatic classification of songs by mood or emotion.

On the other hand, in recent years you cannot talk about music without taking into account the rise in digital streaming music platforms, which offer users to enjoy their music anywhere and anytime. On these platforms, you have pre-set playlists as well as customizable playlists.

Moodsically emerges as the idea to combine the automatic detection of emotions in music and streaming playback.

The rest of the paper is organized as follows: Sect. 2 includes a short survey of works about sentiment analysis and their main drawbacks, with special focus on mood

© Springer International Publishing AG, part of Springer Nature 2019
F. De La Prieta et al. (Eds.): DCAI 2018, AISC 800, pp. 112–119, 2019.
https://doi.org/10.1007/978-3-319-94649-8_14

music classification methods. The proposed tool is described in Sect. 3. Finally, the conclusions are given in Sect. 4.

2 Music and Emotional Model

From the psychological perspective, two main approaches in the area of emotional modeling today are categorical and dimensional. In 1980, Russell published a model of dimensional type [11], which after almost 40 years is still current and constantly revised on emotional regulation and its parameters. It is a model of dimensional type. The circumflex model of affect, see Fig. 1, proposes that all affective states arise from two fundamental neurophysiological systems, one related to *valence* (a pleasure-displeasure continuum) and the other to *arousal*, or levels of excitation. Each emotion can be understood as a linear combination of these two dimensions, or as varying degrees of both valence and arousal. This dimensional model is suitable for emotions recognition experiments because it can locate discrete emotions in its bidimensional space.

Arousal and valence levels represent different aspects but related to the emotional response. According to the relationship between excitement levels and mood, listening to music has a direct effect on the levels of excitation in the state of mind. There is discrepancy between researchers of emotions in music; some of them who argue that music provokes basic emotions, while others deny that this is possible. Others say that music can induce only positive general emotions and negative and others that produces a range of both basic emotions, such as joy, confidence, surprise fear, sadness, aversion, anger or anticipation and complex emotions, for example, optimism, love, submission, fright, disappointment, remorse, contempt or treachery.

2.1 Musical Characteristics and Measure of Emotion

It has been demonstrated in several investigations that manipulating certain characteristics of music can evoke the emotion that we want [1, 3, 5], some of these characteristics are the rhythm, the intensity, the mode, the tempo, the articulation, pitch, the timbre, etc. A brief summary is detailed in Table 1.

Neuropsychology is the discipline that studies the functioning of the brain in order to obtain information about the biological processes of perception and emotion in the human being. With regard to the physiological signals to be taken into account in the recognition of emotions, electrocardiograms have been proposed in the studies together with other signals of the nervous system such as systolic and diastolic blood pressure, dilation of the pupils, respiration, temperature and conductance of the skin. In another study, it was necessary to explore the relationships between several structural characteristics of music, as well as the reports of pleasant experience and activation and different physiological measures (respiratory parameters, skin conductance, or heart rhythm) [2, 10].

In order to systematically evaluate the state of the art of the Music Information Retrieval (MIR) algorithms, the Annual Music Information Retrieval Evaluation eXchange (MIREX), included in 2007 an Audio Mood Classification (AMC) task [9]. Since there, the interest of MIR researchers to classify music by moods is increasing.

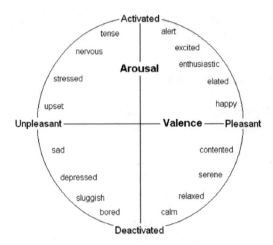

Fig. 1. Emotion Russell model

With the use of artificial intelligence methods, researchers have begun to study how to use computers for emotional modelling [5, 6, 8].

The study of musical emotion is currently a multidisciplinary task, with philosophical, musicological, psychological, computational, biological, anthropological, sociological and even therapeutic considerations. The above studies have found that from any perspective, there is a correlation between music and human emotion. With the big data, the emotional semantic and by neuroscience analysis and processing of large amounts of music data has become a hot topic of research.

2.2 Web Applications with Support for the Emotional Filter in Music Selection

Web applications are beginning to appear as user assistant by including the use of emotions. This challenge supporting a better selection or user-profiling to musical pieces. Web applications are beginning to appear by assistant the use of emotions in some stage to support access to musical pieces. Spotify[1], is undoubtedly the platform for online music reproduction best known to all Internet users. Spotify has predetermined playlists classified according to different moods and activities (Happy, Melancholia, Mood Bosster, ect.). Radioline[2] offers the user the possibility to choose radios that reproduce songs by a type of emotion; it offers eight possibilities: calm, dark, dreamy, energetic, happy, lucky, positive and romantic. Rockola[3] is an interactive music radio in which you can choose music by mood among other options through its graphic search (see Fig. 2). Create playlists based on the following moods: relaxed, sentimental, sad, romantic, animated, happy, melancholy and energetic.

---

[1] https://open.spotify.com/.

[2] http://www.radioline.co/online-radios-podcasts/moods.

[3] http://rockola.fm/.

Table 1. Musical features correlated with discrete emotions in musical expression [10].

Emotion	Musical features
Happiness	Fast tempo, small tempo variability, major mode, simple and consonant harmony, medium-high sound level, small sound. level variability, high pitch, much pitch variability, wide pitch range, ascending pitch, perfect 4th and 5th intervals, rising micro intonation, raised singer's formant, staccato articulation, large articulation variability, smooth and fluent rhythm, bright timbre, fast tone attacks, small timing variability, sharp contrasts between "long" and "short" notes, medium-fast vibrato rate, medium vibrato extent, micro-structural regularity
Sadness	Slow tempo, minor mode, dissonance, low sound level, moderate sound level variability, low pitch, narrow pitch range, descending pitch, "flat" (or falling) intonation, small intervals (e.g., minor 2nd), lowered singer's formant, legato articulation, small articulation variability, dull timbre, slow tone attacks, large timing variability (e.g., rubato), soft contrasts between "long" and "short" notes, pauses, slow vibrato, small vibrato extent, ritardando, micro-structural irregularity
Anger	Fast tempo, small tempo variability, minor mode, atonality, dissonance, high sound level, small loudness variability, high, pitch, small pitch variability, ascending pitch, major 7th and augmented 4th intervals, raised singer's formant, staccato articulation, moderate articulation variability, complex rhythm, sudden rhythmic changes (e.g., syncopations), sharp timbre, spectral noise, fast tone attacks/decays, small timing variability, accents on tonally unstable notes, sharp contrasts between "long" and "short" notes, accelerando, medium-fast vibrato rate, large vibrato extent, micro-structural irregularity
Fear	Fast tempo, large tempo variability, minor mode, dissonance, low sound level, large sound level variability, rapid changes in sound level, high pitch, ascending pitch, wide pitch range, large pitch contrasts, staccato articulation, large articulation variability, jerky rhythms, soft timbre, very large timing variability, pauses, soft tone attacks, fast vibrato rate, small vibrato extent, micro-structural irregularity
Tenderness	Slow tempo, major mode, consonance, medium-low sound level, small sound level variability, low pitch, fairly narrow pitch range, lowered singer's formant, legato articulation, small articulation variability, slow tone attacks, soft timbre, moderate timing variability, soft contrasts between long and short notes, accents on tonally stable notes, medium fast vibrato, small vibrato extent, micro-structural regularity

Allmusic[4] is a web application where you can search for songs by genre, situation of the moment and moods divided into almost three hundred. Once the song were found, you can add the album in which it is published to the user's collection, but it does not allow you to save or play playlists (Fig. 3).

---

[4] https://www.allmusic.com/advanced-search.

Fig. 2. Rockola interface for searches including moods.de Rockola.

Fig. 3. Allmusic interface for searches including moods.

3 Moodsically

Moodsically emerges as a personal music content manager with the idea to combine the automatic detection of emotions along others audio descriptors in music along with streaming playback. The system is implemented using a client-server architecture, the client being any navigator for supporting any devices including mobiles. On the server is the RS, which is responsible for executing the analysing and classification algorithm with the music sent by the client. This analysis and classification is upon the use of Essentia[5] and Gaia[6] library inside the server architecture. The different functionalities processes are carried out by a request of the application by REpresentational State Transfer (REST[7]) services, with which the server itself returns a list of songs that adapts to the profile of the user in the case of playlist tool utilities.

Essentia is an open source library for audio analysis and audio-based musical information retrieval [7]. It contains an extensive collection of reusable algorithms that implement audio input and output function, standard processing of digital signal blocks, statistical characterization of data, and a broad set of musical, spectrum, temporal, tonal and audio descriptors at high level. Essentia is completed with Gaia library linked to Python that implements measures of similarity and classification in the results obtained from the audio analysis, and generates classification models that Essentia can use to calculate high-level description of the music. Among the elements obtained are emotions, tonality, musical style, timbre, whether it is voice or instrumental. It allows the obtaining of high-level descriptors by classifying the low level through SVM [4] with previously trained models.

Moodsically extracts key low-level descriptors: mode, danceability, beats per minute and volume while high-level descriptors as the musical genre, timbre, tonality and emotion (sadness, happiness, party or relaxed) briefly explaining in next section. The extracted metadata are integrated in the user interface, enriching the characteristics of each song (see Fig. 4).

3.1   Extraction of Emotion in Music into the Tool

To classify the songs by emotions, the initial classification stands on extracts the valence and activation probability from song, with 0 to 1 values that determine sadness, happiness, celebration and relaxed moods. The valence (Eq. 1) is calculated by subtracting the probability that the song is sad (*PSad*) to the probability that it is happy (*PHappy*). For arousal (Eq. 2) by subtracting the probability that it is relaxed (*PRelax*) to the probability that it will be a party music (*PParty*). This gives the value of the valence for the X-axis and the activation value for the Y-axis.

$$Valence = PSad - PHappy \qquad (1)$$

$$Arousal = PRelax - PParty \qquad (2)$$

Once the angle and the distance value has been calculated in polar coordinates (Eqs. 3 and 4), the song is classified according to the values based on the 12-segment affective circumference [11, 12].

---

[5] http://essentia.upf.edu/.

[6] http://essentia.upf.edu/documentation/gaia/.

[7] https://en.wikipedia.org/wiki/Representational_state_transfer.

$$\theta = valence/arousal \tag{3}$$

$$r = \sqrt{valence^2 + arousal^2} \tag{4}$$

Fig. 4. Metadata annotation extracted from song

All the music is labeled into the database system with any emotion metadata annotation classified into 12 emotions (happy, exalted, excited, active, angry, frustrated, sad, depressed, bored, calm, relaxed and serene). That allows to connect with several functionalities (recommendation systems, playlist creation/reproduction, etc.) and then with the user interface in the tool, where the emotion is associated with an emoticon (See Fig. 5).

Fig. 5. (Left) Playlist interface – (Right) Playlist generation user tool

4 Conclusions and Future Work

There is a correlation between music and human emotion. The questions of how to give computers the ability to identify and include emotions parameters into human-computer interaction open new challenges facing with artificial intelligence today. The objective of this work has been to develop an application that extracts content information from the songs uploaded by the user, by including emotion and musical characteristics. Moodsically makes available a more accurate management increasing the playlist creation, searching processes or recommender utilities by adding emotional aspects. So far, we only have references of the tools shown in Sect. 2.2. Moreover, none possesses the functionality of the tool shown.

Several are the current open work lines to increase the potential of Moodsically while measure the results in use. One of them by adding a social network in order to propose social based recommendation and other functionalities. Another to include emotional aspects linked to biological data through sensors. Both open works would allow improving the recommendation system with the incorporation of usage data and undiscovered relationships between users.

Acknowledgments. This work was supported by the Spanish Ministry of Economy and Competitiveness and FEDER funds. Project SURF (TIN2015-65515-C4-3-R).

References

1. Kawakami, A., Furukawa, K., Katahira, K., Kamiyama, K., Okanoya, K.: Relations between musical structures and perceived and felt emotions. Music Percept. Interdisc. J. **30**(40), 407–417 (2012)
2. Smeaton, A.F., Rothwell, S.: Biometric responses to music-rich segments in films: the CDVPlex. In: Proceedings of the 2009 Seventh International Workshop on Content-Based Multimedia Indexing, CBMI 2009, pp. 162–168 (2009)
3. Bhat, A.S., Prasad, N.S., Amith, V.S., Murali Mohan, D.: An efficient classification algorithm for music mood detection. In: Western and Hindi Music Using Audio Feature Extraction (2014)
4. Burges, C.J.C.: A tutorial on support vector machines for pattern recognition. Data Mining Knowl. Disc. **2**(2), 121–167 (1998)
5. Laurier, C., Meyers, O., Serra, J., Blech, M., Herrera, P., Serra, X.: Indexing music by mood: design and integration of an automatic content-based annotator. Multimed. Tools Appl. **48**(1), 161–184 (2009)
6. Deng, J.J., et al.: Emotional states associated with music: classification, prediction of changes, and consideration in recommendation. ACM Trans. Interact. Intell. Syst. (TiiS) **5**(1), 4 (2015)
7. Bogdanov, D., Wack, N., Gómez, E., Gulati, S., Herrera, P., Mayor, O., Roma, G., Salamon, J., Zapata, J., Serra, X.: ESSENTIA: an open-source library for sound and music analysis. In: Proceedings of the 21st ACM International Conference on Multimedia (MM 2013), pp. 855–858. ACM, New York (2013)
8. Hu, X., Choi, K., Downie, J.S.: A framework for evaluating multimodal music mood classification. J. Assoc. Inf. Sci. Technol. **68**(2), 273–285 (2017)
9. Hu, X., Downie, J.S., Laurier, C., Bay, M., Ehmann, A.F.: The 2007 mirex audio mood classification task: lessons learned. In: ISMIR 2008 – 9th International Conference on Music Information Retrieval, pp. 462–467 (2008)
10. Juslin, P.N., Laukka, P.: Expression, perception, and induction of musical emotions: a review and a questionnaire study of everyday listening. J. New Music Res. **33**, 217–238 (2004)
11. Russell, J.A.: A circumplex model of affect. J. Pers. Soc. Psychol. **39**(6), 1161–1178 (1980)
12. Yik, J.M., Russell, J.A., Steiger, J.H.: A 12-point circumplex structure of core affect. Emotion **11**(4), 705 (2011)

Arrhythmia Detection Using Convolutional Neural Models

Jorge Torres Ruiz[1(✉)], Julio David Buldain Pérez[2],
and José Ramón Beltrán Blázquez[2]

[1] Accenture, María Zambrano 31, 50018 Zaragoza, Spain
jorge.torres.ruiz93@gmail.com
[2] Department of Electronic and Communication Engineering,
University of Zaragoza, 50018 Zaragoza, Spain
{buldain, jrbelbla}@unizar.es

Abstract. Mostly all works dealing with ECG signal and Convolutional Network approach use 1D CNNs and must train them from scratch, usually applying a signal preprocessing, such as noise reduction, R-peak detection or heartbeat detection. Instead, our approach was focused on demonstrating that effective transfer learning from 2D CNNs can be done using a well-known CNN called AlexNet, that was trained using real images from ImageNet Large Scale Visual Recognition Challenge (ILSVRC) 2012. From any temporal signal, it is possible to generate spectral images (spectrograms) than can be analysed by 2D CNN to do the task of extracting automatic features for the classification stage. In this work, the power spectrogram is generated from a randomly ECG segment, so no conditions of signal extraction are applied. After processing the spectrogram with the CNN, its outputs are used as relevant features to be discriminated by a Multi Layer Perceptron (MLP) which classifies them into arrhythmic or normal rhythm segments. The results obtained are in the 90% accuracy range, as good as the state of the art published with 1D CNNs, confirming that transfer learning is a good strategy to develop decision models in signal and image medical tasks.

Keywords: Neural network · Transfer learning · Electrocardiogram
Arrhythmia detection · Convolutional network · Spectrogram

1 Introduction

An electrocardiogram (ECG) is a medical test that helps doctors to diagnose heart pathologies. This signal is recorded with a series of electrodes placed over the patient's body, typically on his chest, that measure the heart's electrical activity. Over the years, algorithms have tried to help doctors in ECG analysis exploring several ways such as eliminating noise, marking the beat's starting point or calculating the beat variability [1–4]. But, as ECG is a very complex signal, algorithms have problems to produce a real diagnosis. Recently, machine learning has entered in this field to overcome these difficulties. Bibliography shows that neural networks output outstanding results,

© Springer International Publishing AG, part of Springer Nature 2019
F. De La Prieta et al. (Eds.): DCAI 2018, AISC 800, pp. 120–127, 2019.
https://doi.org/10.1007/978-3-319-94649-8_15

reaching over 90% accuracy in arrhythmia detection and, in some particular applications, using deep learning the results increase almost to 100%, as in Kiranyaz et al. [5].

Most studies use electrocardiographic signals and apply some type of preprocessing to train neural networks in ECG analysis. We have tried to explore this classification problem from a different point of view. This paper's main idea was inspired in the research carried by Nguyen and Bui [6]. They tried to classify vocal audio signals using their spectrogram for searching similarities with other previously classified spectrograms. Following this idea, we used a convolutional neural network (CNN) to extract automatically characteristics from spectrograms which have been generated from ECG segments.

Several studies have used CNNs to extract features from ECG signals and evaluate these features for different purposes such heartbeat classification or QRS detection. Rajendra et al. [7] trained a 1D CNN in order to extract characteristics from ECG beat and classified them into five different heartbeat types (non-ectopic, supraventricular ectopic, ventricular ectopic, fusion and unknown). They used wavelets to remove noise and a R-peak detection algorithm to segment the signal into heartbeats for feeding the CNN. The output layer from the CNN was a fully-connected layer with five neurons which determine the heartbeat type from the fed ECG beat. In a similar way, Pyakillya et al. [8] tried to classify heartbeats into four different types (normal sinus rhythm, arrhythmic, other kind of rhythm and very noisy) using a 1D CNN. They preprocessed the signal by applying standard procedures like mean subtraction and fed the CNN with that data, where its last layer was a softmax layer with four outputs. On the other hand, Xiang et al. [9] combined two 1D CNNs in order to automatically detect QRS complex from ECG signals. They fed one CNN with ECG segments, that were preprocessed by averaging and difference operations, to extract coarse-grained features. Another CNN was fed with ECG segments preprocessed by difference operation to extract fine-grained features. The features extracted from both CNNs were concatenated and sent to a MLP which determines if a QRS complex appears in the signal received.

Mostly all works dealing with ECG signal and CNN approach use 1D CNNs and must train them from scratch, usually applying a signal preprocessing, such noise reduction, R-peak detection or heartbeat detection. Instead, our approach was focused on demonstrating that effective transfer learning from 2D CNNs can be done using a well known CNN architecture trained by Krizhevsky et al. [10] called AlexNet. This CNN was trained using real images from ImageNet Large Scale Visual Recognition Challenge (ILSVRC) 2012. ImageNet consists of 1.2 million 256×256 images belonging to 1000 categories. In Hoo-Chang et al. [11] it is proposed that this transfer learning can accelerate substantially the model development when using images in medical decision models. Instead of using the temporal signal, it is possible to generate spectral images (spectrograms) than can be analysed by 2D CNN to do the same task of extracting automatic features for the classification stage. In this work, the power spectrogram is generated from a randomly ECG segment, so no conditions of signal extraction are applied. The phase spectrogram is not included in this analysis and is left to future works to determine if its inclusion is useful or not. The only specification of the spectrogram image is that it must be adjusted to fit with the input of the AlexNet. After the CNN has processed the spectrogram, its outputs are used as relevant features to be discriminated by a Multi Layer Perceptron (MLP) which classifies them into arrhythmic or normal rhythm segment. Fine tuning of the AlexNet was not applied during the learning stage of the MLP.

We have used the MIT-BIH Arrhythmia Database [12] to generate training and validation samples. This database has become one of the most popular databases used in arrhythmia detection or classification. It contains 48 half-hour records where each record presents two electrocardiograms took on different derivations. All electrocardiograms have been revised by independent cardiologists, marking the beginning of each beat, the type of beat and additional information such as rhythm changes or noise artefacts. These characteristics make the database well suited for machine learning.

In the next section, the methodology for generating input data to the models is explained. Section 3 shows the first experiment results, using the outputs of the second convolutional layer of AlexNet as features for a MLP with one hidden layer. Section 4 deals with the results obtained when the MLP has two hidden layers, and Sect. 5 explain a second experiment where the output of the last convolutional layer from the CNN is used as features, showing that results get better as deeper is the representation used from the CNN. Last section makes a discussion and proposes future investigations.

2 Methodology

From each signal in the database, we extracted segments of 2048 samples in both derivations with same starting point randomly generated. For both segments we calculated their spectrogram using a Hamming window of 2048/60 samples and 50% overlapping, without applying a low-pass filter. The frequency range covered up to 360 Hz, which is the maximum frequency present in the ECGs from the database (Fig. 1).

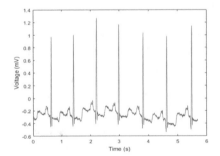 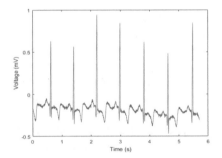

Fig. 1. Left image shows a ECG segment from one signal in MIT-BIH database. The image at the right plots a segment extracted from the same patient's signal but with the second derivation.

The trained convolutional network (AlexNet) chosen for the feature extraction stage requires an image input of size 227 × 227 pixels, for this reason we decided to combine spectrograms from both derivations to generate this input image format, by generating a matrix of 227 × 114 elements from the first spectrogram and a matrix of 227 × 113 elements from the second. We scaled each matrix to range [0–255] in order to store the data as images with one byte per pixel. With both matrixes scaled to the same range, we combined them over the temporal axis saving the result as a grayscale image.

From each electrocardiogram record stored in the database, we have generated 602 images, being 301 from its first half signal and another 301 images from the second half. We have used the first half samples for training and the second ones for validating the MLPs (Fig. 2).

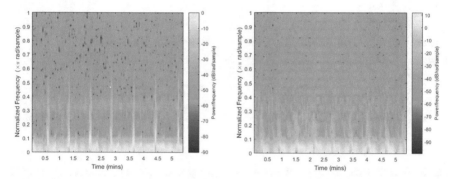

Fig. 2. Both figures show spectrograms generated from signal segments with 2048 samples, where the left one has been taken from a normal rhythm patient and the right one corresponds to an ill patient with a heart pathology. It can be noticed that the heart pathology changes the rhythm that a normal heart would make, having more power across its higher frequencies.

3 First Experiment

As first experiment, the spectrogram images were processed by AlexNet, extracting its activations on the second convolutional layer. This decision was based in the fact that higher layers in a convolutional network tend to present specialized filters for the kind of objects that the images used for training contain. As AlexNet was trained with real world images, we expected to eliminate this filter particularization influence by using low level layers activations as the first trial. So, we used these characteristics to train one MLP that will detect if the segment contains arrhythmic beats. We configured a MLP with one hidden layer and one sigmoid neuron on its output layer, which will output a response in range [0–1], where 0 means that the segment contains arrhythmic beats and 1 that the beats are normal.

We varied the number of hidden neurons over different trials in order to estimate the number of hidden neurons that gets the best classification, trying with 60, 80, 100, 110, 120, 130, 150 and 200 neurons. Finally, we calculated the confusion matrix for each classifier in order to compare their classifying performance, extracting the classification measures: TPR (true positive rate), FNR (false negative rate), FPR (false positive rate), TNR (true negative rate), PPV (possible predictive value), FDR (false discovery rate), FOR (false omission rate), NPV (negative predictive value), and the overall accuracy. As Fig. 3 shows, the classifier with 130 hidden neurons achieves the best results, having an accuracy over 85%.

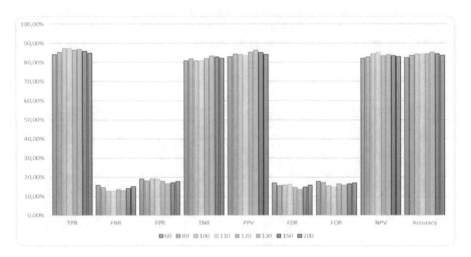

Fig. 3. Overall, the classifiers with one hidden layer give an accuracy between 82% and 85%. The best test results are achieved by the MLP with 130 hidden neurons, reaching an accuracy of 85.3%. Also, the false positive rate and the false negative rate are below 20%.

4   Test for MLP with Two Hidden Layers

After seeing that a MLP with one hidden layer can classify segments with more than 80% accuracy, we tested if a two hidden layer MLP can improve the results. Increasing the number of hidden layers of a MLP allows to generate better decision regions when the problem presents a non-linear behavior. We used a pyramidal architecture with a larger number of neurons in the first hidden layer and a small number in the second, trying with 100-25-1, 120-30-1, 120-50-1 and 300-100-1 architectures. Overall the results increased in a 2% respect to the previous classifiers. The test classification results are shown in Fig. 4.

5   Second Experiment

As second experiment, we tested if extracting the activations on the last convolutional layer from AlexNet can improve the results. This idea was inspired in the research carried by Simonyan and Zisserman [13] where they claim that increasing the number of convolutional layers helps in the image processing. So, we trained the previous MLPs with activations extracted from the last convolutional layer. The test results are shown in Fig. 5. Overall, all the classifiers improved the results that achieved in the

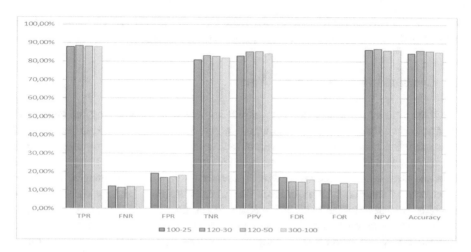

Fig. 4. Results from classifiers with two hidden layers improved the overall results about 1–2%, reaching the MLP with 120 neurons on its first layer and 30 on its last hidden layer an accuracy of 86%.

first experiment. This result is shocking when we consider that the last convolutional layer of the AlexNet includes specialized filters that recognize shapes of real world things like cars, dogs, houses etc. Even with this set of filters, so little adapted to the kind of images present in a spectrogram, increasing the depth of the CNN substantially improves the results of the classification. The best result gives an accuracy around 90%

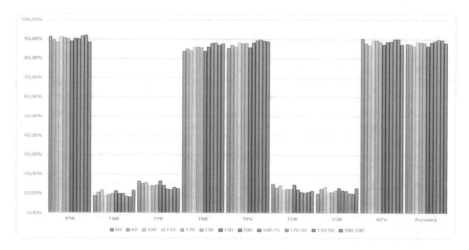

Fig. 5. Results from classifiers trained with activations from the AlexNet's last convolutional layer. The accuracy increased almost 4 points compared to the previous experiment, getting close to 90% accuracy. MLP with two hidden layer and architecture of 120 neurons on its first hidden layer and 30 on its second hidden layer gives an accuracy of 89.9%, classifying the arrhythmic segments with 89.7% precision and the normal segments with 90.1%.

and was obtained for the MLP with two hidden layers, with 120 and 30 neurons, which was trained with activations from the last convolutional layer.

6   Conclusions and Future Work

We tried transfer learning method to explore a new way of analysing the electrocardiograms, avoiding the hand-made signal preprocessing and extracting the relevant characteristics automatically. The results show that our first try can classify ECG segments with a 90% accuracy, which is in the state of the art level. This good result is obtained with a convolutional network generated with real world images, instead of spectrograms images, which confirms that transfer learning strategy can be applied easily as first development stage in medical image classification tasks. As we do not apply any fine-tuning of the CNN, we can think that the classification accuracy could increase a few more points with simple improvements. Our main interest for future research is to generate CNNs, specifically trained for analysing spectrograms of electric signals, that can be used as transferable modules for automatic feature generation in other biomedical fields as electroencephalography and electromyography. Another aspect to explore is if phase coefficients can contribute to get better accuracy, also dealt and processed with specifical CNNs. Another interesting goal is that the exact classification type of the arrhythmia signal could be tried with this methodology.

Acknowledgements. We would like to thank the company Sallén Tech of the Gunnevo group for financing the publication of this work.

References

1. Ho, K.K.L., Moody, G.B., Peng, C.-K.: Predicting survival in heart failure cases and controls using fully automated methods for deriving nonlinear and conventional indices of heart rate dynamics. Circulation **96**, 842–848 (1997)
2. Thakor, N.V., Zhu, Y.-S.: Applications of adaptive filtering to ECG analysis: noise cancellation and arrhythmia detection. IEEE Trans. Biomed. Eng. **38**(8), 785–794 (1991)
3. Antunes, E., Brugada, J., Steurer, G., Andries, E., Brugada, P.: The Differential Diagnosis of a Regular Tachycardia with a Wide QRS Complex on the 12-Lead ECG: Ventricular Tachycardia, Supraventricular Tachycardia with Aberrant Intraventricular Conduction, and Supraventricular Tachycardia with Anterograde Conduction over an Accessory Pathway (1994)
4. Tsipouras, M.G., Fotiadis, D.I., Sideris, D.: An arrhythmia classification system based on the RR-interval signal. Artif. Intell. Med. **33**, 237–250 (2005)
5. Kiranyaz, S., Ince, T., Hamila, R., Gabbouj, M.: Convolutional neural networks for patient-specific ECG classification. In: 37th IEEE Engineering in Medicine and Biology Society Conference (EMBC 2015) (2015)
6. Nguyen, Q.T., Bui, T.D.: Speech classification using SIFT features on spectrogram images. Vietnam J. Comput. Sci. **3**(4), 247–257 (2016)
7. Acharyaa, U.R., Oha, S.L., Hagiwaraa, Y., Tana, J.H., Adama, M., Gertychd, A., Sane, T.R.: A deep convolutional neural network model to classify heartbeats. Comput. Biol. Med. **89**, 389–396 (2017)

8. Pyakillya, B., Kazachenko, N., Mikhailovsky, N.: Deep learning for ECG classification. IOP Conf. Series J. Phys. Conf. Series **913** (2017). 012004

9. Xiang, Y., Lin, Z., Meng, J.: Automatic QRS complex detection using two-level convolutional neural network. BioMed. Eng. OnLine (2018)

10. Krizhevsky, A., Sutskever, I., Hinton, G.E.: ImageNet classification with deep convolutional neural networks. In: NIPS (2012)

11. Hoo-Chang, S., Roth, H.R., Gao, M., Le, L., Ziyue, X., Nogues, I., Yao, J., Mollura, D., Summers, R.M.: Deep convolutional neural networks for computer-aided detection: CNN architectures, dataset characteristics and transfer learning. IEEE Trans. Med. Imaging **35**(5), 1285–1298 (2016)

12. MIT-BIH Arrhythmia Database [Internet]. Harvard-MIT Division of Health Sciences and Technology (1980). https://www.physionet.org/physiobank/database/mitdb/. Accessed Feb 2018

13. Simonyan, K., Zisserman, A.: Very deep convolutional networks for large-scale image recognition. In: ICLR (2015)

A Novel Sentence Vector Generation Method Based on Autoencoder and Bi-directional LSTM

Kiyohito Fukuda$^{(\boxtimes)}$, Naoki Mori, and Keinosuke Matsumoto

Osaka Prefecture University,
1-1 Gakuen-cho, Nakaku, Sakai, Osaka 599-8531, Japan
fukuda@ss.cs.osakafu-u.ac.jp

Abstract. Recently, dramatic performance improvement in computing has enabled a breakthrough in machine learning technologies. Against this background, generating distributed representation of discrete symbols such as natural languages and images has attracted considerable interest. In the field of natural language processing, word2vec, a method to generate distributed representations of words is well known and its effectiveness well reported. However, an effective method to generate the distributed representation of sentences and documents has not yet been reported.

In this study, we propose a method of generating the distributed representation of sentences by using an autoencoder based on bi-directional long short-term memory (BiLSTM). To obtain the information and findings that necessary to generate effective representations, the computational experiments are carried out.

Keywords: Sentence vector · Bi-directional long short-term memory
Autoencoder

1 Introduction

Recently, dramatic performance improvement in computing has enabled a breakthrough in machine learning technology. Against this background, generating distributed representation of discrete symbols such as natural languages [1] and images [2] has attracted considerable interest. By generating distributed representations of discrete symbols, we can define similarity in meaning and grammar between those symbols as the distance between them. Therefore, such representations can be expected to be applied to various tasks that need to take into account the semantic relationships.

In the field of natural language processing, word2vec [3,4], a method to generate distributed representations of words is well known and its effectiveness well reported. Word2vec allows us to deal with semantic relationship between words. However, although there have been lots of studies on the distributed representation of sentences and documents [5,6], an effective method to generate such

F. De La Prieta et al. (Eds.): DCAI 2018, AISC 800, pp. 128–135, 2019.
https://doi.org/10.1007/978-3-319-94649-8_16

representation has not yet been reported. Thus, to analyze the information of the text such as novels and journal articles in units of sentences is very difficult problem.

In this study, with a view towards realizing generation of an effective distributed representation of sentences, we propose a method of generating the distributed representation by using an autoencoder [7] based on long short-term memory (LSTM) [8,9]. To obtain the information and findings that necessary to generate effective representations, the computational experiments are carried out.

Hereinafter, we refer to the distributed representations of words as word vectors, those of sentences as sentence vectors, and those of documents as document vectors.

2   Related Work

2.1   Word2Vec

Word2vec [3,4] is the most famous word vector generation method. It acquires the meaning of words by learning whether a word in sentence can be exchanged for the other word. It generate word vector by training neural networks to reconstruct words from context. Word vectors are weights of trained neural networks.

There are two types of models in word2vec, namely continuous bag-of-words (CBOW) model and skip-gram model. In CBOW model, word2vec learns to minimize the sum of error rate when predicting context words from the target word. On the other hand, it learns to minimize the sum of error rate when predicting the target word from context words in skip-gram model.

2.2   Doc2Vec

Doc2vec [10] is document vector generation method expanding the concept of word2vec to document. There are also two types of learning models in doc2vec, namely distributed bag-of-words (DBOW) model and distributed memory (DM) model. DBOW model is the learning model that corresponding to skip-gram model in word2vec. In DBOW model, doc2vec learns to predict word vector of each word contained in document from document vector. On the other hand, DM model is the learning model that corresponding to CBOW model in word2vec. In DM model, it learns to predict the word vector of focus word in document from word vectors of context words and document vector.

3   Proposed Method

In this study, we propose the sentence vector generation method by using an autoencoder based on LSTM. Figure 1 shows the outline of the proposed method.

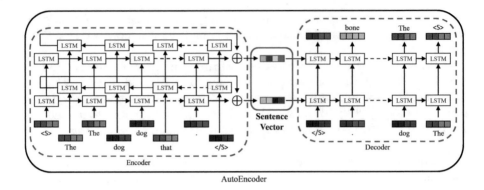

Fig. 1. Outline of the proposed method

3.1 Encoder and Decoder

In the proposed method, the encoder-decoder model [11] confirmed to be effective in the field of natural language processing is used. Encoder-decoder model is used as autoencoder by making the input of encoder and the output of decoder same. We think that intermediate representation connecting the encoder and decoder can be considered as the features of sentence. In addition, we construct the networks of encoder-decoder model by using LSTM because the proposed method has to deal with time-series data such as sentences.

The networks of the proposed method are constructed in multi layer structure. We expect that different information can be saved in intermediate representation in different layers of the encoder. We use a bi-directional LSTM (BiLSTM) [12] in the architecture of the encoder. By using it, the proposed method can consider the information on the whole input sequence at each time.

3.2 Data Form of Input and Output

Input and output in the proposed method are word vectors of each word obtained by morphological analysis of an input sentence. The algorithms of input and output data generation are as follows:

1. By performing morphological analysis on input sentence s, a sequence of N words $w_1 w_2 \cdots w_N$ is obtained.
2. The symbol representing beginning of sentence w_S is added to the beginning of word sequence. In addition, the symbol representing end of sentence w_E is added to the end of it. As a result, the word sequence size changes to $N + 2$.
3. The word vector $\boldsymbol{v}_w$ for each word in $w_S w_1 w_2 \cdots w_N w_E$ is obtained by the pre-trained word2vec.
4. Input set of the forward LSTM in the encoder represented by $\mathcal{X}_f$ is defined as

$$\mathcal{X}_f = \{\boldsymbol{v}_{w_S}, \boldsymbol{v}_{w_1}, \cdots, \boldsymbol{v}_{w_N}\} = \{\boldsymbol{v}_0, \boldsymbol{v}_1, \cdots, \boldsymbol{v}_N\}. \tag{1}$$

5. Input set of the backward LSTM in the encoder represented by $\mathcal{X}_b$ and input set in the decoder represented by $\mathcal{Y}_i$ are defined as

$$\mathcal{X}_b = \mathcal{Y}_i = \{\boldsymbol{v}_{w_E}, \boldsymbol{v}_{w_N}, \cdots, \boldsymbol{v}_{w_1}\} = \{\boldsymbol{v}_{N+1}, \boldsymbol{v}_N, \cdots, \boldsymbol{v}_1\}. \tag{2}$$

6. Output set in the decoder represented by $\mathcal{Y}_o$ is defined as

$$\mathcal{Y}_o = \{\boldsymbol{v}_{w_N}, \boldsymbol{v}_{w_{N-1}}, \cdots, \boldsymbol{v}_{w_S}\} = \{\boldsymbol{v}_N, \boldsymbol{v}_{N-1}, \cdots, \boldsymbol{v}_0\}. \tag{3}$$

Here, the size of input and output set are $|\mathcal{X}_f| = |\mathcal{X}_b| = |\mathcal{Y}_i| = |\mathcal{Y}_o| = N + 1$.

3.3 Learning and Generation of Sentence Vector

The proposed method is trained to output the same data as input data in encoder from decoder. The algorithms of model learning are as follows:

1. A sentence s is extracted from sentences for training.
2. Set $i = 1$.
3. By performing the operation explained in Sect. 3.2 to s, training data $\mathcal{X}_f, \mathcal{X}_b, \mathcal{Y}_i, \mathcal{Y}_o$ are generated.
4. The ith element of $\mathcal{X}_f$ represented by $\boldsymbol{x}_i^f$ is input in the forward LSTM of the encoder.
5. The ith element of $\mathcal{X}_b$ represented by $\boldsymbol{x}_i^b$ is input in the backward LSTM of the encoder.
6. i is updated to $i = i + 1$. After that, if $i \leq N + 1$, it returns to 4.
7. The sum of the hidden state vectors in the forward LSTM layers and the backward LSTM layers are inherited to the first hidden state vectors in the decoder.
8. Set $j = 1$.
9. The jth element of $\mathcal{Y}_i$ represented by $\boldsymbol{y}_j^i$ is input in the decoder. The output in the decoder at that time is defined as $\boldsymbol{o}_j^i$, and the error between $\boldsymbol{o}_j^i$ and the jth element of $\mathcal{Y}_o$ represented by $\boldsymbol{y}_j^o$ is defined as $L\left(\boldsymbol{o}_j^i, \boldsymbol{y}_j^o\right)$.
10. j is updated to $j = j + 1$. After that, if $j \leq N + 1$, it returns to 9.
11. By using back propagation, the model is trained to minimize the sum of errors in the decoder represented by $\sum_{j=1}^{N+1} L\left(\boldsymbol{o}_j^i, \boldsymbol{y}_j^o\right)$.

In addition, the algorithms of sentence vector generation are as follows:

1. By performing the operation explained in Sect. 3.2 to the sentence that wants to generate sentence vector represented by s, input data $\mathcal{X}_f, \mathcal{X}_b$ are generated.
2. Set $i = 1$.
3. The ith element of $\mathcal{X}_f$ represented by $\boldsymbol{x}_i^f$ is input in the forward LSTM of the encoder.
4. The ith element of $\mathcal{X}_b$ represented by $\boldsymbol{x}_i^b$ is input in the backward LSTM of the encoder.
5. i is updated to $i = i + 1$. After that, if $i \leq N + 1$, it returns to 3.
6. The sum of the hidden state vectors in the forward LSTM layers and the backward LSTM layers represented by $\boldsymbol{h}_s$ is obtained as the sentence vector of s.

4   Experiments

In this study, we compare the similarities between sentences calculated using sentence vectors generated by the proposed method and existing method to obtain the information and findings that is necessary to generate the effective sentence vectors. We use doc2vec as the existing method in this experiment.

Note that these experiments are carried out by using Japanese sentences because our research goal is to generate sentence vector for Japanese and automatically generate Japanese novels.

4.1   Advance Preparation

In the proposed method, word vectors are needed as the input data. In this paper, we used word2vec as a word vector generation method because it was confirmed to be effective and easy to implement. Table 1 shows the settings of word2vec. It was implemented with chainer which is a flexible framework for neural networks. The parameters that are not listed in Table 1 were default parameters. Japanese wikipedia text and 4566 novels contributed to Japanese website called "*Shōsetsuka ni narō*" [13] were used as the training data for word2vec. These novels were ranked top 100 in each genre during each period. The size of the training data was about 5.0 GB. The words whose frequency is less than or equal to the threshold were defined as unknown words.

Table 1. Settings of word2vec

Word2vec model	Skip-gram
Training method	Negative sampling
Window size	3
Vector size	200
Sampling size	5
Epochs	10
Batch size	5000
Optimizer	Adam
Learning rate α	0.000005
Threshold	5
Vocabulary	559676

4.2   Experimental Procedures

The experimental procedures are as follows:

1. 2 novels of 2 genres called "different world (love)" and "real world (love)" are obtained from "*Shōsetsuka ni narō*". Sentences of more than n_{min} words and less than n_{max} words not containing character name are randomly extracted.

2. Synonymous sentence set $\mathcal{S}$ is generated by replacing parts of sentences extracted in 1 with synonyms or synonymous expression.
3. 2 novels of genres named "different world (love)" and "real world (love)" are obtained from "*Shōsetsuka ni narō*". These novels are different from novels obtained in 1.
4. Candidate sentence set $\mathcal{C}$ is defined as the novels obtained in 1 and 3.
5. Sentence vectors of candidate sentence set and synonymous sentence set are generated by the proposed method and existing method.
6. The cosine similarities between synonymous sentence set and candidate sentence set are calculated with sentence vectors of 5. Count the number of synonymous sentences whose similarities with original sentences of them in candidate sentence set are within the top n. It is defined as N_c.
7. The accuracies of the proposed method and existing method represented by a are defined as

$$a = \frac{N_c}{|\mathcal{S}|} \tag{4}$$

4.3 Experimental Conditions

Table 2 shows the experimental conditions of the proposed method and doc2vec. The former were implemented with chainer and the latter were implemented with

Table 2. Experimental conditions

$n_{\min}$	10
$n_{\max}$	23
Training data size	8758720 sentences
Encoder architecture	2 layers (BLSTM, BLSTM)
Encoder unit size	(200, 200)
Decoder architecture	2 layers (LSTM, LSTM)
Decoder unit size	(200, 200)
Decoder loss function	Mean squared error
Epochs	30
Batch size	1000
Optimizer	Adam
Learning rate α	0.00001
Doc2vec model	DM
Window size	3
Vector size	200
Threshold	5
Candidate sentence set size	6787
Synonymous sentence set size	100

gensim. The parameters that are not listed in Table 2 were default parameters. Sentences of more than $n_{\min}$ words and less than $n_{\max}$ words that decompose 4566 novels contributed to "*Shōsetsuka ni narō*" by the period were used as the training data. Conversational sentences were removed from the training data.

4.4   Experimental Results

Table 3 shows the results of experiments. The accuracies of the proposed method's first and second layer are higher than those of existing method in all n in Table 3. This suggests that the proposed method can generate sentence vectors that can define the similarity in meaning. We thought that the proposed method could consider the word meaning because of inputting word vectors.

When we qualitatively analyzed original sentences and synonymous sentences that have high similarities, it was found that the different sentence information could be saved in the different layers of the proposed method. The sentence information on vocabulary and grammar can be saved in the proposed method's first layer and the abstract information such as sentence meaning can be saved in the proposed method's second layer. It was also found that the accuracies of the proposed method's second layer were higher than first layer. These suggest that the meaning of sentence is more important than grammar of it to quantitatively deal with similarity between sentences.

On the other hand, the accuracies of doc2vec are not high. This suggests that doc2vec is inappropriate for generating sentence vector because sentence has fewer words than document.

Table 3. Experimental results

	$a\ (N_{\mathrm{c}})$				
Top n sentences	$n=1$	$n=2$	$n=3$	$n=4$	$n=5$
First layer (proposed method)	0.90 (90)	0.90 (90)	0.92 (92)	0.93 (93)	0.93 (93)
Second layer (proposed method)	**0.91** (91)	**0.94** (94)	**0.94** (94)	**0.95** (95)	**0.96** (96)
Existing method (doc2vec)	0.72 (72)	0.77 (77)	0.78 (78)	0.80 (80)	0.81 (81)

5   Conclusion

In this study, we proposed the sentence vector generation method by using autoencoder based on LSTM. To obtain the information and findings that is necessary to generate the effective ones, computational experiments were carried out. The result of experiments was that the proposed method could generate sentence vectors because it could save different information on sentence by using multi layer networks based on BiLSTM.

In future work, we plan to carry out the comparison experiments with sentence vector generation method other than doc2vec such as skip-thought vector [14]. More detailed analysis of sentence vectors generated by the proposed method is also an important future work in order to investigate what kind of sentence information is saved in them. In our proposed method, no special technique to deal with Japanese sentences is not used though they have special features such as that Japanese is commutative between phrases and it is non-commutative within phrases. Therefore, we have to propose a novel method of generating distributed representations of Japanese sentences. In addition, we have to consider methods for analyzing novels in units of sentences and for automatically generating sentences based on sentence vectors as a first step of our final goal, which is an automatic novel generation.

Acknowledgements. This work was supported by JSPS KAKENHI Grant, Grant-in-Aid for Scientific Research(C), 26330282.

References

1. Pennington, J., Socher, R., Manning, C.D.: Glove: global vectors for word representation. In: EMNLP, vol. 14, pp. 1532–1543 (2014)
2. Vinyals, O., Toshev, A., Bengio, S., Erhan, D.: Show and tell: a neural image caption generator. In: Computer Vision and Pattern Recognition (2015)
3. Mikolov, T., Chen, K., Corrado, G., Dean, J.: Efficient estimation of word representations in vector space. CoRR, abs/1301.3781 (2013)
4. Mikolov, T., Sutskever, I., Chen, K., Corrado, G.S., Dean, J.: Distributed representations of words and phrases and their compositionality. In: Burges, C.J.C., Bottou, L., Welling, M., Ghahramani, Z., Weinberger, K.Q. (eds.) Advances in Neural Information Processing Systems, vol. 26, pp. 3111–3119. Curran Associates, Inc. (2013)
5. Logeswaran, L., Lee, H.: An efficient framework for learning sentence representations. In: International Conference on Learning Representations (2018)
6. Ponti, E.M., Vulic, I., Korhonen, A.: Decoding sentiment from distributed representations of sentences. CoRR, abs/1705.06369 (2017)
7. Hinton, G.E., Salakhutdinov, R.R.: Reducing the dimensionality of data with neural networks. Science **313**(5786), 504–507 (2006)
8. Gers, F.A., Schmidhuber, J.A., Cummins, F.A.: Learning to forget: continual prediction with LSTM. Neural Comput. **12**(10), 2451–2471 (2000)
9. Hochreiter, S., Schmidhuber, J.: Long short-term memory. Neural Comput. **9**, 1735–1780 (1997)
10. Le, Q.V., Mikolov, T.: Distributed representations of sentences and documents. CoRR, abs/1405.4053 (2014)
11. Sutskever, I., Vinyals, O., Le, Q.V.: Sequence to sequence learning with neural networks. In: Proceedings of the 27th International Conference on Neural Information Processing Systems, NIPS 2014, vol. 2, pp. 3104–3112. MIT Press, Cambridge (2014)
12. Schuster, M., Paliwal, K.K.: Bidirectional recurrent neural networks. Trans. Sig. Proc. **45**(11), 2673–2681 (1997)
13. Shōsetsuka ni narō. https://syosetu.com/
14. Kiros, R., Zhu, Y., Salakhutdinov, R., Zemel, R.S., Torralba, A., Urtasun, R., Fidler, S.: Skip-thought vectors. arXiv preprint arXiv:1506.06726 (2015)

Recognizing the Order of Four-Scene Comics by Evolutionary Deep Learning

Saya Fujino$^{(\boxtimes)}$, Naoki Mori, and Keinosuke Matsumoto

Graduate School of Engineering, Osaka Prefecture University,
1-1 Gakuencho, Nakaku, Sakai, Osaka 599-8531, Japan
fujino@ss.cs.osakafu-u.ac.jp

Abstract. In recent years, comic analysis has become an attractive research topic in the field of artificial intelligence. In this study, we focused on the four-scene comics and applied deep convolutional neural networks (DCNNs) to the data for understanding the order structure. The tuning of the DCNN hyperparameters requires considerable effort. To solve this problem, we propose a novel method called evolutionary deep learning (evoDL) by means of genetic algorithms. The effectiveness of evoDL is confirmed by an experiment conducted to identify structural problems in actual four-scene comics.

Keywords: Four-scene comics · Comics structure
Deep convolutional neural network · Genetic algorithm
Hyperparameter

1 Introduction

The use of artificial intelligence to enable computers to understand comics is one of the most important and interesting topics in the field of comic engineering. Many studies about comic analysis based on image recognition have been conducted as background research for the development of deep learning. For understanding the content in comics, one of the authors has proposed a method for detecting the order of a four-scene comic structure with the help of deep convolutional neural networks (DCNNs) [1]. This is more difficult than simple detection of characters but is more important for better understanding of comic content. We believe that the most interesting features of comics are the story and intentions of authors.

In this study, we focused on four-scene comics and applied DCNNs to the those data to understand the order structure. We applied evolutionary deep learning (evoDL) [2,3] to tune the DCNN hyperparameters. To this end, a novel network structure with three paths of DCNN has been introduced into the evoDL.

To demonstrate the effectiveness of evoDL, computer simulations are performed taking the example of a real four-scene comic.

2   Related Works

Many studies, such as those reported in [4,5], have been performed in the field of character detection, which is a very significant aspect of comic engineering. However, some studies have just applied conventional face detection methods to comic images. We have applied the DCNN to recognize anime storyboards [2] and for sketch recognition [3]. However, these studies do not take into account the story or the content. Studies about generating four-scene comics have been reported in [6,7] The method of detecting the order of scenes in four-scene comics using DCNN has been proposed [8] for various scene combination. Although this is a novel approach to analyze comics, the accuracy is not high; the best result obtained was 67% for the recognition task between (scenes 1, 2) and (scenes 3, 4).

The study combining the genetic algorithm (GA) with DCNNs is reported in [9]. Although many other studies have also been conducted [10], none of these have applied evolving DCNNs to analyze four-scene comics.

3   Four-Scene Comics

The four-scene comic is the famous comic strip format in which scene panels appear ordered from top to bottom. The format is used for short story contents. The four-scene comics have a structure known as Kishotenketsu, representing a famous Japanese Kanji character.

1. Ki: The first scene, which introduces the story.
2. Sho: The second scene, which follows or explains the story of the first scene.
3. Ten: The third scene, which has an unexpected twist or an accident.
4. Ketsu: The final scene, which concludes the story.

The aforementioned structure implies that there exists a certain relationship among the different scene images and a correlation between the scene images and order.

4   Deep Convolutional Neural Network

DCNN is one of the deep learning methods widely used in the field of image recognition. In this study, we used the architecture of AlexNet [1] as DCNN.

4.1   Three-Path DCNN

We define a model with a three-path DCNN. Figure 1 shows the model. Recently, although the multipath models have been studied in [11], no research has applied the multipath model to four-scene comic analysis. The model can extract features by using multiple paths with functional specialization. Using this approach, we deal with the unique problem of classification of the two halves of four-scene comics to obtain the specific tendency for continuities of each half scene for proper recognition.

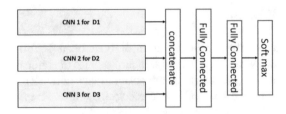

Fig. 1. Three-path DCNN model

In our experiments, two images A and B corresponding to a certain scene in the four-scene comics are used as input data. The sizes of A and B are 227×157, both grayscale with 255 gradations. Let the numerical tensor of A and B be X^A and X^B respectively. In the case of grayscale, A and B become matrices. We made three types of input data for the three-path DCNN. The first data **D1** is obtained by concatenating X^A and X^B in the vertical direction. The data sizes of **D1** become 227×314. **D1** data represents a natural appearance of two scenes in four-scene comics. Next, the data Y obtained by a function $\phi(X^A, X^B)$ is utilized. The reason for using Y is the hypothesis that humans recognize the continuous images in four-scene comics based on a certain correlation between the two images. In this study, we adopted addition $(X^A + X^B)$ and subtraction $(X^A - X^B)$ as ϕ. Other types of functions such as the Hadamard product, $\max(x, y)$ or $\min(x, y)$ can also be used. The dataset created by $X^A + X^B$ is denoted as **D2**, while that created by $X^A - X^B$ is set to **D3**.

Figure 2 shows the classes of this experiment. We denote the first-half scene as class1 and the latter-half scene as class2.

Fig. 2. The classes of this experiment

5   Evolutionary Deep Learning

We propose evoDL [2] as the novel evolutionary approach for finding the fine set of hyperparameters for the DCNN model.

5.1   Representation

Table 1 shows the details of alleles of the genotype in GA.

Table 1. The genes of individuals in GA

Design variables	Allele
The number of filters (NF)	32, 48, 64
Filter Size (FS)	3, 5, 7
Pooling Size (PS)	3, 5, 7
Pooling Type (PT)	0(not use), 1(Max), 2(Average)
NL	512,1024
Activation function (Ac)	0(not use), 1(ReLU), 2(leaky ReLU)

NL: The number of nodes in fully connected layer 1

In this study, we also observed that a smaller network is suitable if the performance is identical in terms of utilization of computer resources. The following three restrictions are introduced into our network setting:

The number of filters: The total number of filters is fixed at 352. We searched for the number of filters only in convolutional layers 1, 2 and 4, 5. When we decide the number of filters using evoDL, the number of filters in convolutional layer 3 is decided uniquely.

The number of nodes in fully connected layers: In this study, we used three fully connected linear layers before output. The number of nodes of the last fully connected layer is fixed by the number of classes. For the number of nodes of the first fully connected N1 and the second fully connected N2, we used the restriction, $N1 \times N2 = 2^{18}$. This restriction is equivalent fixing the number of weights between the first and second fully connected layers.

The first and second pooling: In past study [8], the pooling layers generate both the responses-normalization layers, so we set the restriction that must be set pooling 1 and pooling 2 and selected both PTs for fine tuning.

Figure 3 shows an example of the chromosome of an individual. Each string in the locus is related to the strings presented in Table 1. The basic structure of the DCNN is fixed according to AlexNet. In Fig. 3, the seven genes, namely "Ac" represent the type of the activate functions that are set just after five convolutional layers and two fully connected layers. If "Ac" is 0, no activate function is used, if it is 1, the ReLU function is set after the target layer, if it is 2, the leaky ReLU is set after there. A different pooling method is used based on the value of PT. If PT = 1, maximum pooling is used and average pooling is used if PT = 2. On the contrary, no pooling method is used if PT = 0.

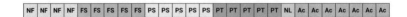

Fig. 3. An example of genotype

5.2   Fitness Function

In evoDL, the fitness function $F(s)$ of the GA is defined as the sum of the accuracy in k-fold cross-validation of training data.

6   Experiments

Here, we show the experimental results of evolving a DCNN model using evoDL.

6.1   Target Problem and evoDL Setting

In this experiment, we focus on the transition of four-scene comics. The datasets were prepared by a Japanese four-scene comics called "Compeito! 1" [12] authored by FUJINO HARUKA, which has 188 stories. The datasets **D1–D3** shown in Sect. 4.1 are used. The size of the **D1** image is 227×314 pixels and that of the **D2, D3** images is 227×157 pixels. In this experiment, the DCNN model is evolved to recognize 2 classes between the first-half scenes (class1) and the latter half scenes (class2), the hypothesis that there exists a certain correlation between class1 and class2.

We use 264 images as training data in evolution. In the GA parts of evoDL, the fitness in Sect. 5.2 is calculated by using the results of a two-fold cross-validation. Each individual is translated into a DCNN model and trained. After this, we get this DCNN accuracy and loss. Because of the two-fold cross-validation, we repeat this 2 times and obtain the fitness value by using the method explained in Sect. 5.2. In contrast, the test data has 112 images. We evaluated the final elite DCNN model by using this test data.

The hyperparameters of one path in three-path DCNNs are obtained using evoDL. Next, each DCNN path based on **D1–D3** is trained independently. After finishing three DCNN training stages by hyperparameters obtained using evoDL, all intermediate layers of the three DCNNs before the fully connected layers are concatenated and connected to new fully connected layers. Next, the weights of the convolution layers are fixed and only a new fully connected layer is trained by three paths.

6.2   Experimental Conditions

We used a flexible framework for deep learning library Keras [13] in a DCNN part, and Java for a GA part in evoDL. Table 2 shows the setting of GA. Batch size is fixed at 20, and we used Adam as the optimizer. The learning rate is set to 2×10^{-6}. The whole process of evoDL evolution for one DCNN in the three-path model is as follows:

1. We obtained the best hyperparameters of each DCNN model from an elite individual of GA in the final generation.
2. DCNN models are trained with obtained hyperparameters. In this step, the maximum number of epochs is 2000, and all training data are used because cross-validation is unnecessary.

Table 2. Setting of GA in evoDL

Generation size	20
Population size	20
Chromosome length	27
Crossover type	Uniform
Crossover rate	1.0
Mutation rate of each locus	$\frac{1}{L}$ (L is chromosome length)
Selection	Tournament selection
Tournament size	2
Elitism	True

6.3   Results and Discussion

Figs. 4, 5 and 6 show the elite individuals searched by evoDL and the DCNN models obtained from these individuals. From Figs. 4, 5 and 6, we can observe that the filter sizes of convolution layers 1, 2 are the same as that of layers 7 and 3, respectively. On several loci, such as PT of the first layer and activation function of the first and second layer, the genes are same in all individuals. In particular, the first activation functions used all leaky ReLU in each DCNN because the input data has many negative values. The basic features of input data are the same because the original images are the same. We can assume that the structure of DCNN obtained using evoDL is reasonable.

Moreover, the numbers of nodes in fully connected layers 1 and 2 are set to 1024 and 256 for a three-path model. Table 3 shows the accuracy for each elite DCNN model after 2000 epochs on training stage. The accuracy is particularly low in case of the **D3** dataset that contains the continuities of the two scenes. However, it does not demonstrate a substantial difference between class1 and class2. Table 4 describes the accuracies for three-path model after 500, 1000,

1500, and 2000 epochs of training. The amount of information in **D3** is less than that in **D1** and **D2**. However, **D3** contains the important information. Hence, the three-path model delivers the best accuracy. Because each DCNN can extract different features, the three-path model can obtain better accuracy as compared to other DCNNs. Moreover, the accuracy of the test data is higher than the accuracy in [8], which is 0.67. In case the classification of four-scene comics is done by only image information, humans can only recognize the images from the **D1** dataset. However, our results indicate that images of **D2** and **D3** are also useful for the computer. These results indicate that a more effective ϕ, shown in Sect. 4.1, may exist for recognition of four-scene comics.

When we feed **D2** and **D3** to the DCNN models, the DCNN models can find the similarity between the scenes by superimposition of data (**D2**) and the continuity of scenes obtained by subtraction data (**D3**) more clearly. Those results show that DCNN can recognize the order of four-scene comics by understanding a certain part of manga by a human-like viewpoint.

Fig. 4. The elite individual for **D1**

Fig. 5. The elite individual for **D2**

Fig. 6. The elite individual for **D3**

Table 3. Results of each DCNN to test data

Model name	Accuracy
DCNN for **D1**	0.70
DCNN for **D2**	0.73
DCNN for **D3**	0.64

Table 4. Results of three-path DCNN to test data

Epoch	Accuracy 1
500	0.77
1000	**0.79**
1500	0.78
2000	0.77

7   Conclusion

In this study, we proposed a novel method, evoDL, to obtain DCNN models for four-scene comic recognition from GA. Using AlexNet as the basic DCNN model, evoDL was used to tune hyperparameters of the DCNN within the limitation of network size. Moreover, we prepare three types of datasets and feed them to the three-path DCNN model. We can confirm the effectiveness of three-path DCNN model by evoDL. In a future work, we intend to extend evoDL to various datasets such as animes or other four-scene comics using the three-path model.

Acknowledgments. A part of this work was supported by JSPS KAKENHI Grant, Grant-in-Aid for Scientific Research(C), 26330282. A part of this work was supported by LEAVE A NEST CO., LTD. I would like to thank FUJINO HARUKA for providing her comic book as the dataset. The authors would like to acknowledge the helpful discussions with Dr. Miki Ueno of Toyohashi University of Technology, Japan.

References

1. Krizhevsky, A., Sutskever, I., Hinton, G.E.: Imagenet classification with deep convolutional neural networks. In: Bartlett, P., Pereira, F.C.N., Burges, C.J.C., Bottou, L., Weinberger, K.Q. (eds.) Advances in Neural Information Processing Systems, vol. 25, pp. 1106–1114 (2012)
2. Fujino, S., Hatanaka, T., Mori, N., Matsumoto, K.: The evolutionary deep learning based on deep convolutional neural network for the anime storyboard recognition. In: 14th International Conference Distributed Computing and Artificial Intelligence, DCAI 2017, Porto, Portugal, 21–23 June 2017, pp. 278–285 (2017)
3. Fujino, S., Mori, N., Matsumoto, K.: Deep convolutional networks for human sketches by means of the evolutionary deep learning. In: Joint 17th World Congress of International Fuzzy Systems Association and 9th International Conference on Soft Computing and Intelligent Systems, IFSA-SCIS 2017, Otsu, Japan, 27-30 June 2017, pp. 1–5 (2017)
4. Takayama, K., Johan, H., Nishita, T.: Face detection and face recognition of cartoon characters using feature extraction. In: Image, Electronics and Visual Computing Workshop, p. 48 (2012)
5. Burie, J.-C., Nguyen, N.-V., Rigaud, C.: Comic characters detection using deep learning. In: 2nd International Workshop on coMics ANalysis, Processing and Understanding (MANPU) (2017)
6. Matsumoto, K., Ueno, M., Mori, N.: 2-scene comic creating system based on the distribution of picture state transition. In: Advances in Intelligent Systems and Computing, vol. 290, pp. 459–467 (2014)
7. Ueno, M.: Computational interpretation of comic scenes. In: Advances in Intelligent Systems and Computing, vol. 474, pp. 387–393 (2016)
8. Ueno, M., Mori, N., Suenaga, T., Isahara, H.: Estimation of structure of four-scene comics by convolutional neural networks. In: Proceedings of the 1st International Workshop on coMics ANalysis, Processing and Understanding, p. 9. ACM (2016)
9. Yunming, P., Zhining, Y.: The genetic convolutional neural network model based on random sample. Int. J. u- e-Serv. Sci. Technol. **8**(11), 317–326 (2015)

10. Suganuma, M., Shirakawa, S., Nagao, T.: A genetic programming approach to designing convolutional neural network architectures. In: Proceedings of the Genetic and Evolutionary Computation Conference, pp. 497–504. ACM (2017)
11. Wang, M.: Multi-path convolutional neural networks for complex image classification. CoRR, abs/1506.04701 (2015)
12. Fujino, H.: Compeito ! 1 (Confetti ! 1). Houbunsha (2007)
13. Chollet, F., et al.: Keras (2015). https://github.com/keras-team/keras

A Big Data Platform for Industrial Enterprise Asset Value Enablers

Alda Canito$^{(\boxtimes)}$ ⓘ, Marta Fernandes ⓘ, Luís Conceição ⓘ,
Isabel Praça ⓘ, and Goreti Marreiros ⓘ

GECAD - Research Group on Intelligent Engineering and Computing
for Advanced Innovation and Development,
Polytechnic of Porto, Porto, Portugal
{alrfc,mmdaf,lmdsc,icp,mgt}@isep.ipp.pt

Abstract. The growing ubiquity of IoT, along with bigger steps towards full digitalization in the manufacturing industry, makes it easier to constantly monitor equipment activity and implement predictive maintenance approaches. Big Data solutions are best suited to process the large amounts of data generated through monitorization – additionally, they also allow for processing of unstructured data, such as documents used in not fully-digitalized processes. This paper describes the creation of a small Hadoop cluster, without high-availability, its integration in the InValue architecture and the processes through which it was populated with historical data from a relational warehouse. The degree of parallelization on the data ingestion tasks and its effect on performance were evaluated for the different kinds of datasets that are currently being used for batch data processing.

Keywords: Predictive maintenance · Big data architecture · Hadoop ecosystem
Data ingestion · Data processing

1 Introduction

A company's maintenance policy plays a significant role in its ability to provide quality products and services, while keeping high production rates. For this, machine uptime must be kept as high as possible; identifying and fixing problems before machines reach the point of failure. This way, preventive actions can be adopted, and machine downtime reduced, while at the same time avoiding compromising the quality of the end product. To reach valuable insights about machine status and behaviour there is a need to acquire and process as much data as possible from the machines' activities, their environment, their processing rates, etc.

The growing adoption of IoT technologies [1] brought a new range, variety and scope of data. The machines themselves often facilitate data concerning their activities via some API or framework, which can be accessed over gateways. Additional information, not made available by the machine, can be captured through the installation of sensors, which have become much more accessible and easy to interconnect with other systems.

© Springer International Publishing AG, part of Springer Nature 2019
F. De La Prieta et al. (Eds.): DCAI 2018, AISC 800, pp. 145–154, 2019.
https://doi.org/10.1007/978-3-319-94649-8_18

A predictive maintenance system aims to maximize the interval between interventions on the machines, while minimizing unscheduled repairs; to do so, regular monitorization of the equipment's condition is required [2]. Insights and knowledge can be extracted from the available data [3]; however, traditional data analysis techniques are not able to deal with the massive scale of the information involved and, therefore, new technologies and processes are required [3, 4]. Big Data solutions offer a way to deal with large amounts of heterogeneous data and were traditionally applied to human-generated data and used for sales prediction [5, 6], to mine user relationships and recommendation systems [7], among others. A trend of applying Big Data solutions to industrial, machine-generated data, has started to emerge as of recently. However, the digitalization process of the small and medium-sized companies of this industry sector has only just begun, with a large part of the information still being kept in paper documents and/or manually introduced into digital systems. Moreover, this information is frequently not interconnected in a global, factory-wide solution.

Our proposed platform, the Industrial Enterprise Asset Value Enablers (InValue) [8], currently being implemented, aims to provide a platform that facilitates the shift from traditional maintenance approaches to more proactive ones, employing state-of-the-art Big Data analytics to gather insights from machine and sensor-generated data. The platform addresses all stages the data must go through, from the acquisition, to storage, data processing and, finally, its delivery to end users.

This paper presents the set up of a Hadoop cluster for Big Data processing under the InValue platform [9]. This cluster is part of the Data Processing module and in charge of processing large amounts of data in order to generate alerts, diagnostics, statistics and suggest predictive actions, among other roles. Particular focus is given to the cluster's internal architecture, how it integrates with InValue's architecture, the services added to it, and the data ingestion and storage processes.

This document is organized as follows: (1) Introduction, wherein the theme and motivations were presented; (2) The InValue Platform, which contextualizes the scenario; in section (3), Big Data Architecture, the inner architecture of the Big Data solution is presented, along with a description of the services it needs; (4) Batch Data Ingestion, clarifying the data population process and, finally, (5) Conclusions.

2   The InValue Platform

The InValue platform was designed to facilitate the implementation of Predictive Maintenance approaches and is currently installed in a metallurgical company, specialized in custom precision parts production, which has just recently begun its digitalization process. The platform is comprised of three layers: (1) Data acquisition, (2) Data Processing and (3) Information delivery (Fig. 1). The architecture is described in further detail in [9].

The Data Acquisition module is able to monitor machines of different ages and technologies. Machine-generated data (either production data or provided by the machine's own sensors) is made available through a bus with a described protocol; further information of interest concerning the machine's status that is not provided by the machines themselves is captured by a number of sensors previously installed with

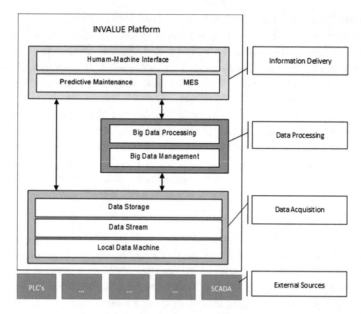

Fig. 1. InValue system's architecture [9].

Table 1. Data acquisition rates

Data	Acquisition rate
Machine-generated	0.2 Hz
Temperature (sensor)	1 Hz
Noise (sensor)	100 Hz
Vibration (sensor)	100 Hz
Manufacturing Orders	Every 2 weeks (batch)
Machine Logs	Every 2 weeks (batch)

this purpose in mind. Additionally, relevant information can be provided by external systems, such as production management software, local SCADAs, etc. Table 1, shows current data acquisition rates.

So far, the Data Acquisition module captures circa 1.6 GB a day, having captured more than 230 GB over the course of 7 months. This information describes several features of the equipment, such as machining coordinates, tool in use, coolant level, motion time, vibration, noise inside the machine, etc. The manufacturing orders are stored as text files containing information about expected vs actual production times, quantities, which machines and employees were assigned to each task, etc., and are manually uploaded to the platform every 2 weeks. Additionally, the machine logs are also manually uploaded with the same frequency. These are also text files, containing information about the internal status of the machines, such as uptimes and downtimes, which buttons were pressed and error logs, among others.

The Data Processing module is responsible both for pre-processing the data and for employing Machine Learning and Data Mining techniques with the purpose of identifying components that might be approaching failure, diagnosing failures, and proposing possible corrective measures. This information is delivered by the Information Delivery module not only to the end users it concerns, but also to the Manufacturing Execution Systems (MES) in use.

The data to be processed is not only substantial but also subject to errors, highly heterogeneous and often incomplete. Furthermore, the different sources provide data at different speeds and with different complexity levels: these characteristics make it harder for traditional data analysis approaches to process and extract value from the data and thus the usage of Big Data technologies becomes a necessity.

3 Big Data Architecture

Big Data commonly refers to large volumes of data, which are frequently incomplete, inconsistent and noisy, as is the case of the data generated by machines and sensors in the manufacturing industry. Using Big Data technologies for predictive maintenance, while a relatively novel approach, is already a hot research topic, with the number of new publications directly concerning them increasing every year [10]. InValue's Data Acquisition module captures large amounts of data at great speeds; the challenges that arise from the need to process this data can be tackled by the use of distributed architectures. Apache Hadoop [11] is the first *de facto* mainstream framework for Big Data processing: providing its own distributed file system (HDFS: Hadoop Distributed File System) and programming framework (MapReduce). Its distributed nature allows it to perform computationally heavy jobs over unstructured data, being particularly suited for our batch processing needs [12, 13].

Cloudera's distribution of Apache Hadoop is fairly popular and eases the processes involved with installation, analysis and data management, making them easier and more reliable: it was therefore chosen to answer InValue's Big Data processing requirements. Hadoop works by splitting files into blocks and distributing them through a computer cluster. Similarly, its computing strategy (MapReduce) works by splitting jobs into several smaller tasks, which run in parallel nodes. To satisfy the processing needs of the data extracted from InValue's relational warehouse, a small cluster (without high availability) comprised of four nodes, was set up. This particular set up is meant for development/testing purposes; a higher number of nodes and a different distribution of responsibilities would be required for an industrial environment. Considering the necessity of performing SQL queries in a distributed environment for data analytics and the need to schedule diverse tasks, the following services were added: Hive, Hue, Yarn, Sqoop, Oozie, Zookeeper and Impala. Figure 2, below, presents a simplified view of the distribution of roles and services through the nodes.

According to [14], a small cluster without high-availability should assign roles to three types of hosts: Master hosts, Utility/Edge hosts and Worker hosts. However, due to the very small size of InValue's cluster, comprised only of four (4) hosts, the responsibilities of the Master host and the Utility/Edge Hosts are aggregated in Node 1, with the exception of the Secondary NameNode, which is delegated to Node 4. As

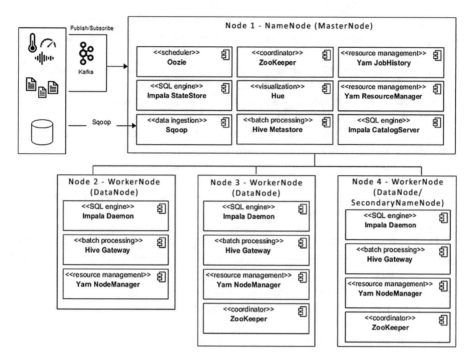

Fig. 2. The InValue's Hadoop cluster

such, Node 1 is the NameNode and ResourceManager/JobHistoryServer, responsible for splitting data and assigning tasks through the DataNodes/NodeManagers (Nodes 2 through 4). The Worker hosts include worker responsibilities for Impala, Hive and Zookeeper.

Apache Hive [15] performs long-running, batch-oriented jobs in the form of SQL-like queries. The data is stored in Tables, which are serialized and stored in HDFS directories. Furthermore, Tables can be partitioned in subdirectories (Partitions) and files (Buckets). A Hive MetaStore, installed in the Master host, contains schemes and data useful for query optimization and works as a catalogue. The Worker hosts contain Hive Gateways.

Apache Yarn [16] organizes the workload by job/application and distributes/schedules it through the worker nodes accordingly. The Master host serves as the global ResourceManager.

Apache Sqoop [17] was installed on the Master host and works as a data extractor, converting large datasets from relational database formats into HDFS and vice-versa. It is used to extract information from InValue's relational warehouse and flush it into the Big Data platform, making it ready for use. It converts commands to MapReduce tasks and uses YARN's framework to import/export data in a fault-tolerant, parallel way. Additionally, it allows for incremental loads, making it very useful for extracting data that has been added to the warehouse between executions.

Apache Oozie [18] was installed on the Master host and is a workflow scheduler, responsible for managing Apache Hadoop jobs in a multi-tenant, secure, scalable and reliable fashion.

Apache Zookeeper [19] is present in Nodes 1, 3 and 4 and is the main coordination service with the goal of relieving applications from being concerned about common problems faced by distributed systems such as dead-locks and race conditions. To do so, it exposes a set of primitives that can be used by client applications.

Apache Impala [20] works as a SQL engine that enables using familiar syntax to query data stored in HDFS in a read-mostly fashion. Not only it is compatible with the majority of popular file-formats (defaulting to Parquet), but it also uses the standard Hadoop components (it integrates with the Hive Metastore, for instance) and distributes its daemon processes through the same machines as the rest of the Hadoop infrastructure to enhance its performance. As such, Impala's Catalog and StateStore were installed in the Master host, while the daemons are installed in the Worker hosts.

Finally, Apache Kafka [21], serves as the communication broker between InValue's Data Acquisition module and the cluster, employing the Publish/Subscribe pattern. Kafka is installed in the Data Acquisition module, providing streams for the upper layer's consumption: it transmits the data provided by the sensors and the machines in real-time as a stream that is consumed both by the Data Processing layer and by a service that populates InValue's relational database for warehousing purposes. This database can also be queried by the Data Processing layer, as will be described below.

4 Batch Data Ingestion

The InValue platform's Big Data Analytics are twofold: (a) batch processing for predictive model generation and (b) stream processing for real-time statistics and alerts. These two different scenarios are depicted in Fig. 3, below:

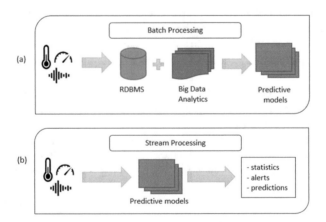

Fig. 3. (a) Bach processing and (b) Stream processing in the InValue platform

In an initial approach, the data will be extracted from the relational database and processed in batch in order to generate a number of predictive models. These models will be used in a second, future phase, wherein they will be executed against data streams to generate statistics, alerts and predictions in real-time. This section describes the data ingestion operations necessary for the batch processing phase.

The information gathered by the Data Acquisition module is stored in a SQL Server database and represents a total of 47 features. A feature extraction process was applied to these in order to obtain a subset of 32 features [22] to flush into HDFS, through Sqoop, in a preliminary approach.

Sqoop transforms commands into MapReduce tasks and uses YARN to import/export data, supporting parallelism on top of fault-tolerance. It is possible to import all tables from a database using a single command: all tables are required to have a primary key that can be used to split the table's contents across a number of mappers. InValue's SQLServer database, however, was designed with read/write performance in mind and does not feature primary keys for any of its tables. Given this scenario, tables were uploaded to the HDFS server separately. 'Timestamp' was selected as the column to split by, considering its values are unique within the same table and are replicated similarly across all tables.

When it comes to splitting tasks through the nodes, [23] suggests splitting according to the total number of available worker processor cores. [17], on the other hand, suggests setting the number of mappers to a low number and slowly incrementing it while evaluating how performance is affected.

There isn't a single heuristic to define the optimum number of mappers to use on a Sqoop extraction, as many factors can influence the outcome. These include, but are not limited to: the dataset size and type, hardware used at the database server, concurrent requests being made to the database, number of nodes in the cluster and their processing power, whether primary keys are defined and indexed or not, etc. Increasing the number of mappers does not necessarily produce better results [24] and the best solution often has to be discovered empirically. It was decided to initially follow the approach stated in [17], and then gradually increase the number of mappers until the maximum suggested in [23]. Given that InValue's Hadoop Cluster is comprised of 3 worker nodes, with 8 cores each, the maximum number of mappers would, theoretically, be 24. Yet, not all 24 cores can reasonably be expected to be available (e.g. consider that daemon processes are running on all nodes). This is one factor that must be accounted for when setting up the usage share of CPU (virtual core) that can be assigned to each MapReduce task. In InValue's cluster, the number of virtual cores is set to 8: meaning that even if the number of mappers is manually set to a higher number, only eight of them will run in parallel, while the remaining will run sequentially as soon as possible.

To assess the number of mappers best suited for our cluster, several runs for each different number of mappers considered were made, calculating the average and standard-deviation for each in order to evaluate performance. The differences in performance according to the number of mappers are visible in Table 2. These values describe the ingestion of one month of machine information, captured at a frequency of 0.2 Hz and corresponding to 73666 rows (approx. 3.1 MB). As described above, several factors come into play when it comes to performance and therefore some

Table 2. Effect of the number of mappers in data retrieval time (seconds) for 0.2 Hz

	M = 2	M = 3	M = 4	M = 8
AVG	18,91	19,09	18,68	25,42
σ	1,63	0,77	1,12	2,90

variance in the results was to be expected; however, a lower number of mappers seemed to generate better average results, with 4 mappers, on average, achieving better transfer rates; indicating that even though more virtual cores were available, more mappers do not necessarily translate to better performance; for smaller datasets, the parallelization gains do not outweigh the costs.

A similar approach was taken for the tables containing data gathered at a higher acquisition rate. In the first experiment, the best results were obtained with between 3 and 4 mappers and, therefore, 3 was selected as the starting point. Similarly, it was slowly increased and the results are shown in Table 3, below:

Table 3. Effect of the number of mappers in data retrieval time (seconds) for 100 Hz

	M = 3	M = 4	M = 6	M = 8
AVG	728,66	675,51	620,16	625,00
σ	80,97	47,76	35,39	54,02

These values correspond to one month of information gathered at a frequency of 100 Hz and comprised of 201849310 rows (approx. 7.5 GB).

The influence of the dataset size can clearly be seen in this second test, with a lower number of mappers no longer correlating with higher performance. For this dataset in particular, more parallelization led to better performance, but only to a certain extent: peaking at around 6 mappers, after which, increasing the number of mappers was no longer beneficial.

5   Conclusions

This paper described the deployment of a development/testing Hadoop cluster for the InValue's proposed architecture. The Cloudera distribution of Apache Hadoop was chosen for this scenario given its popularity, community support, ease of use and available online documentation, tackling many known installation issues.

Because of the processing power required to analyse the existing historical data, a small cluster of 4 nodes, without high-availability, was set up and the following services were installed: Hive, Hue, Yarn, Sqoop, Oozie, Zookeeper and Impala. In a preliminary phase, a subset of the tables was imported with SQOOP from InValue's relational SQLServer database. The data transfer process benefits greatly from parallelism and an evaluation of the best number of mappers to use in the extraction process was made. As it is known that several factors come into play when it comes to data transference, with the data volume being particularly important, different experiments were made for datasets with different acquisition rates. Satisfying levels of performance were achieved and will be applied to future data transfers.

The real-time needs of the platform were not addressed in this phase. Future work will have to deal with not only fast-arriving data but also support unstructured data (e.g. text documents). As such, new services will have to be installed in the cluster – of which Cassandra's NoSQL-based solution and Apache Spark's real-time processing are good starting points -, which could affect performance and require modifications to the cluster's architecture.

Acknowledgments. The present work has been developed under the EUREKA - ITEA2 Project INVALUE (ITEA-13015), INVALUE Project (ANI|P2020 17990), and has received funding from FEDER Funds through NORTE2020 program and from National Funds through FCT under the project UID/EEA/00760/2013.

References

1. Chui, M., Loffler, M., Robert, R.: The internet of things. Mckinsey Q. (2) (2010)
2. Mobley, R.K.: An Introduction to Predictive Maintenance. Elsevier Science, New York (2002)
3. McAfee, A., Brynjolfsson, E.: Big data: the management revolution. Harvard Bus. Rev. **90** (10), 60–68 (2012)
4. Cohen, J., Dolan, B., Dunlap, M., Hellerstein, J.M., Welton, C.: MAD skills: new analysis practices for big data. Proc. VLDB Endowment **2**(2), 1481–1492 (2009)
5. Al-Noukari, M., Al-Hussan, W.: Using data mining techniques for predicting future car market demand; DCX case study. In: 3rd International Conference on Information and Communication Technologies: From Theory to Applications, ICTTA 2008, pp. 1–5. IEEE (2008)
6. Chon, S.H., Slaney, M., Berger, J.: Predicting success from music sales data: a statistical and adaptive approach. In: Proceedings of the 1st ACM Workshop on Audio and Music Computing Multimedia, pp. 83–88. ACM, October 2006
7. Martens, D., Provost, F., Clark, J., de Fortuny, E.J.: Mining massive fine-grained behavior data to improve predictive analytics. MIS Q. **40**(4) (2016)
8. InValuePt. InValuePT - Home (2017). http://www.invalue.com.pt/. Accessed 01 Feb 2018
9. Canito, A., et al.: An architecture for proactive maintenance in the machinery industry. In: International Symposium on Ambient Intelligence. Springer (2017)
10. O'Donovan, P., Leahy, K., Bruton, K., O'Sullivan, D.T.: Big data in manufacturing: a systematic mapping study. J. Big Data **2**(1), 20 (2015)
11. The Apache Software Foundation. Welcome to Apache Hadoop (2018). http://hadoop. apache.org/. Accessed 25 Jan 2018
12. IBM: What is the Hadoop Distributed File System (HDFS)? https://www-01.ibm.com/ software/data/infosphere/hadoop/hdfs/
13. Borthakur, D.: HDFS Architecture Guide (2013). https://hadoop.apache.org/docs/r1.2.1/ hdfs_design.html. Accessed 05 Feb 2018
14. Cloudera. Cluster Hosts and Role Assignments (2018). https://www.cloudera.com/ documentation/enterprise/latest/topics/cm_ig_host_allocations.html. Accessed 05 Feb 2018

15. Thusoo, A., Sen Sarma, J., Jain, N., Shao, Z., Chakka, P., Anthony, S., Liu, H., Wyckoff, P., Murthy, R.: Hive: a warehousing solution over a map-reduce framework. Proc. VLDB Endowment **2**(2), 1626–1629 (2009)
16. Vavilapalli, V.K., Murthy, A.C., Douglas, C., Agarwal, S., Konar, M., Evans, R., Graves, T., Lowe, J., Shah, H., Seth, S., Saha, B.: Apache hadoop yarn: yet another resource negotiator. In: Proceedings of the 4th Annual Symposium on Cloud Computing, p. 5. ACM, October 2013
17. Ting, K., Cecho, J.J.: Apache Sqoop Cookbook. O'Reilly Media, Sebastopol (2013)
18. Islam, M., Huang, A.K., Battisha, M., Chiang, M., Srinivasan, S., Peters, C., Neumann, A., Abdelnur, A.: Oozie: towards a scalable workflow management system for hadoop. In: Proceedings of the 1st ACM SIGMOD Workshop on Scalable Workflow Execution Engines and Technologies, p. 4. ACM, May 2012
19. Apache ZooKeeper: What is zookeeper (2014). http://zookeeper.apache.org. Accessed 01 Feb 2018
20. Bittorf, M.K.A.B.V., Bobrovytsky, T., Erickson, C.C.A.C.J., Hecht, M.G.D., Kuff, M.J.I.J. L., Leblang, D.K.A., Robinson, N.L.I.P.H., Rus, D.R.S., Wanderman, J.R.D.T.S., Yoder, M. M.: Impala: a modern, open-source SQL engine for Hadoop. In: Proceedings of the 7th Biennial Conference on Innovative Data Systems Research (2015)
21. Garg, N.: Apache Kafka. Packt Publishing Ltd. (2013)
22. Fernandes, M., Canito, A., Bolón, V., Conceição, L., Praça, I., Marreiros, G.: Predictive Maintenance in the Metallurgical Industry: data analysis and feature selection. In: World Conference on Information Systems and Technologies, pp. 478–489. Springer, Cham (2018)
23. White, T.: Hadoop: The Definitive Guide. O'Reilly Media, Inc. (2012)
24. Groover, M., Malaska, T., Seidman, J., Saphira, G.: Hadoop Application Architectures: Designing Real-World Big Data Applications. O'Reilly Media, Inc. (2015)

Estimating the Purpose of Discard in Mahjong to Support Learning for Beginners

Miki Ueno[✉], Daiki Hayakawa, and Hitoshi Isahara

Toyohashi University of Technology,
1-1 Hibarigaoka, Tempaku-cho, Toyohashi, Aichi 441-8580, Japan
ueno@imc.tut.ac.jp

Abstract. It is always difficult for beginners to learn rules of a new game. A game is classified depending on various aspects, for instance, one with perfect or imperfect information. Because of developing the computer program for a game with perfect information, it is significant to focus on a game with imperfect information as the main target for computational research. Especially, Mahjong is one of popular games with imperfect information. From the aspect of the game informatics, we focus on providing a support system for human players. In order to support mahjong beginners, we constructed a system that displays hints and estimates the players' purpose of discard based on the support vector machine.

Keywords: Estimating purpose of discard
Learning support for beginners · Mahjong
Agame with imperfect information

1 Introduction

A game is played by clear rules and is based on victory or defeat; so, it is good subject for computer science. A game is classified depending on various aspects, for example, the number of players, or the total scores, etc. A game is also classified as one with perfect or imperfect information based on whether all kinds of information about the action and states of game are provided. In the 1950s, computational research on chess revealed that chess is a game with perfect information. In the artificial intelligence fields, an increasing amount of research is now being conducted on *Shogi* and *Igo*, which are games with perfect information. In a game with perfect information, computers as players have begun to defeat human players. In *igo*, there are numerous states to be considered, and hence, Igo is a good example of not only a game with perfect information but also one with imperfect information, based on statistical methods.

AlphaGo [1,2], the computer program developed by Google DeepMind, defeated professional human *Igo* players in 2015 without handicap. Therefore, it

© Springer International Publishing AG, part of Springer Nature 2019
F. De La Prieta et al. (Eds.): DCAI 2018, AISC 800, pp. 155–163, 2019.
https://doi.org/10.1007/978-3-319-94649-8_19

is significant to focus on a game with imperfect information as the main target for computational research.

In games with imperfect information [3], current research on *Werewolf* is popular in the artificial intelligence field, but *Mahjong* over the years is one of the major examples of a game with imperfect information because of the number of players involved and the various rules. Mahjong is popular, especially in East Asia, and its rules[1] are similar to that of *Seven Bridge* in a game series of *Gin rummy*, which is one of the most famous card games in the world.

In game informatics, the aim of research can be roughly divided into two types: one, to make strong computer players, and two, to provide a support system for human players. Focused on the latter type, there are several studies on the support system for a game with perfect information [4]. In research on mahjong, some studies describe methods introduced for the support system estimating states [5,6] that require safety give-up [7] or one more tile for completion by the opponent player [8]. However, there is less research than that required for a game with perfect information. All the previous studies provide hints of action to be taken but do not provide the reason for taking the action.

Therefore, it is an important issue which parts of complex rules as research target and how to define the problem on computers.

The aim of this research is to support learning the state of the mahjong game by constructing a system that displays hints of sets[2] of a hand and estimating the purpose of user action in a user turn based on a support vector machine (SVM).

2 Basic Concept

2.1 Basic Mahjong Rule for Our Research

The game pieces and scoring of Mahjong slightly differs depending on regional variations. The aim of mahjong is to make sets as well as get the highest value in all of players. In addition to get winning, players to make sets as earlier as possible and steal scores to beat opponents player. In this paragraph, basic mahjong rule for this research is introduced as follows.

Types of Pieces

- 136 pieces[3], which contains 34 distinct kinds of piece; four of each kind.
  - Three suits, which run from one to nine.
  - Four directional tiles
  - Three Cardinal tiles
- The objective of the game is to put together a complete set which contains four sets of threes either three of a kind of the same suit (or *pung*) or a sequence of the same suit and a pair, for total of 14 pieces.

---

[1] https://www.mahjongtime.com/chinese-official-mahjong-rules.html.
[2] To make several sets for winning.
[3] In this research, we ignore 8 pieces of *the flower and season tiles* from the 144 pieces.

Flow of the Game

1. Each player starts with 13 tiles.
2. With each turn, a player picks up a 14th tile, and then discards one tile placed face up on the center of the table.
3. At this point, other players can choose to pass, take the tile to complete a set (pung, *chow* or *kong*) or to declare a win.
4. The first player who completes the set of 14 tiles wins the hand.

2.2 Preliminary Experiment

To investigate the kinds of support required by mahjong beginners, a questionnaire survey is conducted on five subjects. The following supports are obtained.

- Indicate candidate sets based on the present player's situation
- Indicate discard when an opponent player has only one more tile for completion, and when a player does not have discard safety
- Indicate player's hand that has more ways of completion
- Explain after the play, several candidates' discard actions during their turn of play

In addition, we obtained the history of web-search by the wearable device [9] on mahjong beginner in order to find out what kinds of information they required. Referring to these surveys, the mahjong beginner needs information about how to make sets. Thus, our research aim is to support making sets.

3 Constructing Hints for the Sets System

In this section, we construct the system that can display hints for making sets.

3.1 Setting up the System

To make the rules simple, we focused only on one mahjong player under the condition that *shantensuu*[4] must not be increased.

This system considers player's hand and following three kinds of information.

- Whether *ready hand*[5] or not
- Self-draw tile
- Kinds and the number of claim of a tile

On the other hand, the system ignores *kong* and *dora* to avoid complex situations in this section. *Dora* is a bonus tile randomly selected at the beginning of a game or *kong*. The rule for *dora* originated from Japan.

---

[4] Required turn for a player to be out.
[5] Player declares his/her hand when requiring one more tile for completion as information for opponent players.

3.2 Algorithm

To display several hints, the following three algorithms were constructed.

1. To judge whether the hand is complete or not
2. To show existing sets in a player's hand
3. To indicate possible sets in future based on the present hand

Three algorithms were combined to display variable hints for mahjong beginners. The flow of this system is as follows.

1. User input 14 tiles, of which 13 tiles are in the player's hand and one is a self-draw tile. The system judges whether the player's hand is complete or not.
2. If the user's hand is complete, the system displays the sets.
3. If the user's hand is incomplete, the system displays hints based on the player's present hand.

The system displays two major types of possible sets, the maximum scores one and the minimum required tiles for completion, as hints.

3.3 Experiment 1

To confirm the effectiveness of the hint system for beginners, an experiment is carried out. A subject(in his twenties, male) plays mahjong 10 times each with/without this system. For a half experiment, a subject can use this system anytime.

Table 1 shows the result of the experiment with or without the system. The rate of completion on using this system is 20% higher than that when not using this system. On the other hand, the rate of *waiting*[6] and the average score of using system is lower than that of not using it.

Table 1. The result of 10 games w./w.o use

	Rate of completion	Rate of waiting	The average score
With the hint system	30%	60%	3800 points
Without the system; only ref. web page	10%	30%	3000 points

---

[6] Player needs one more tile for completion.

4   Estimation of Suitable Purpose for Player's Discard in the Present State

The hint system is helpful for beginners to notice two major types of sets that can be made and select one tile to discard based on the present situation. However, the problem lies in deciding which tile to select and which to discard and the reason for this. Because the purpose of discard varies in Mahjong, this part is difficult for a beginner to understand.

Therefore, by knowing the reason, the method of estimating suitable selection with purpose, based on the database of *expert play* derived from machine learning, and displaying it is useful. In this section, we show how to estimate a suitable reason for a player's discard with the present state of hand by using a support vector machine.

Although users can have multiple purposes at the same time, to make rules simple, each situation is assigned to one of three; two major purpose and the "another"; using the algorithm mentioned in Sect. 3.

4.1   Defining the Purpose of Player's Discard

The two types of majority purpose of player's discard are defined as follows.

Max when the player is out to gain the maximum score
min when the player needs minimum tiles for completion

For the experiment, these types are used as class label. If a certain state is regarded as both Max and min, Max has high priority, and therefore, that state is assigned to the max label. If a certain state is regarded as neither Max nor min, the state is assigned to the "another" label.

4.2   Target Data Construction

The following game playing logs were obtained from Tenho [10] which is a popular on-line mahjong website.

Target period Jan. 1st, 2016–Dec. 31st, 2016.
Total numbers of games 48874 (only four-players games, only expert stage)
Total numbers of states of games 463374 (completion: 389721,   drawn game:73653)

For our research, we exclude states of the drawn game and therefore we use 389721 states. To make the rules simple, states including *kong* are removed from the target data for our research.

4.3   Two Types of Purpose for Estimation

We focused on the n-th state of player's discard and describe two kinds of purposes.

Discard purpose in a certain turn. The purpose of the player's discard on the n-th turn

Transition of discard purpose. A purpose transition between the n and $(n-1)$-th turns.

Table 2 shows the numbers of transition or no-transitions of the discard purpose in the target data. Table 3 shows the numbers of transitions categorized into six types.

Table 2. Total numbers of transition or no-transition in the target data

Num. of states	Transition	No-transition	Not subject (including *kong*)
389621	168920	194450	26261

Table 3. Number of states of each transition in target data

Max→min	Max→another	min→Max	min→another	another→Max	another→min
34075	77343	5138	2288	44358	5718

5   Experiment 2

In this experiment, we estimate two types of purpose as shown in 4.3.

5.1   Feature Vectors

We estimate the purpose of player's discard by using a SVM. For the SVM, we use two types of binary feature vectors as below.

Existing features. There are 22320-dimensional features according to existing research [8]. The existing features are not for estimating player's discard but for estimating *waiting*. The existing features refer to the basic features of mahjong.

Extended features. There are 204-dimensional of extended features, defined by ourselves, comprising the player's hand and *dora*. Table 4 shows the elements of the extended features.

Table 4. Extended features for n-th state

Features	Num. of dimension
Kinds and num of player's hand	$34 \times 5 = 170$
Kinds of *Dora*	34

5.2 Experimental Settings and Result

The number of states for the experiment is 198, randomly selected an equivalent number of states of six types as shown in Sect. 4: three kinds of classes from the view of n-th state; and six kinds from the view of the $n - 1$ and the n-th state transition.

Thirty-eights states are used for the grid-search to define γ and C of RBF kernel of SVM and 160 states are used for the 10-fold cross validation to calculate accuracy.

Table 5. Result of estimating the purpose of player's discard on a certain state

Kinds of Features	Dimension of features	Average accuracy
Existing features	22320	32.6%
Extended features	204	42.8%

Table 6. Result of estimating the transition purpose of player's discard

Kinds of Features	Dimension of features	Average accuracy
Existing features	44640	31.4%
Extended features	408	39.2%

Table 5 shows the result of the purpose on a certain state. The accuracy using only the extended features is almost 9% higher than the chance rate. Figure 1 shows the result in form of a confusion matrix. The figure indicates that the another class is easiest to classify.

Table 6 shows the result of the purpose transition. Those feature dimensions are twice as purpose on a certain scene.

We combine the hint system as shown in Sect. 3 and the method of estimation in order to support selected hints for a real-time play.

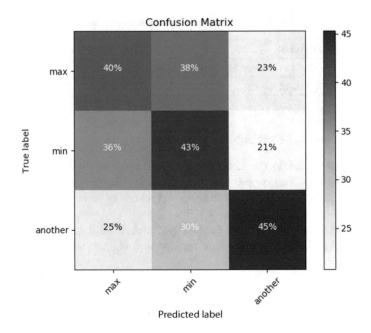

Fig. 1. The confusion matrix of extended feature for a certain purpose

6 Conclusion

In order to support mahjong beginners, we constructed a system that displays
hints and estimated the player's purpose of discard based on the database of
expert play. After each experiment, we introduce estimating method to the hint
system to show more helpful information.

Following topics can be used to conduct further related work on the subject:

– Consider more ways of displaying hints
– Select features for SVM
– Estimate multiple purpose of discard

Acknowledgments. This research is partially supported by DAIKO FOUNDATION
and JSPS KAKENHI Grant Number 17K17809.

References

1. David, S., et al.: Mastering the game of go with deep neural networks and tree
   search. Nature **529**, 484–503 (2016)
2. David, S., et al.: Mastering the game of Go without human knowledge. Nature
   550, 354–359 (2017)
3. Bowling, M., et al.: A demonstration of the polaris poker system. In: Proceedings
   of 8th International Conference on Autonomous Agents and Multiagent Systems,
   vol. 2, pp. 1391–1392 (2009)

4. Taeshi, I.: A learning support system by using the future position, SIG Technical Reports, vol. 7, 2017-GI-38 (2017)
5. Miki, A., Miwa, M., Chikayama, T.: Learning to rank moves in mahjong using SVM with tree kernels. In: Proceedings of Game Programing Workshop 2008, vol. 11, pp. 60–66 (2008)
6. Yamamoto, Y., Hoki, K.: Performance analysis of Mahjong evaluation functions represented by primitive features as tile positions and scores. In: Proceedings of 9th Entertainment and Computing Symposium (2015)
7. Mizukami, N., et al.: Adapting one-player mahjong players to four-player mahjong by recognizing folding situations. In: Proceedings of Game Programing Workshop 2013 (2013)
8. Mizukami, N., Tsuruoka, Y.: Building computer mahjong players by modeling opponent players using game records and a Monte Carlo method. In: Proceedings of Game Programing Workshop 2014, pp. 48–55 (2014)
9. Ueno, M., Morishita, M., Isahara, H.: Artificial curation for creating learners manual based on data semantics and user personality. AISC, pp. 247–253. Springer, Cham (2018). https://doi.org/10.1007/978-3-319-62410-5_30
10. Tenho. http://tenhou.net/

A Web-Based Micro-service Architecture for Comparing Parallel Implementations of Dissimilarity Measures

Daniel-Stiven Valencia-Hernández[ID], Ana-Lorena Uribe-Hurtado[(✉)][ID], and Mauricio Orozco-Alzate[ID]

Facultad de Administración, Departamento de Informática y Computación,
Grupo de Ambientes Inteligentes Adaptativos - GAIA,
Universidad Nacional de Colombia - Sede Manizales,
km 7 vía al Magdalena, Manizales 170003, Colombia
{dsvalenciah,alhurtadou,morozcoa}@unal.edu.co

Abstract. The performance of an application can be significantly improved by using parallelization, as well as by defining micro-services which allow the distribution of the work into several independent tasks. In this paper, we show how a micro-service architecture can be used for developing an efficient and flexible application for the nearest neighbor classification problem. Several dissimilarity measures are compared, in terms of both accuracy and computational time, for sequential as well parallel executions. In addition, a web-based interface was developed in order to facilitate the interaction with the user and easily monitoring the progress of the experiments.

Keywords: Micro-services · Dissimilarity measures
Multi-core implementation

1 Introduction

Classifying objects in a fast, efficient and automatic way is a very important issue in applications such as industry and medicine [7]. Examples in those areas include, in the industrial case, discarding or accepting manufactured objects in a production chain or identifying diseases in the case of health care. The automatic classification is typically performed by using an algorithm —the so-called classifier— which is designed according to the information from a collection of samples equipped with class labels. Once deployed, the classifier assigns labels to new incoming and unlabeled samples which are typically represented as images or other digital signals derived from them, e.g. histograms or spectra. Among all the available classification algorithms, the k nearest neighbor rule (kNN) is a

---

A.-L. Uribe-Hurtado—Estudiante del Doctorado en Ingeniería, Industria y Organizaciones - Universidad Nacional de Colombia - Sede Manizales.

straightforward and well-known classifier that uses a dissimilarity measure, typically a distance, as criterion to assign class labels: when $k = 1$, unlabeled objects are assigned with the class label of the most similar object from the collection of labeled examples. According to [6], kNN is one of the most popular algorithms in data mining for object classification. The selection of an appropriate dissimilarity measure is the only free parameter involved in the design of this classifier if $k = 1$; therefore, kNN is very often used in comparative studies for finding which dissimilarity measures are more efficient and give better classification accuracies for particular applications and time-constrained automation scenarios.

In many cases, the above-mentioned representations of the objects —e.g. raw images or their processed versions such as histograms and spectra— are organized as long vectors whose i-th entries must be compared against entries from a bunch of other vectors in either a bin-to-bin or a cross-bin [9] fashion, according to the dissimilarity measure that has been chosen. Moreover, the collection of labeled examples for the comparisons might be a fairly large dataset. All these factors may contribute to turning both, the computation of the dissimilarity measure and the estimation of the classification performance, computationally expensive when the algorithm must perform hundreds or even thousands of comparisons. It is, therefore, desirable that these comparisons are made as fast as possible such that the industrial production is maximized or, in cases of simulation, to make the experiment reproducible in a computationally acceptable time or, similarly, in a medical case, to make prompt decisions that allow an early identification of a medical condition.

Efficiency and classification accuracy are both targets to be optimized in simulations as well as in real-world applications. A solution for the first one is the parallelization of the algorithms such that they are scalable over multi-core architectures, allowing researchers and practitioners to execute them on a chosen number of cores. Practitioners, however, are often not experts on parallel computation and, therefore, not able to dig into the details of the implementations. A graphical user interface (GUI) would help them to easily monitor the distribution of the computation load as well as to track the progress of the results in order to make informed decisions about the best configuration for their automatic classification systems.

As a solution motivated by the above-mentioned reasons, we present in this paper a web-based application —including a monitoring GUI— that benefits from (i) Python libraries for parallel processing that help to accelerate the dissimilarity measure comparisons, (ii) a micro-service architecture to distribute the application in different works, in a multi-core architecture, such that the classifier is independently executed from the monitoring task and its web-based (GUI) reports. In particular, the contribution of this work is consist in the integration, under the same micro-service architecture of three processes: (i) the user interface micro-service (UI_{ms}), which deploys the graphical user interface; (ii) the hardware monitoring micro-service (HM_{ms}), which monitors the CPU core and the CPU frequency usages; and (iii) the kNN classifier micro-service

(kNN$_{ms}$), that performs the parallel classification of the object. Further details are given in Sect. 2.

For the sake of illustration, we test the classifier with several classical distances to compare histograms, either invoked from *Scikit-learn* or, alternatively, implemented by us from scratch; namely: Minkowski, Manhattan, Euclidean, Chebyshev, Canberra, Hamming, Cosine, City-block, Bray-Curtis and Correlation. As exemplar applications, 17 spectral datasets[1] from PRTools [3] were taken into account; but results for only four of them are shown due to space constrains. Those datasets cover a variety of industrial and medical applications.

The remaining of the paper is organized as follows: Sect. 2 explains the micro-service architecture and how the application is launched, monitored and deployed. Section 3 shows the experiments and results obtained with the application. Finally, our conclusions are presented in Sect. 4.

2 Micro-service Architecture and Application Deployment

According to [1], the implementation of an application under the micro-service architecture allows to separate its functionalities such that the application turns more flexible and gets a fast response to changes when maintenance to the different processes of the software is made. Micro-services are often used in transactional applications, particularly for those implementing basic operations such as *create*, *read*, *update* and *delete* (CRUD) over a relational database. We aim to show that such an architecture is suitable to be used for other types of applications, such as simulation and object classification, as well as to illustrate that its strength lies in the possibility of the application components to work independently from each other.

The micro-service architecture allows us to distribute the application in three independent processes: (i) UI$_{ms}$ corresponds to the GUI that interacts with the end user and allows loading the file to be processed, selecting the number of cores according to the machine architecture and choosing the programming language to execute the classifier in parallel, see Fig. 2; (ii) kNN$_{ms}$ that launches the kNN classifier with different dissimilarity measures; and (iii) HM$_{ms}$ which is in charge of monitoring the frequency and loading of the cores. The interaction and description of the micro-services shown in Fig. 1 are the following:

- HM$_{ms}$: depends on the `psutil` Python library that permits to obtain the frequency and load charge of the cores. This information is obtained from the hardware of the actual machine, by receiving a `get` request from UI$_{ms}$ at a predefined rate and, afterwards, sending the values via the `get` response.
- kNN$_{ms}$: implements the kNN classification algorithm. Even though this classifier is typically used with the Euclidean distance, we selected a collection of the most common dissimilarity measures used in classification [2,4,10] (see Table 1) in order to test which of them has the best accuracy and the best

---

[1] Available at: http://www.37steps.com/prhtml/prdisdata/specdata.html.

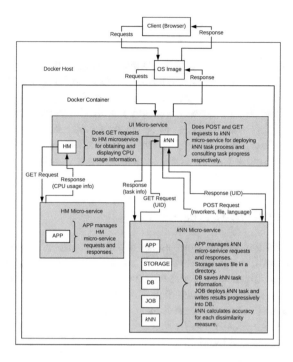

Fig. 1. Micro-service architecture for the nearest neighbor classification.

computational processing time when applying kNN under a leave-one-out test [11] for the performance estimation. This micro-service was tested with 17 datasets from PRTools but, due to space constraints, results for only four of them are reported below. They correspond to real-world classification problems from medical and industrial domains. Internally, kNN$_{ms}$ requires the Scikit-learn, Scipy and Numpy libraries to compute the dissimilarities doing calls to the Scikit-learn measures[2]. Additionally, the dissimilarity measures are launched over the machine in a parallel way by using Scikit-learn. We used the traditional Single Instruction Multiple Data (SIMD) Flynn's taxonomy [5] to split the test vector to be classified and send different data portions over different cores to process the comparisons of the vector data. Each vector is partitioned into blocks and launched over the 32 cores.

- UI$_{ms}$: corresponds to a friendly user interface which provides the required tools for visualizing the results of kNN$_{ms}$, as well as for showing the current usage of hardware resources as reported by HM$_{ms}$. UI$_{ms}$ is composed by the HM and kNN internal processes; the first one sends a request to HM$_{ms}$ to obtain the percentage of usage of each core and CPU frequency and, the

---

[2] http://scikit-learn.org/stable/modules/generated/sklearn.metrics.pairwise.distance_metrics.html and https://docs.scipy.org/doc/scipy/reference/spatial.distance.html.

second one, sends a request to kNN_{ms}, to obtain partial and total results from the classifier.

The micro-services were deployed using Docker[3]. It installs all the package dependencies that each micro-services needs, regardless of the operating system (OS) packages. The Docker platform creates its own OS image and installs Python, JavaScript and the HTTP server version required by the application implemented on it. HM_{ms} and kNN_{ms} were implemented in Python and UI_{ms} in JavaScript using the rendering engine *React* to monitor and deploy the information as close as possible to real time, allowing the end user to see the progress of the process.

3   Experiments and Results

Results were obtained on a Dell server with two Intel® Xeon® E7-4820 @ 2.00 GHz processors, each one with 8 cores. Each core implements multi-threading (2 threads per core), allowing to send up to 32 threads at the same time, 64 GB in RAM, with hard disk SATA DELL drivers, running Ubuntu Server and Docker 17.09.1-ce version as the micro-service platform. The micro-service web-based application provides a friendly visualization and interaction, such that the user can see and load data without requiring previous knowledge about the operating system or having to implement scripts in a shell to get results.

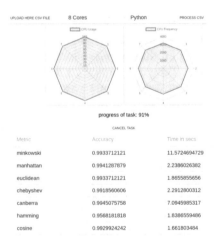

Fig. 2. Graphic user interface deployed by UI_{ms} for 8 cores

Figure 2 illustrates how the user can load his data and track the progress and behavior of the CPU at a predefined refreshing rate. Moreover, if several

---

[3] https://docs.docker.com/engine/docker-overview/#docker-registries.

researchers participate in the development of the application, each one can implement his code independently, requiring only the definition of how the processes communicate in order to achieve effective interaction among the micro-services. Leave-one-out classification results and average elapsed time (in seconds) for five executions of four representative datasets, are shown in Table 1—for 1 and 32 cores, respectively. Those results corresponds to the ones reported by HM_{ms}, which receives the information from kNN_{ms}. The lowest accuracy values are underlined and the highest ones are highlighted in boldface. A size ranking of datasets is shown in brackets, *aviris_clear* is the largest dataset with 21015 instances, 200 features and 17 classes. The smallest one is *tecator* with 215 instances, 96 features and 2 classes. The speed-up was computed with the basic formula $S = \frac{T_1}{T_N}$, where T_1 and T_N are the elapsed times with 1 and 32 threads respectively. Analyzing the values reported in Table 1, as well as in Figs. 3 and 4, the following results can be highlighted: Cosine $\left(1 - \frac{\sum_{i=1}^{n} x_i y_i}{\left(\sqrt{\sum_{i=1}^{m} x_i^2} \sqrt{\sum_{i=1}^{m} y_i^2} \right)} \right)$, when executed on a single core, is the cheapest dissimilarity measure to be computed with all the datasets. Its speed-up (0.024) is, consequently, the smallest one since the parallelization turns more costly than the sequential execution

Table 1. Accuracy, average elapsed time of one core (Time 1) and 32 cores (Time 32) in seconds and speed-up, for five executions of kNN_{ms} using one and 32 cores

Dissimilarity measures	*aviris_clear* (1)				*plastic2* (2)			
	Acc.	Time 1	Time 32	Speed-up	Acc.	Time 1	Time 32	Speed-up
1 Minkowski	0.634	5205.158	443.880	**11.726**	**0.997**	478.341	42.062	**11.372**
2 Manhattan	0.794	261.819	24.745	10.581	**0.997**	8.969	2.765	3.243
3 Euclidean	0.706	16.282	27.516	0.592	**0.997**	1.377	4.253	0.324
4 Chebyshev	0.506	145.896	28.994	5.032	0.996	8.796	3.071	2.864
5 Canberra	**0.804**	501.338	76.740	6.533	**0.997**	42.461	7.231	5.872
6 Hamming	0.741	115.159	25.320	4.548	0.956	4.908	2.779	1.766
7 Cosine	0.683	13.320	26.738	0.498	0.996	1.004	4.331	0.232
8 City-block	0.794	262.121	24.879	10.536	**0.997**	8.943	2.762	3.238
9 Bray-Curtis	0.793	154.035	27.701	5.561	**0.997**	7.507	2.914	2.576
10 Correlation	0.665	104.954	24.948	4.207	0.996	4.375	2.816	1.554
Dissimilarity measures	*lung_autofl_2* (3)				*tecator* (4)			
	Acc.	Time 1	Time 32	Speed-up	Acc.	Time 1	Time 32	Speed-up
1 Minkowski	0.870	11.78	1.330	8.862	**0.981**	0.286	0.196	1.461
2 Manhattan	0.864	0.205	0.326	0.628	**0.981**	0.006	0.196	0.032
3 Euclidean	0.866	0.045	0.348	0.130	0.977	0.006	0.197	0.031
4 Chebyshev	0.862	0.215	0.395	0.543	0.967	0.008	0.195	0.043
5 Canberra	**0.886**	1.085	0.495	2.194	0.963	0.030	0.196	0.153
6 Hamming	0.743	0.113	0.333	0.338	0.767	0.007	0.198	0.035
7 Cosine	0.878	0.029	0.414	0.070	0.977	0.005	0.196	0.024
8 City-block	0.864	0.214	0.332	0.643	**0.981**	0.010	0.198	0.048
9 Bray-Curtis	0.865	0.167	0.412	0.405	**0.981**	0.008	0.196	0.042
10 Correlation	0.872	0.110	0.437	0.252	0.977	0.006	0.197	0.032

itself. In contrast, Minkowski $\left(\sum_{i=1}^{n}(x_i - y_i)^p\right)^{\frac{1}{p}}$, for $p = 3$, obtains the highest speed-up: the parallel execution is 11.726 times faster than its sequential counterpart, see Fig. 5. The reported accuracy, for each dissimilarity measure, is the same for both the sequential execution and the multi-core one consequently, we report it only once; see Table 1. Remember that we performed a leave-one-out test which, therefore, is deterministic. All the cores execute the same algorithm because we use a SIMD strategy for the parallelization. The Hamming dissimilarity measure exhibits the lowest accuracy, as expected since this measure is particularly suitable for binary data instead of quantitative data. In Table 1, we can see that dissimilarities derived from Manhattan ($L_1 = \sum_{i=1}^{n} |x_i - y_i|$) and Euclidean $\left(L_2 = \sqrt{\sum_{i=1}^{n}(x_i - y_i)^2}\right)$ are the most accurate ones. Canberra —a distance introduced by Lance-Williams [4, 10]— is a weighted version of L_1 and Minkowski [8], which, in turn, it is a generalization of both L_1 and L_2. The Bray-Curtis (Bray-Curtis) dissimilarity measure is an exception since it is not derived from L_1 and L_2.

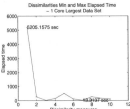 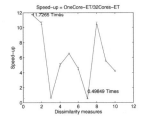

Fig. 3. Elapsed time for one core

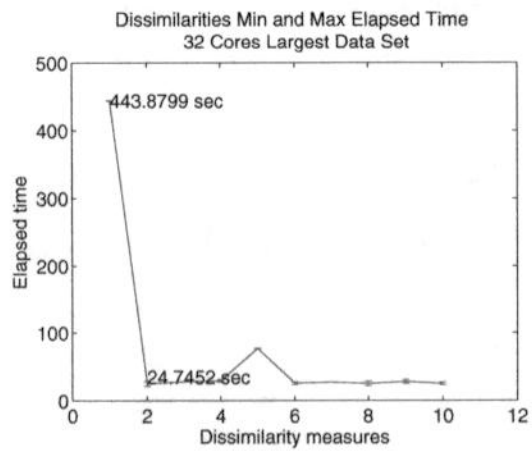

Fig. 4. Elapsed time for 32 cores

Fig. 5. Speed-up

4   Conclusions

The implementation of a web-based micro-service application allowed us to provide a user-friendly interface, as well as a fast implementation, for the comparison of several dissimilarity measures. With such an application, the researcher, depending on his strengths, can concentrate on the development of a part of the code without interfering with the other components of the application. In addition, we showed that parallelizing the application is an option to obtain results in a shorter computational time, getting speed-ups of up to 10 times faster with the parallel version with respect to the sequential one when using the largest dataset. In spite that the deployment of the micro-services has a cost, it is minimal —in this application— in comparison with the cost of the classification task. However, measuring the former one is of interest as future work. It is important to note that, even though Cosine is the fastest measure, it is not the one with the best accuracy. Several bin-to-bin dissimilarity measures were taken into account in this paper, all of them having an $\mathcal{O}(n)$ computational complexity. We experimentally examined two issues: first, for which of them the kNN algorithm presents the best accuracy and, second, how significant the reduction of the computational time is when using the parallel version over the multi-core architecture. As future work, we suggest to test the application with dissimilarity measures of a different type in order to perform the same comparisons, in terms of accuracy and computational efficiency but, alternatively, on many-core architectures (GPU).

Acknowledgments. The authors acknowledge support to attend DCAI'18 provided by Facultad de Administración, Universidad Nacional de Colombia - Sede Manizales (UNAL) and GAIA research group. Anonymous reviewers are acknowledged as well as Oscar David Arbeláez-Echeverri for his support in the development of the application.

References

1. Ciavotta, M., Alge, M., Menato, S., Rovere, D., Pedrazzoli, P.: A microservice-based middleware for the digital factory. Procedia Manufact. **11**, 931 – 938 (2017). 27th International Conference on Flexible Automation and Intelligent Manufacturing, FAIM 2017, 27-30 June 2017, Modena, Italy (2017)
2. Cui, M., Cui, J., Li, H.: Dimensionality reduction for histogram features: a distance-adaptive approach. Neurocomputing **173**, 181–195 (2016)
3. Duin, R.P.W., Juszczak, P., de Ridder, D., Paclik, P., Pekalska, E., Tax, D.M.J.: PRTools4: a matlab toolbox for pattern recognition (2007)
4. Florindo, J.B., Bruno, O.M.: Texture classification using non-euclidean Minkowski dilation. Phys. A **493**, 189–202 (2018)
5. Flynn, M.J.: Some computer organizations and their effectiveness. IEEE Trans. Comput. **C-21**, 948–960 (1972)
6. Gutiérrez, P.D., Lastra, M., Bacardit, J., Benítez, J.M., Herrera, F.: GPU-SME-kNN: scalable and memory efficient kNN and lazy learning using GPUs. Inf. Sci. **373**, 165–182 (2016)
7. Jain, A.K., Duin, R.P.W.: Pattern recognition. In: Gregory, R.L. (ed.) The Oxford Companion to the Mind, 2 edn., pp. 698–703. Oxford University Press, Oxford (2004)
8. Santana, F., Santiago, R.: A generalized distance based on a generalized triangle inequality. Inf. Sci. **345**, 106–115 (2016)
9. Swaminathan, M., Yadav, P.K., Piloto, O., Sjöblom, T., Cheong, I.: A new distance measure for non-identical data with application to image classification. Pattern Recogn. **63**, 384–396 (2017)
10. Todeschini, R., Ballabio, D., Consonni, V., Grisoni, F.: A new concept of higher-order similarity and the role of distance/similarity measures in local classification methods. Chemometr. Intell. Lab. Syst. **157**, 50–57 (2016)
11. Wong, T.T.: Performance evaluation of classification algorithms by k-fold and leave-one-out cross validation. Pattern Recogn. **48**(9), 2839–2846 (2015)

A Parallel Application of Matheuristics in Data Envelopment Analysis

Martín González[1(✉)], Jose J. López-Espín[1(✉)], Juan Aparicio[1(✉)], and Domingo Giménez[2]

[1] Center of Operations Research, Miguel Hernández University of Elche, Alicante, Spain
martingonzes@gmail.com, jlopez@umh.es, j.aparicio@umh.es
[2] Department of Computer Science and Systems, University of Murcia, Murcia, Spain

Abstract. Data Envelopment Analysis (DEA) is a non-parametric methodology for estimating technical efficiency and benchmarking. In general, it is desirable that DEA generates the efficient closest targets as benchmarks for each assessed unit. This may be achieved through the application of the Principle of Least Action. However, the mathematical models associated with this principle are based fundamentally on combinatorial NP-hard problems, difficult to be solved. For this reason, this paper uses a parallel matheuristic algorithm, where metaheuristics and exact methods work together to find optimal solutions. Several parallel schemes are used in the algorithm, being possible for them to be configured at different stages of the algorithm. The main intention is to divide the number of problems to be evaluated in equal groups, so that they are resolved in different threads. The DEA problems to be evaluated in this paper are independent of each other, an indispensable requirement for this algorithm. In addition, taking into account that the main algorithm uses exact methods to solve the mathematical problems, different optimization software has been evaluated to compare their performance when executed in parallel. The method is competitive with exact methods, obtaining fitness close to the optimum with low computational time.

1 Introduction

Data Envelopment Analysis (DEA) is a mathematical programming, non-parametric technique commonly used to measure the relative performance of a set of homogeneous processing units, which use several inputs to produce several outputs. These operating units are usually called Decision Making Units (DMUs) in recognition of their autonomy in setting their input and output levels. Thanks to being a non-parametric technique, DEA does not need to suppose a particular functional form for the production function, technical efficiency may be easily evaluated with multiple inputs and outputs and it also produces relevant benchmarking information from a managerial point of view. In particular,

© Springer International Publishing AG, part of Springer Nature 2019
F. De La Prieta et al. (Eds.): DCAI 2018, AISC 800, pp. 172–179, 2019.
https://doi.org/10.1007/978-3-319-94649-8_21

DEA provides both input and output efficient targets, the coordinates of the projection point on the estimated efficient frontier, and represents levels of operation that can make the corresponding inefficient DMU perform efficiently.

Traditional DEA measures maximize the total technical effort associated with the evaluated unit in order to reach the efficient frontier. Instead, it seems more natural to assume that inefficient DMUs apply a Principle of Least Action, a well-known law in physics, with the aim of being technically efficient. Otherwise, inefficient units would need to make an extra effort, decreasing inputs and/or increasing outputs, to reach the frontier. The application of this 'natural' Principle of Least Action is linked to the determination of the closest targets on the efficient frontier of the corresponding DEA production possibility set. This drawback of traditional DEA measures has aroused increasing interest among researchers to develop new models capable of yielding achievable targets. Examples are the papers by Briec and Lesourd [2], Pastor and Aparicio [3], Aparicio and Pastor [1,4–6] and Aparicio et al. [7].

The application of the Principle of Least Action has been recently studied from a metaheuristic perspective (Benavente et al. [8], López-Espín et al. [9] and González et al. [10]). In [8,9] heuristics were used to generate valid solutions for a subset of restrictions of the problem, while in [10] all the constraints are incorporated, the heuristics are improved, and new ones are developed, thereby generating initial populations of solutions that satisfy all constraints.

Our paper takes up where González et al. [10] left off in the application of metaheuristics to the approach in [1]. The improvement of previous heuristics for the generation of valid solutions is a possible option, but greatly limits the search for valid solutions for large problem sizes, because when the number of variables grows, the number of valid solutions decreases. Exact methods can also be used to solve these problems. The main drawback of these methods is the great amount of time needed to solve a NP-hard problem. When the problem grows, the number of possible combinations between variables increases exponentially.

The contributions of this work include the development of a parallel algorithm that belongs to the class of hybrid metaheuristics [11]. New parallel features have been included in the matheuristic developed in [12]. The algorithm developed is focused on the need to solve multiple simultaneous models. This is due to the DEA problem that concerns us, in which numerous models must be analyzed for each DMU evaluated. The aim is to separate the number of DMUs to be evaluated in the most efficient way in the different available threads. For this, message-passing (MPI) and shared memory (OpenMP) programming have been considered.

The remainder of the paper is organized as follows. In Sect. 2, a brief introduction to the main notions associated with Data Envelopment Analysis is presented, and existing approaches for determining closest targets are outlined. The working problem is also presented in this section. The parallel algorithm used to generate and improve valid solutions is studied in Sect. 3. In Sect. 4, the results of some experiments are summarized. Section 5 concludes the paper and outlines some possible lines of research.

2   Data Envelopment Analysis and the Problem to Be Solved

DEA involves the use of mathematical programming to construct a non-parametric piecewise surface over the data in the input-output space. Technical efficiency measures associated with the performance of each DMU are then calculated relative to this surface, as a distance from it.

Before solving the mathematical programming model, we introduce some notations. Let us assume that data on m inputs and s outputs for n DMUs are observed. For the j-th DMU, these are represented by $x_{ij} \geq 0$, $i = 1, \ldots, m$, and $y_{rj} \geq 0$, $r = 1, \ldots, s$.

One of the models that can be solved by applying the Principle of Least Action in DEA is that by [1]:

$$\max \left\{ \beta_k - \frac{1}{m} \sum_{i=1}^{m} \frac{t_{ik}^-}{x_{ik}} \right\}$$

s.t.

$$
\begin{aligned}
\beta_k + \frac{1}{s} \sum_{r=1}^{s} \frac{t_{rk}^+}{y_{rk}} &= 1 & (c.1) \\
-\beta_k x_{ik} + \sum_{j=1}^{n} \alpha_{jk} x_{ij} + t_{ik}^- &= 0 \ \forall i & (c.2) \\
-\beta_k y_{rk} + \sum_{j=1}^{n} \alpha_{jk} y_{rj} - t_{rk}^+ &= 0 \ \forall r & (c.3) \\
-\sum_{i=1}^{m} \nu_{ik} x_{ij} + \sum_{r=1}^{s} \mu_{rk} y_{rj} + d_{jk} &= 0 \ \forall j & (c.4) \\
\nu_{ik} &\geq 1 \ \forall i & (c.5) \\
\mu_{rk} &\geq 1 \ \forall r & (c.6) \\
d_{jk} &\leq M b_{jk} \ \forall j & (c.7) \\
\alpha_{jk} &\leq M(1 - b_{jk}) \ \forall j & (c.8) \\
b_{jk} &= 0, 1 \ \forall j & (c.9) \\
\beta_k &\geq 0 & (c.10) \\
t_{ik}^- &\geq 0 \ \forall i & (c.11) \\
t_{rk}^+ &\geq 0 \ \forall r & (c.12) \\
d_{jk} &\geq 0 \ \forall j & (c.13) \\
\alpha_{jk} &\geq 0 \ \forall j & (c.14)
\end{aligned}
\tag{1}
$$

The definition and interpretation of the decision variables and constraints of the model 1 can be found in [1].

One weakness of the approach in model 1 is that it uses a "big M" in (c.7) and (c.8). These constraints allow us to link d_{jk} to α_{jk} by means of the binary variable b_{jk}. The value of M can be calculated if and only if all the facets that define the DEA technology are previously determined. Unfortunately, the identification of all these facets is a combinatorial NP-hard problem. This weakness will be overcome in the new approach introduced here, since the new methodology does not need to resort to a big M to obtain the desired result.

3   Parallel Algorithm

In order to improve the performance of the algorithm developed in [12], it has been parallelized at different levels. Some of the objectives to parallelize the algorithm are related to the difficulty of solving problems such as the one proposed in Sect. 2, where a lot of computing time is taken to find satisfactory solutions. So, the proposed parallelization models are intended to reduce computation time to solve these models and improve the fitness of the solutions obtained. For this, both shared memory (OpenMP) and message-passing (MPI) schemes have been used, since it is possible to use them either separately or in combination. The metaheuristic parallelised in this scheme is based on generating initial solutions to the given model, and improving them with the intention of finding satisfactory solutions. For this, the following scheme is followed: Initialization, Improvement, Selection, Crossing and Diversification. Each parallelization models tries to improve the solution quality or computational time:

- OpenMP: Shared memory parallelization functions have been included within the algorithm's own functions, making them faster. Mainly, these improvements have been included in the initialization function, where it is possible to distribute the generation of solutions between the different threads, and in the improvement and crossing functions, where these tasks can be divided into smaller functions. This model of parallelism has been introduced with the intention of improving the internal loops of the main metaheuristic algorithm. Therefore, in all parts of the algorithm where many models must be evaluated (initialization, improvement, crossing), whenever the evaluation of the models are independent of each other, this type of parallelism is introduced.
- MPI: Message-passing functions have been included over and above metaheuristics. These functions optimize the computing time and the fitness of the final solutions. This level of parallelism is found in a higher level than the shared memoria scheme. In this way, the number of models to be evaluated are divided in equal parts in the different threads. Thus, the computation time is decreased as the number of cores increases. If it is not possible to divide all the DMUs into equal groups, the remaining units will be assigned randomly to the different cores.

The OpenMP scheme can also be used at the same level as the MPI scheme. A comparison between these two schemes is developed in the results of the experiment. This comparison is made to decide where implement each of the schemes. Figure 1 shows how the algorithm works and all the possible configurations.

In the literature, there are several mathematical methods that can solve both mixed integer linear programming (MILP) and linear programming (LP) problems. In the algorithm developed in this paper, a MILP-based decomposition is used to divide the main DEA problem, which is difficult to solve, into smaller LP-type problems that are easier to solve. In this regard, an exact method able of optimally solving numerous LP problems is needed. For this task, two exact methods that work optimally are evaluated to measure their performance in combination with our parallel algorithm. The software packages used are CPLEX and GUROBI.

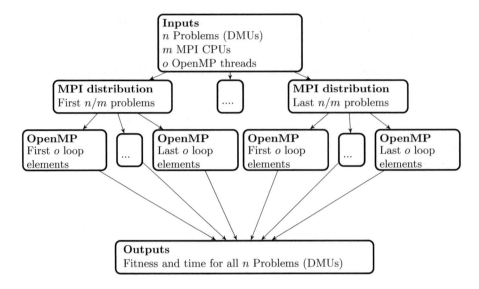

Fig. 1. Generation of metaheuristics and structure of the algorithm.

4   Experimental Results

Experiments were conducted to analyze the effectiveness of the parallel scheme developed. Two different exact methods are compared in terms of fitness and time, working in parallel. The performance of each exact method is also evaluated. For all the experiments, the IBM ILOG CPLEX Optimization Studio (CPLEX) and GUROBI are used. The system used in the experiments is a AMD Phenom II X6 1075T CPU (hexacore) at 3 GHz with 16 GBytes of RAM, private L1 and L2 caches of 64 KBytes and 512 KBytes respectively, and a L3 cache of 6 MBytes shared by all cores. For all the experiments, a standard problem has been taken with the following dimensions: 3 inputs, 2 outputs and 50 DMUs. Regarding the data, in our simulations the m inputs and s outputs of each of the n DMUs are generated at random but taking into account that the production function that governs the production situation is the well-known Cobb-Douglas function [13].

First, we analize the behavior of the different models of parallelism in the main algorithm will be studied. For this, the two available paradigms (MPI and OpenMP) will be used to perform the same function: divide the DMUs to be evaluated in equal groups, thus creating parallel executions of the algorithm, and therefore, of the metaheuristics and exact methods. In this evaluation, while comparing the parallel models, the different exact methods used in this paper are also compared: CPLEX and GUROBI. Therefore, there is going to be a comparison between how the MPI model works with CPLEX and GUROBI and how the OpenMP model also works with CPLEX and GUROBI.

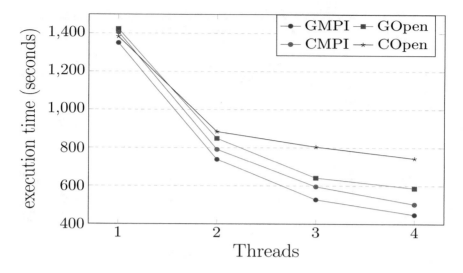

Fig. 2. Comparison of the execution time (in seconds) using different threads with the same problem, for each optimizer and different parallel algorithms: GUROBI with MPI (GMPI) and OpenMP (GOpen), and CPLEX with MPI (CMPI) and OpenMP (COpen).

Figure 2 shows how, as the number of CPUs increases, the computation time decreases. However, it can be seen how, when the number of CPUs grows, the improvement in the computational time decreases, becoming lower each time. This is due to the fact that, depending on the number of DMUs to be evaluated, from certain divisions, the resulting DMU groups have practically the same number of elements, therefore the computation time is smaller, but similar. In addition, this figure also shows that the exact methods chosen work well in parallel, being possible to execute multiple instances of them at the same time. It must be taken into account that the files in which the problems to be solved are written must have different names to avoid any confusion during the execution. This experiment shows that the MPI message-passing programming model works better than the OpenMP shared memory model in this first step of the algorithm. Therefore reaffirming what was stated in Sect. 3.

Another important parameter to analyze, is the objective value achieved by each optimizer. For this, it is necessary to emphasize that the initial generation of solutions is done in a random way, so that the initial solutions generated by an optimizer are not the same as for the other, since the executions are independent of each other. To evaluate the results, 20 executions have been made for each optimizer, making 5 for each set of threads. The results shown in Fig. 3 show the average values of all the executions for each set of threads. Analyzing the obtained values, it can be affirmed that, after performing several experiments, the CPLEX optimizer obtains better fitness values in all the sets. As mentioned before, the generation of initial solutions is randomly created and is independent

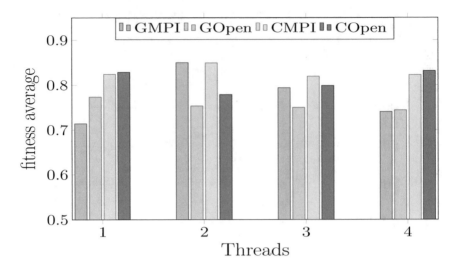

Fig. 3. Objective values obtained for each optimizer and different parallel algorithms: GUROBI with MPI (GMPI) and OpenMP (GOpen), and CPLEX with MPI (CMPI) and OpenMP (COpen). The values are the average of 5 executions.

for each optimizer. Even so, after a statistical study, CPLEX obtained better fitness values in this experiment.

5   Conclusions and Future Works

The application of the Principle of Least Action in DEA is a topic of relevance in recent DEA literature. However, it is well-known that from a computational point of view, this has usually been tackled with inadequate approaches, associated with combinatorial NP-hard problems.

Parallel algorithms are good solutions for solving these kind of problems. This is because a high number of independent problems must be solved, and those problems can be divided and solved by different threads. Furthermore, several optimized software packages can be used, but not all of them work well in parallel. To improve performance, is necessary to include optimizers that can be executed in different instances. For that, CPLEX and GUROBI are tools for these tasks. The parallel algorithm proposed in these paper works in an optimal way with both of these optimizers and with different parallel paradigms.

For future work, we propose the use of other free optimizers, and to check their performance compared with the most powerful optimizers in the market. In addition, it is also desirable to incorporate improvements in parallelism, so that the internal functions of the metaheuristic algorithm can be executed more quickly and accurately.

Acknowledgements. J. Aparicio and M. González thank the financial support from the Spanish 'Ministerio de Economía, Industria y Competitividad' (MINECO), the 'Agencia Estatal de Investigacion' and the 'Fondo Europeo de Desarrollo Regional' under grant MTM2016-79765-P (AEI/FEDER, UE).

References

1. Aparicio, J., Ruiz, J.L., Sirvent, I.: Closest targets and minimum distance to the Pareto-efficient frontier in DEA. J. Prod. Anal. **28**, 209–218 (2007)
2. Briec, W., Lesourd, J.B.: Metric distance function and profit: some duality results. J. Optim. Theory Appl. **101**(1), 15–33 (1999)
3. Pastor, J.T., Aparicio, J.: The relevance of DEA benchmarking information and the least-distance measure: comment. Math. Comput. Modell. **52**, 397–399 (2010)
4. Aparicio, J., Pastor, J.T.: A well-defined efficiency measure for dealing with closest targets in DEA. Appl. Math. Comput. **219**, 9142–9154 (2013)
5. Aparicio, J., Pastor, J.T.: Closest targets and strong monotonicity on the strongly efficient frontier in DEA. Omega **44**, 51–57 (2014)
6. Aparicio, J., Pastor, J.T.: On how to properly calculate the Euclidean distance-based measure in DEA. Optimization **63**(3), 421–432 (2014)
7. Aparicio, J., Mahlberg, B., Pastor, J.T., Sahoo, B.K.: Decomposing technical inefficiency using the principle of least action. Eur. J. Oper. Res. **239**, 776–785 (2014)
8. Benavente, C., López-Espín, J.J., Aparicio, J., Pastor, J.T., Giménez, D.: Closest targets, benchmarking and data envelopment analysis: a heuristic algorithm to obtain valid solutions for the shortest projection problem. In: 11th International Conference on Applied Computing (2014)
9. López-Espín, J.J., Aparicio, J., Giménez, D., Pastor, J.T.: Benchmarking and data envelopment analysis. An approach based on metaheuristics. In: Proceedings of the International Conference on Computational Science, ICCS 2014, Cairns, Queensland, Australia, 10–12 June 2014, pp. 390–399 (2014)
10. Gónzalez, M., López-Espín, J.J., Aparicio, J., Giménez, D., Pastor, J.T.: Using genetic algorithms for maximizing technical efficiency in data envelopment analysis. In: Proceedings of the International Conference on Computational Science, ICCS 2015, Reykjavík, Iceland, 01–03 June 2015, vol. 51, pp. 374–383 (2015)
11. Talbi, E.-G.: Hybrid Metaheuristics. SCI, vol. 434. Springer, Germany (2013)
12. González, M., López Espín, J.J., Aparicio, J., Giménez, D., Talbi, E.: A parameterized scheme of metaheuristics with exact methods for determining the principle of least action in data envelopment analysis. In: Program of the 2017 IEEE Congress on Evolutionary Computation (2017)
13. Cobb, C.W., Douglas, P.H.: A theory of production. Am. Econ. Rev. **18**(1), 139–165 (1928)

Prediction Market Index by Combining Financial Time-Series Forecasting and Sentiment Analysis Using Soft Computing

Dinesh Kumar Saini$^{(\boxtimes)}$, Kashif Zia, and Eimad Abusham

FCIT, Sohar University, Sohar, Oman
{dinesh,kzia,eabusham}@soharuni.edu.om

Abstract. In recent years, a lot of research is focusing on predicting real-world outcomes using Social networks data (for example, Twitter Data). Sentiment Analysis of the twitter data thus has become one of the key aspects of making predictions involving human sentiments. Stock market movements are very sensitive and it affects investment of the investors because of this prediction is the main interest of the researchers. Soft computing approaches and nature-inspired computing has a lot of potential in predicting the market movement. In this paper, soft computing techniques are used to predict market trends using sentiments extracted from market data. The results indicate that by selecting suitable neural networks architecture and selecting suitable regression coefficients can improve the overall accuracy and correlation of the predictions. Stock market information people use for investment decisions. Forecasting must be accurate otherwise it will not be effective in the decision. There are techniques like trend based classification, adaptive indicators selection and market trading signals are used in forecasting.

Keywords: Prediction · Market index · Financial · Time series
Sentiment Analysis · Soft computing

1 Introduction

This document constitutes Springer's guidelines for the preparation of proceedings papers. Soft computing approaches like analytical and statistical modeling, neural networks, behavioral financial modeling, machine learning technologies are used to predict the sentiment of the market. The market moves and investors emotions and sentiments need to be analyzed on real-time online transactional data and that to be on the very fast basis. While processing data on a real-time basis, linear or nonlinear modeling approaches can be used. Autoregressive Integrated Moving Average (ARIMA) model which was suggested long ago by Box and Jenkins in 1970 is still used in linear computing. It uses time series concept. Most of the world stock markets behave perfectly linear it is not evident, which is coming from market-specific domain knowledge of the market and residual variance of expected return and original return is reasonably high.

Market trends and financial forecasting is a quite challenging task in current time because of non-linearity in the market trends. Heteroscedasticity and Autoregressive

© Springer International Publishing AG, part of Springer Nature 2019
F. De La Prieta et al. (Eds.): DCAI 2018, AISC 800, pp. 180–187, 2019.
https://doi.org/10.1007/978-3-319-94649-8_22

Conditional Heteroscedasticity (ACH) are two popular techniques used in financial data forecast. In recent time soft computing techniques like Artificial Neural Networks (ANN), Genetic Algorithms, Machine Learning, are used for prediction and forecasting.

NIFTY is the share market of India and we took some sample dataset from the historical data and try to predict sentiment for the market move. In the paper, we used Alchemy API and artificial neural network for prediction. Figure 1 shows the framework.

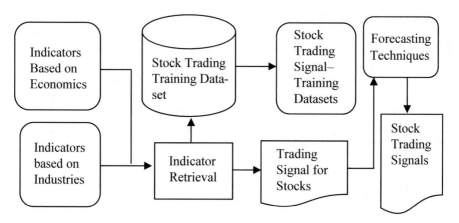

Fig. 1. Stock market forecasting mechanism on, town/city, and country.

2 Proposed Framework

The Alchemy API as an open source platform which can be used for natural language processing from social networking data like twitter. Alchemy API can extract, analyze sentiment and tags etc. Alchemy is used for performing Sentiment Analysis on the Twitter data set after reducing it to reduced sequential dataset all-purpose proposed by Chen and Lazer (2013).

3 Structuring Your Paper

Each tweet was classified as positive, negative or neutral. This was followed by creating a sentiment score for each day scaled from −1 to 1. Each tweet is given a weight" +1 for the positive tweet, −1 for the negative tweet and 0 for neutral. Sentiment score for a day is then calculated as follows:

$$Sentiment\ Score = \frac{\sum_i w_i}{N}$$

Where, wi is the weight for an ith tweet for the day and N is the total number of tweets on the given day. For the analysis, we attempted to explore any relationship between the previous 4 days of sentiment scores and the percentage change in the market index values from the previous day.

4   Length of Papers

Artificial Neural Networks seemed to be a natural choice for the same given its reliability and precision in discovering the non-linear hidden relationship between the input and output data. The following lists the ANN parameters used: Input and Output Variables: Selecting input and output variables was trivial: 4 input variables corresponding to the previous four days of sentiment scores, 1 output variable corresponding to the percentage change in the NIFTY 50 index value from the previous day. Figure 2, it is shown how ANN layers are working.

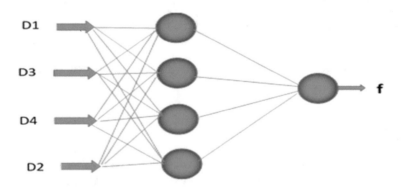

Fig. 2. Artificial neural network

A number of hidden layers: We also were quite content to use only one hidden layer for the network given only about 100 records of data.

Data Partition: The dataset was partitioned as follows:

- 59% of the training dataset
- 16% of validation dataset
- 25% of the testing dataset from 17th October to 29th November

Training Algorithm, number of neurons in the hidden layer and the Transfer Functions: The critical parameters remaining were the training algorithm to be used and number of neurons in the first hidden layer. Empirical observations have suggested that Levenberg Marquardt seems to outperform other training algorithms for a smaller dataset consisting of few hundreds of records. So we chose, Levenverg Marquardt Algorithm as our training Algorithm and experimented with a number of hidden units and transfer functions of hidden layers to find out the optimal configuration for the given data. The transfer function was selected as linear for output layer (only one output node) and experimentation was performed with both Logistic and hyperbolic tangent functions for the hidden layer. For each transfer function, we noted the best configuration and compared the results for both.

Comparison Criteria for the Results: A number of statistical performance parameters were used to compare various neural network architectures based on their performance on the testing dataset.

Mean ARE is Absolute Relative Error, which signify the neural network prediction and training quality. Actual and desired values difference is computed to get this index. The size of the network also matters; small networks have better quality. The mean of the ARE gives us a reasonable estimate of the accuracy of a particular neural network architecture.

AIC: Akaike Information Criterion (AIC) is proposed by Hu (2007). Networks are analyzed with weights which are hidden units. Akaike information criteria help in analyzing fitness criteria for architecture search. Hidden units increase the cost in the network but still, these are preferred. Network errors affect the result so it needs to reduce in hidden units. Neural network depends on an optimal number of weights.

Correlation: Network output and actual values are related and this relationship needs to be measured using a statistical measure called correlation. The range of correlation varies from −1 to +1. The relationship is measured on the scale from −1 to +1, closer to 1 the stronger to the relationship. If the relationships are closer to −1, then the relationship will be negative liner relationship. If the value of correlation is 0 then the relationship is no linear relationship.

Fitness: the test is used for testing the network, errors must be in less quantity, fitness help in calculating inverse mean for network errors.

R-Squared is used for forecasting accuracy of the models, it uses mean for the records. Values are used for indicating the model accuracy if the value is 1, the model is better and if 0 then model is not a good model.

5 Forecasting Time Series Using Artificial Neural Networks

Two approaches are used liner vs nonlinear. Moving average, exponential smoothing and time series regression is used in linear. ARIMA is also one of the mechanisms used for linear forecasting. In ANN is using nonlinear forecasting as shown in Fig. 3.

The financial market is nonlinear which is now established by financial analysts and researchers.

Autoregressive Conditional Heteroskedasticity (ACH) and General Autoregressive Conditional Heteroskedasticity (GACH) models support nonlinearity and are used in prediction of stock markets.

Input and Output Variables: The input variables selected were the lagged values of the closing prices of NIFTY 50 Index values. However, with very less lag between

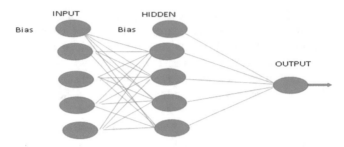

Fig. 3. Feedforward ANN

each input variable, the correlation between them increases and this leads to overfitting over the training data. Similar is the case with large lag by the input variables which leads to under fitting. So we resorted to a compromise; i.e. input variables are various lagged structures with different lag bw each (lag of 1 or two trading days). We have experimented with various possible values for lagged structures: 5, 10, 15 or 20 input variables and finally selected the Architecture which produces best results over the testing data set). The output layer consists of just one output neuron similar to stage1.

A Number of Hidden Layers: Only one hidden layer is used for Stage 2 as well.

Data Partition: Data partition was performed similarly to the previous stage. The testing data set includes records from 1st week of January to December 31, about 25 percent of the entire dataset.

Training Algorithms, Number of Neurons in the Hidden Layer and Transfer Functions: Many Training Algorithms were Experimented with for the Given Time Series: Quasi-Newton, Levenberg-Marquardt, Batch-Back Propagation, Online Back-Propagation, Quick Propagation and Conjugate Gradient Descent. Each Training algorithm was experimented both with Logistic and Hyperbolic Tangent Transfer functions for the hidden layer. The linear transfer function was used for the output layer. A number of architectures were tried and tested with a different number of neurons in the hidden layer for each training Algorithm and corresponding Transfer functions and the best Architecture was noted for every combination of training algorithm and the transfer function.

Comparison Criteria for the Results: This is similar to the description for Stage1.

6   Results

A snapshot of the dataset used for training the neural network during Stage-1 is shown below. Here SS stands for the Sentiment Score. SS for Date-1 is the Sentiment Score for the previous date and so on as shown in Fig. 4.

After experimenting with both a number of hidden units and the transfer functions for the hidden layer, 4-9-1 Architecture with a Logistic transfer function for the hidden layer was selected. The selected Architecture was then queried to predict the percentage

SS for Date-1	SS for Date-2	SS for Date-3	SS for Date-4	%Change from Date-1
-0.301	0.107	0.174	-0.411	0.220
-0.281	-0.034	-0.035	0.028	-0.582
0.244	-0.282	-0.034	-0.035	-0.186
-0.173	0.284	0.339	0.244	3.160
-0.521	-0.224	0.284	0.339	0.568
0.726	-0.521	-0.224	0.284	4.275
-0.636	0.719	-0.521	-0.224	-7.211
-0.558	-0.664	0.719	-0.521	0.862
0.761	-0.559	-0.664	0.719	2.328
0.711	0.759	-0.559	-0.664	1.843
-0.861	0.180	0.759	0.710	-3.143
-0.311	-0.864	0.180	0.710	-0.341
0.427	0.067	0.114	0.096	2.328
-0.543	0.421	0.067	0.114	-1.528
-0.499	-0.545	0.421	0.067	0.858
0.776	-0.499	-0.545	0.421	3.376
-0.699	-0.070	-0.110	0.772	-4.957
0.529	-0.710	-0.070	-0.110	0.185
0.600	0.529	-0.710	-0.070	-1.758
-0.379	0.598	0.529	-0.710	2.933
-0.815	-0.391	0.598	0.529	-3.161
0.475	0.055	0.105	-0.818	1.079
0.283	0.474	0.055	0.105	1.417
-0.482	0.278	0.474	0.055	2.177

Fig. 4. Dataset used for training the neural network during stage-1

changes for the testing dataset i.e. from Oct 17 to Nov 29. The percentage changes were then converted to the NIFTY index values. These index values were then used to note down the performance of the selected architecture w.r.t NIFTY index values (instead of the % changes; Since we want to combine the results of Stages 1 and 2, so we need to keep consistency in parameters and keep them with w.r.t actual index values only), shown in Fig. 5.

B NIFTY 50 Index Stage1	C Actual NIFTY 50 Index
5245.977	5240.500
5215.451	5227.750
5205.734	5215.700
5370.228	5282.550
5400.751	5336.700
5631.644	5338.000
5225.544	5322.950
5270.602	5320.400
5393.315	5347.900
5492.718	5380.350
5320.089	5362.950
5301.949	5366.300
5425.385	5421.000
5342.485	5412.850
5388.297	5415.350
5570.210	5386.700
5294.072	5350.250
5303.847	5334.600
5210.628	5287.800
5363.435	5315.050
5193.910	5258.500
5249.973	5253.750
5324.358	5274.000
5440.268	5225.700
5232.479	5238.400

Fig. 5. Nifty values computed using the trained architecture 4-9-1 vs. actual ones.

With respect to the Index values, the selected Architecture gives a correlation of 0.819, Mean ARE as 0.0142 and R-Squared as 0.6708. On the testing dataset, as shown in Fig. 6.

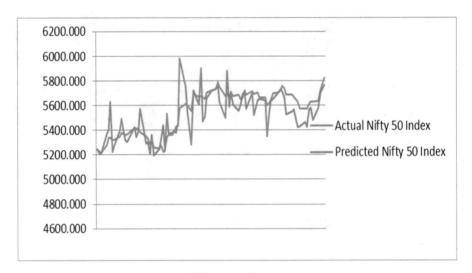

Fig. 6. The plot of actual vs. predicted (by the 4-9-1 arch.) nifty 50 values from aug–nov

The following figure presents a plot of Actual vs. Predicted Nifty 50 Index values over the testing Dataset (Oct 17–Nov 29, for the Stage1, as shown in Fig. 7.

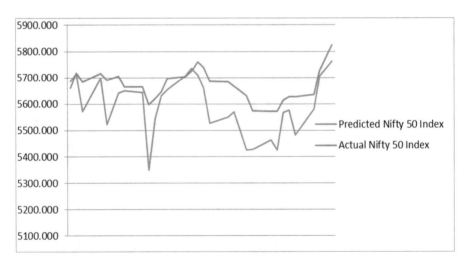

Fig. 7. The plot of actual vs. predicted

Nifty 5o Index values over the testing Dataset Description: For this Stage, we used the historic data of NIFTY 50 Index values consisting of about 2000 records from NSE (National Stock Exchange of India) official website (www.nseindia.com).

7 Conclusion

The major objective of the paper is to combine the following two approaches of predicting NIFTY index 50 values. Artificial Neural Networks with input variables as previous four days of Sentiment Scores and the output variable as the percentage change from the last day closing prices of the NIFTY 50 Index. For calculating Sentiment Score for a day, Sentiment Analysis was performed on the Twitter data for the day. Artificial Neural Networks applied to the Historic data of the closing prices of the NIFTY 50 Index i.e. Financial Time Series forecasting using ANN using simple Multiple Linear Regression and see if the accuracy of the predictions improves. For the same, we used a testing dataset of the closing prices of one and a half month from October 17 to November 29, to validate our hypothesis. The results indicate that by selecting suitable neural networks architecture for each of the above-stated methods and selecting suitable regression coefficients can improve the overall accuracy and correlation of the predictions.

References

Chen, R., Lazer, M.: Sentiment analysis of twitter feeds for the prediction of stock market movement. Stanf. Educ. (2013). http://cs229.stanford.edu/proj2011/ChenLazer-SentimentAnalysisOfTwitterFeedsForThePredictionOfStockMarketMovement.pdf

Hu, S.: Akaike information criterion. Cent. Res. Sci. Comput. **93**, 1–20 (2007). http://www4.ncsu.edu/~shu3/Presentation/AIC_2012.pdf

Relaxation Method of Convolutional Neural Networks for Natural Language Processing

Ryo Iwasaki$^{(\boxtimes)}$, Taku Hasegawa, Naoki Mori, and Keinosuke Matsumoto

Osaka Prefecture University, 1-1 Gakuen-cho, Naka-ku, Sakai, Osaka, Japan
iwasaki@ss.cs.osakafu-u.ac.jp

Abstract. Deep learning has developed into one of the most powerful methods in the machine learning field. In particular, convolutional neural networks (CNNs) have been applied not only to image recognition tasks but also to natural language processing (NLP). To reuse older deep learning models, transfer learning techniques have been widely used in the image recognition field. However, there has been little research on transfer learning in NLP. In this paper, we propose a novel transfer learning model based on a relaxation method of CNNs for NLP. The effectiveness of the proposed method is verified using computer simulations, taking a film review score recognition task as an example.

Keywords: Deep learning · Convolutional neural network
Relaxation · Natural language processings

1 Introduction

Recently, many deep learning models have been proposed. In image recognition tasks in the field of computer vision (CV), transfer learning techniques have been reported as improving performance [1].

Convolutional neural networks (CNNs) have been applied to the field of natural language processing (NLP), because the convolution filters of a CNN can capture local features of input well. In NLP research, word vectors are very important for improving performance. To obtain pre-trained word vectors, Word2Vec [2] has been used. A transfer learning method often used in NLP tasks is pre-trained word embedding.

In this paper, we propose a novel transfer learning method based on a relaxation method of CNNs for NLP. In our method, the CNN is trained on a relaxation problem first. Next, we transfer the weights of the model trained on the relaxation problem to the model trained on the original problem as initial values. The effectiveness of the proposed method is confirmed by taking computer simulations of film review recognition tasks using Yoon Kim's CNN model [3] as an example. In this paper, we show only the results for film review recognition tasks with an NLP model. However, we believe that our method is so flexible that our transfer method can be used when a CNN model other than Yoon Kim's is used or when the model is tested on other than film review recognition tasks.

© Springer International Publishing AG, part of Springer Nature 2019
F. De La Prieta et al. (Eds.): DCAI 2018, AISC 800, pp. 188–195, 2019.
https://doi.org/10.1007/978-3-319-94649-8_23

2   Related Work

The CNN has convolution filters to capture local features and has been shown to be effective in the field of NLP [3,4]. CNN models in NLP are used to deal with sequential data. Besides long short-term memory (LSTM) and recurrent neural networks (RNNs), CNN has been used in NLP in recent years [4,5]. CNN models have fewer parameters than LSTM models, and one CNN model has been reported to equal the best of other methods in performance on NLP tasks [6]. Character-level CNNs have achieved results that have equaled the best or have been competitive [7]. One NLP model uses CNN as a part of an encoder-decoder model for generating a general-purpose vector representation of posts and comments on Twitter [8,9].

Releasing the large public image repository ImageNet [10] had an impact on transfer learning in the field of CV. Adapting knowledge makes it possible to build a model having good generalized performance. The potential to transfer knowledge can benefit a wide range of applications in CV, NLP, and other fields. Transfer learning methods prevent models from overfitting. Various surveys have been conducted on transfer learning [11].

Transfer learning techniques have been reported to improve CNN performance in CV [1]. To address various CV tasks, models are frequently not trained from scratch but fine-tuned by pre-training them on other datasets [12]. In general, the initial values of CNN models affect their performance. Fine-tuning first became popular in CV because CNN models are often used in CV. Pre-trained networks were reported to perform better at various tasks in image processing [13,14]. Some studies have shown potentiality for transfer learning in the field of NLP [15].

One of the transfer learning methods often used in NLP tasks is pre-trained word embedding. To obtain pre-trained word vectors, Word2Vec [2] is often used. Although the word vectors obtained from Word2Vec do not have positive or negative meanings, it has been reported that the vectors obtain such meanings after training on a text classification task [3]. Given this information, transferring the weights of the word embedding layer should give more-effective knowledge to the receiving model. Some papers on NLP report and discuss the results of fine-tuning [12,16].

3   Proposed Method

We propose a method which combines the concept of transfer learning with a relaxation method. An important point regarding the relaxation problem is that it must be easier than the original problem. We obtain a CNN model trained on the relaxation problem. Next, on the original problem we train a neural network that has been initialized with the weights of the pre-trained model. We define our proposed method in detail below.

We define an original problem as follows:

$$X_0 = \{x_1, x_2, \cdots, x_i, \cdots, x_n\}, \quad x \in \mathcal{X}_0 \tag{1}$$

$$Y_0 = \{t_1, t_2, \cdots, t_i, \cdots, t_n\}, \quad t \in \mathcal{T}_0 \tag{2}$$

where x_i is data i and t_i is a label of x_i. $|\mathcal{T}_0|$ is represented as the variety of labels in Y_0. If the target problem is a film review with five score rankings, $|\mathcal{T}_0| = 5$.

Here, we make P data set sequences X_0, X_1, X_2, $\cdots$, X_P, where $X_{i+1} \subseteq X_i$. We regard X_i as a new dataset and define the new label $Y_i = \{t_j\}$, $t_j \in \mathcal{T}_i$. If $\mathcal{T}_{i+1} \subset \mathcal{T}_i$, we define a problem represented by $\{X_{i+1}, Y_{i+1}\}$ as a relaxation problem derived from the problem represented by $\{X_i, Y_i\}$. $\mathcal{T}_{i+1} \subset \mathcal{T}_i$ implies that the number of label types in reduced. The weights of the model trained on $\{X_{i+1}, Y_{i+1}\}$ are transferred to the model to be trained on $\{X_i, Y_i\}$. M_i denoted a model trained on $\{X_i, Y_i\}$.

When we have N epochs for total training, we arrange $\alpha_i N$ epochs to train the model M_i under the following condition.

$$\alpha_0 + \alpha_1 + \cdots + \alpha_P \cdots = 1$$

In the proposed method, M_P is trained first, and M_{P-1}, initialized with the weight of M_P, is trained as the next step. We repeat this procedure until we obtain M_0. If $\mathcal{T}_{i+1} \subset \mathcal{T}_i$, we have to increase the number of output connections, because the number of label types increases. There are several ways to initialize connections for new labels, such as random initialization or using weight averages.

The definition of a relaxation problem and the way of creating $\{X_i\}$ can be extended, for example by using a multi-path system.

4   Experimental Setup

To test our transfer method, we experimented with the simplest conditions. As shown in Table 1, we set values to the parameters in this paper.

Table 1. Parameters of our approach

Dataset name	P	N
MR	1	30
SST	1	30

4.1   Dataset

To test the model, we used two benchmarks: one from Movie Reviews (MR)[17][1] and one from Stanford Sentimental Treebank (SST)[18].[2] Table 2 presents information on these datasets. The reason for using film review datasets is that the

---

[1] https://www.cs.cornell.edu/people/pabo/movie-review-data/.

[2] http://nlp.stanford.edu/sentiment/.

film review datasets we use have rating data, but our method could be used with other datasets.

In this paper, we set 1 to P when we train the model on the above datasets. Thus, we obtain two problems from each dataset.

MR has a number of datasets, and we used scale data written by Denni Schwartz, James Berardinelli, Scott Renshaw and Steve Rhodes. From these datasets, we obtained two problems X_0 and X_1 with $\mid \mathcal{T}_0 \mid = 5$ and $\mid \mathcal{T}_1 \mid = 2$. $\mathcal{T}_0$ consists of five labels: $[0, 0.2]$, $(0.2, 0.4]$, $(0.4, 0.6]$, $(0.6, 0.8]$ and $(0.8, 1.0]$. $\mathcal{T}_1$ of two labels: $[0, 0.5]$ and $(0.5, 1.0]$.

From SST, we obtained two datasets X_0' and X_1' with $\mid \mathcal{T}_0' \mid = 5$ and $\mid \mathcal{T}_1' \mid = 2$. $\mathcal{T}_0'$ consists of five labels: $[0, 0.2]$, $(0.2, 0.4]$, $(0.4, 0.6]$, $(0.6, 0.8]$ and $(0.8, 1.0]$. $\mathcal{T}_1'$ consists of two labels: $[0, 0.4]$ and $(0.6, 1.0]$.

Table 2. Dataset Information. $\mid \mathcal{T} \mid$: The number of classes. $\mid X \mid$:Dataset size. $|V|$:Vocabulary size. $|V_{\text{pre}}|$: Size of words included in the Word2Vec model. Test: Test set size (10-fold means that 10-fold cross-validation was used).

Data	$\mid \mathcal{T} \mid$	$\mid X \mid$	$|V|$	$|V_{\text{pre}}|$	α	Test
X_0 (MR)	5	5006	55449	31423	$\frac{2}{3}$	10-fold
X_1 (MR)	2	5006	55449	31423	$\frac{1}{3}$	10-fold
X_0' (SST)	5	11855	21705	18867	$\frac{2}{3}$	2210
X_1' (SST)	2	9633	19403	17024	$\frac{1}{3}$	1823

4.2 Model

We used Yoon Kim's CNN model [3]. The reason for using Yoon Kim's model is that Yoon Kim shows the performance of his model and he evaluates the model on datasets which include those we used in our experiment. The CNN model structure was not changed at all. We used only the single channel model, not the multi-channel. Pre-trained word vectors were obtained from a Word2Vec model trained on 100 billion words from Google News.[3] Table 3 shows information about the models used. We used six models, and our approach was applied to three of them. In Table 3, RL indicates whether we used our Relaxation Method. ELV indicates whether we used word vectors obtained from Word2Vec as the initial values of the embedded layer in the model. EL indicates whether we trained the embedded layer in the model. Random means that the weights of the output layer in the model are initialized randomly. Average means that we transferred the weights from nodes associated with the negative class to very negative and negative, from positive to positive and very positive, and the neutral weights are the mean values of the negatives and positives.

---

[3] https://code.google.com/p/word2vec/.

Table 3. Used model information.

Model Name	Legend Name	RL	ELV	EL	How to initialize output layer
CNN-rand	(A)	×	×	○	Random
CNN-static	(B)	×	○	×	Random
CNN-non-static	(C)	×	○	○	Random
CNN-rand-relax	(D)	○	×	○	Average
CNN-static-relax	(E)	○	○	×	Average
CNN-non-static-relax	(F)	○	○	○	Average

5   Experimental Results

We tested the model on two datasets. Figures 1 and 3 show the results of the test on MR, and Figs. 2 and 4 show the results of the test on SST. The X axes in Figs. 1 and 2 represent the iteration. The X axes in Figs. 3 and 4 represent legends of the models. The Y axes in Figs. 1, 2, 3 and 4 represent the accuracy. Table 4 shows the average accuracy of the results.

Table 4 shows that on the MR datasets two of our models performed better than other models do. Moreover, as shown in Fig. 3, results from our models with 10-fold cross-validation had a narrower range of values. It appears that results stabilize through training on an easy dataset, using our transfer learning method, before the model is used on an actual target dataset.

Although our best model was as good as Yoon Kim's best model on SST, as shown in Table 2 X_1' had less data than X_0'. When we trained models on SST, we also used phrase data. The amount of data in the SST experiment is more than

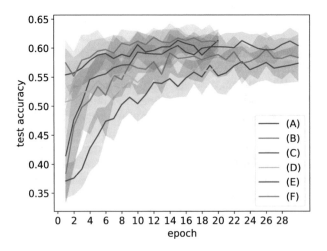

Fig. 1. Changes in accuracy of MR

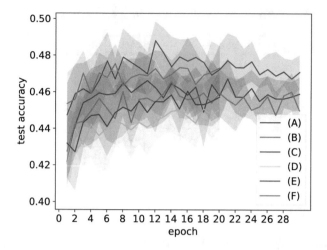

Fig. 2. Changes in accuracy of SST

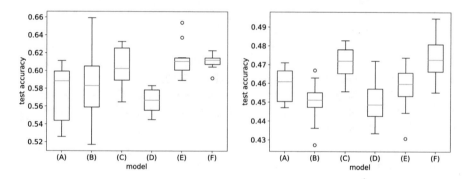

Fig. 3. Box-and-whisker plot of MR   **Fig. 4.** Box-and-whisker plot of SST

is shown in the Table: $\mid X_0' \mid$ had twice as much as $\mid X_1' \mid$, in fact. Moreover, our models for the relaxation problems had fewer parameters than did the original problems. Thus, we confirmed that models to which our method was applied were trained faster without any loss in performance.

You should take care of overfitting relaxation problems if you use our method. Interestingly, Table 4 shows that the performance of CNN-rand-relax on both datasets was worse than CNN-rand. It is assumed that performance of models become worse when models have less knowledge, for example, models do not have pre-trained word vectors.

Table 4. Result of experiments accuracy.

Model name	MR	SST
CNN-rand	0.574	0.459
CNN-static	0.583	0.450
CNN-non-static	0.604	0.471
CNN-rand-relax	0.566	0.451
CNN-static-relax	**0.613**	0.457
CNN-non-static-relax	0.611	**0.473**

6   Conclusion

In this paper, we proposed a relaxation method for transfer learning in NLP. The 10-fold cross-validation results of our models with the relaxation method showed a narrower variety of values. In other words, the results were stabilized. We confirmed that we were able to train a model faster by using the relaxation method. In this paper, we experimented under $P = 1$ and showed only the result of the film review recognition tasks with Yoon Kim's model.

Future research will involve changing the parameters of various test models. To prove that our method is effective under other experimental conditions, we will also evaluate various models using various datasets.

A part of this work was supported by JSPS KAKENHI Grant, Grant-in-Aid for Scientific Research(C), 26330282 and by JSPS KAKENHI Grant, Grant-in-Aid for JSPS Fellows, 16J10941.

References

1. Girshick, R.B., Donahue, J., Darrell, T., Malik, J.: Rich feature hierarchies for accurate object detection and semantic segmentation. CoRR, abs/1311.2524 (2013)
2. Mikolov, T., Sutskever, I., Chen, K., Corrado, G.S., Dean, J.: Distributed representations of words and phrases and their compositionality. In: Burges, C.J.C., Bottou, L., Welling, M., Ghahramani, Z., Weinberger, K.Q. (eds.) Advances in Neural Information Processing Systems, vol. 26, pp. 3111–3119. Curran Associates Inc. (2013)
3. Kim, Y.: Convolutional neural networks for sentence classification. CoRR, abs/1408.5882 (2014)
4. Kalchbrenner, N., Grefenstette, E., Blunsom, P.: A convolutional neural network for modelling sentences. CoRR, abs/1404.2188 (2014)
5. dos Santos, C., Gatti, M.: Deep convolutional neural networks for sentiment analysis of short texts. In: Proceedings of COLING 2014, the 25th International Conference on Computational Linguistics: Technical Papers, Dublin, Ireland, pp. 69–78. Dublin City University and Association for Computational Linguistics, August 2014
6. Dauphin, Y.N., Fan, A., Auli, M., Grangier, D.: Language modeling with gated convolutional networks. CoRR, abs/1612.08083 (2016)

7. Zhang, X., Zhao, J.J., LeCun, Y.: Character-level convolutional networks for text classification. CoRR, abs/1509.01626 (2015)
8. Vosoughi, S., Vijayaraghavan, P., Roy, D.: Tweet2vec: learning tweet embeddings using character-level CNN-LSTM encoder-decoder. CoRR, abs/1607.07514 (2016)
9. Cook, D.J., Feuz, K.D., Krishnan, N.C.: Transfer learning for activity recognition: a survey. Knowl. Inf. Syst. 36, 537–556 (2013)
10. Russakovsky, O., Deng, J., Su, H., Krause, J., Satheesh, S., Ma, S., Huang, Z., Karpathy, A., Khosla, A., Bernstein, M.S., Berg, A.C., Li, F.-F.: Imagenet large scale visual recognition challenge. CoRR, abs/1409.0575 (2014)
11. Pan, S.J., Yang, Q.: A survey on transfer learning. IEEE Trans. Knowl. Data Eng. 22(10), 1345–1359, October 2010
12. Howard, J., Ruder, S.: Fine-tuned language models for text classification. ArXiv e-prints, January 2018
13. Agrawal, P., Girshick, R.B., Malik, J.: Analyzing the performance of multilayer neural networks for object recognition. CoRR, abs/1407.1610 (2014)
14. Razavian, A.S., Azizpour, H., Sullivan, J., Carlsson, S.: CNN features off-the-shelf: an astounding baseline for recognition. CoRR, abs/1403.6382 (2014)
15. Yosinski, J., Clune, J., Bengio, Y., Lipson, H.: How transferable are features in deep neural networks? CoRR, abs/1411.1792 (2014)
16. Mou, L., Meng, Z., Yan, R., Li, G., Xu, Y., Zhang, L., Jin, Z.: How transferable are neural networks in NLP applications? CoRR, abs/1603.06111 (2016)
17. Pang, B., Lee, L.: Seeing stars: exploiting class relationships for sentiment categorization with respect to rating scales. CoRR, abs/cs/0506075 (2005)
18. Socher, R., Perelygin, A., Wu, J., Chuang, J., Manning, C.D., Ng, A., Potts, C.: Recursive deep models for semantic compositionality over a sentiment treebank. In: Proceedings of the 2013 Conference on Empirical Methods in Natural Language Processing, pp. 1631–1642, Seattle, Washington, USA. Association for Computational Linguistics, October 2013

Simple and Linear Bids in Multi-agent Daily Electricity Markets: A Preliminary Report

Hugo Algarvio[1,2($\boxtimes$)], Fernando Lopes[2], and João Santana[1]

[1] Instituto Superior Técnico, Universidade de Lisboa, INESC-ID, Lisbon, Portugal
{hugo.algarvio,jsantana}@tecnico.ulisboa.pt
[2] LNEG-National Research Institute, Est. do Paço do Lumiar 22, Lisbon, Portugal
fernando.lopes@lneg.pt

Abstract. Variable generation (VG) has several unique characteristics compared to those of traditional thermal and hydro-power plants, notably significant fixed capital costs, but near-zero or zero variable production costs. Increasing the penetration of VG tend to reduce energy prices over time, increase the occurrence of zero or negatively priced periods, and reduce the cleared energy levels of existing plants. This paper presents an overview of an agent-based system, called MATREM, to simulate electricity markets. Special attention is devoted to a case study that aims at analyzing the behavior of a simulated day-ahead market in situations with increasing levels of wind generation, and also comparing market schedules and prices in situations involving either simple and linear bids.

Keywords: Electricity market · Day-ahead market
Simple and linear bids · Wind power · MATREM system

1 Introduction

Market forces drive currently the price of electricity and reduce the net cost through increased competition. Several market models have been considered to achieve the key objectives of ensuring a secure operation and facilitating an economical operation, notably electricity pools and bilateral transactions [1,2]. A pool, or market exchange, involves basically a specific form of auction, where participants send bids to sell and buy electricity, for a certain period of time, to a market operator, that analyzes the bids and calculates a market price that must be followed by all participants—the market-clearing price.

Tailored long-term bilateral contracts are very flexible since the negotiating parties can specify their own contract terms, independent of the market operator. Such contracts consist essentially in direct negotiations of energy prices,

H. Algarvio and F. Lopes—This work was performed under grant PD/BD/105 863/2014 (H. Algarvio).

quantities, time of delivery, duration, among other possible issues, between two parties. Market participants often enter into bilateral contracts to hedge against pool price volatility [3].

The last decades have seen the raising of a consent concerning global warming and sustainability. The European Union (EU) is undoubtedly one of the regions in the world with a strong commitment towards a sustainable society. There are ambitious goals for the next years and decades. Several European countries are required to comply with the respective 20-20-20 targets [4]. Accordingly, renewable generation or variable generation (VG), such as wind and photovoltaic (PV), has increased significantly in recent years.

Variable generation has several unique characteristics compared to those of traditional power plants, notably: increases the variability and uncertainty of the net load (load minus VG), has significant fixed capital costs but near-zero or zero variable production costs, and has unique diurnal and seasonal patterns. Thus, VG is the lowest-variable-cost resource, is normally operated at its maximum available capacity limit, and would operate below this level only when the marginal cost is zero or negative (i.e., VG is the marginal resource). Together, these characteristics can significantly influence the performance and outcomes of wholesale markets. In particular, large penetrations of variable generation may reduce market prices due to their near-zero, zero, or negative bids, increase price volatility because of their increased volatility, and reduce the capacity utilization of conventional resources (but see, e.g., [5]).

The potential impacts of large penetrations of variable generation should be analyzed to determine if existing market designs are still effective. As electricity markets continue to evolve, there is a growing need for advanced modeling approaches that simulate how market participants may act and react to changes in the financial and regulatory environments in which they operate. Accordingly, an ongoing work is looking at using software agents to help manage both the complexity of wholesale energy markets and the unique challenges of bilateral contracting of electricity. In particular, we are developing an agent-based simulation tool enabling market participants to submit their hourly offers to a day-ahead energy market. On the basis of the marginal pricing theory—that is, system marginal pricing (SMP) or locational marginal pricing (LMP)—the system simulates a day-ahead market, by accepting or rejecting the offers submitted and establishing the dispatch schedule for a specific day of operation (this is typically defined at noon of the day before the day under consideration). The tool also enables market participants to negotiate the terms and conditions of bilateral contracts.

The remainder of the paper is structured as follows. Section 2 presents an overview of the agent-based simulation tool, called MATREM (for Multi-Agent TRading in Electricity Markets). It also describes the type of offers (simple and linear) that can be used by the agents participating in the simulated day-ahead market. Section 3 describes a case study aiming at: (i) illustrating how the tool operates in typical trading scenarios, (ii) analysing the market behavior in situations with different levels of wind generation, and mainly (iii) analyzing the market outcomes when considering either simple or linear offers. Finally, Sect. 4 states the conclusions and outlines some avenues for future work.

2   The MATREM System: An Overview

MATREM allows the user to conduct a wide range of simulations regarding the behavior of energy markets under a variety of conditions.[1] The agents are essentially computer systems capable of flexible action and able to interact, when appropriate, with other agents to meet their design objectives. They have a knowledge base to store declarative knowledge. Specifically, the knowledge base contains the beliefs an agent has about itself, about the environment (i.e., a simulated electricity market), and about the other agents operating in the environment. Also, the agents have an internal data store to record their (top-level) achievement goals (e.g., "maximize-profit"or "calculate-market-clearing-price"). They are able to combine information about their goals with the contents of their knowledge base to select actions that further specific goals. To this end, they are equipped with a generic model of individual behavior.

The agents are currently being developed using the JAVA Agent Development Framework (JADE)—an agent-oriented platform offering a framework for multi-agent system development, and fully integrated with the JAVA programming language. JADE is an open source platform for peer-to-peer agent based applications [8]. Agent communication is done by sending and receiving messages. The communication paradigm involves the following main tasks: a sender agent (S) prepares and sends a message to a receiver agent (R), (i) the JADE run-time posts the message in the message queue of agent R, (iii) the agent R is notified about the receipt of the new message, and (iii) the agent R gets the new message from the message queue and processes it.

The agents may represent generating companies (GenCos), retailers (Retail-Cos), aggregators, large and small consumers, market operators and system operators. GenCos may sell electrical energy either to organized markets or directly to retailers and other market participants through bilateral contracts. Retail-Cos buy electricity in wholesale markets and re-sell it to customers in retail markets. Aggregators are entities that support groups of end-use customers in trading electrical energy. Large consumers can take an active role in the market by buying electrical energy in the pool or by signing bilateral contracts. Small consumers buy energy from retailers and possibly other market participants.

The system supports both pool trading and bilateral trades or contracts. A day-ahead market (DAM) sells energy to RetailCos and buys energy from GenCos in advance of time when the energy is produced and consumed (see, e.g., [9]). Bilateral trades are defined by privately negotiated bilateral contracts that can be either physical or financial obligations. MATREM also supports a futures market for trading standardized bilateral contracts, as well as a marketplace for negotiating the details of tailored (or customized) long-term bilateral contracts (see, e.g., [10]). A more detailed description of pool trading of electricity follows.

---

[1] See [6] for a detailed description of the MATREM system and [7] for its classification according to a number of specific dimensions related to both liberalized electricity markets and intelligent agents.

2.1   The Day-Ahead Market

The day-ahead market clears to meet bid-in load demand for an entire day, one day in advance. The pricing mechanism is founded on the marginal pricing theory. Under system marginal pricing (SMP), generating companies compete to supply demand by submitting bids in the form of price and quantity pairs. These bids are ranked in increasing order of price, leading to a supply curve. Similarly, retailers and possibly other market participants submit offers to buy certain amounts of energy at specific prices. These purchase offers are ranked in order of decreasing price, leading to a demand curve. The market-clearing price (or system marginal price) is defined by the intersection of the supply curve with the cumulative demand curve. This price is determined on an hourly basis and applied to all generators uniformly, regardless of their bids or location. Generators are instructed to produce the amount of energy corresponding to their accepted bids and buyers are informed of the amount of energy that they are allowed to draw from the system. The bids and offers are referred to as *simple offers* and include the price p to buy/sell electricity and the respective quantity q—that is, $so = (p, q)$.

Locational marginal pricing (LMP) is a more complex variation of marginal pricing—as in SMP, the system collects both generation and load purchase bids, and then determines the optimal generation dispatch. However, the system runs now an optimal power flow procedure that defines the energy price at each bus of the network—that is, the marginal cost depends on the location where the electrical energy is produced or consumed.

Conceptually, the price p is now a function that depends on the traded quantity q. In the case of a supply-side agent, this (linear) function takes the following form:

$$p = p_0 + \delta \cdot q, \qquad p \in [p_0, p_{max}] \qquad (1)$$

where p_0 is the initial price, δ represents an agent's elasticity to the market price, and p_{max} is the maximum price. If the traded quantity q is equal to the minimum quantity q_{min}, the price p is the minimum price p_0. Otherwise, if q is equal to the maximum quantity q_{max}, p is the maximum price p_{max}. For a demand-side agent, the price function takes the following form:

$$p = p_{max} - \delta \cdot q, \quad p \in [p_0, p_{max}] \qquad (2)$$

Market participants set now the initial price p_{max} for the quantity q_{min} and the minimum price p_0 for the maximum quantity q_{max}. The bids and offers are referred to as *linear offers*.

3   The Case-Study

The case-study includes two different parts. In the first part, the agents are only allowed to make simple offers. However, in the second part, the agents can make linear offers. The market-clearing price is obtained by considering either the SMP or the LMP algorithms. Both parts include two scenarios: a scenario of low wind generation and a scenario of high wind generation.

Table 1. Key characteristics of producer agents

Agent	Technology	Power (MW)	Marginal Price (€/MWh)	P_0 (€/MWh)	P_{max} (€/MWh)	Slope (%)
Thermal 1	Coal	628	45.00	42.75	47.25	0.7
WindPower	Wind	1500	0.00	0.00	0.00	0.0
CCGT 1	Gas	830	54.50	52.16	57.65	0.7
CCGT 2	Gas	990	56.12	53.29	58.89	0.6
CCGT 3	Gas	1176	57.90	54.53	60.27	0.5

3.1 System Marginal Pricing: Simple Offers

Table 1 summarizes the main characteristics of the five producer agents considered in the study. For simplicity, the agents prepare offers by considering the marginal price and the nominal power (except the agent referred to as "Wind-Power"). The retailer agents prepare different offers depending on both the demand and the strategic price. Figure 1 depicts the simple offers of all retailer agents.

As noted earlier, we consider two scenarios involving different levels of wind generation. Table 2 presents the simulation results for both scenarios. Figure 2 depicts the market-clearing prices for each of the 24 h of the two days under consideration.

By analyzing Table 2, we can conclude that large penetrations of wind benefit the demand-side agents (but not the traditional supply-side agents, i.e., the disptachable power plants). Specifically, the demand-side agents can buy energy at low market prices, reducing the profit of the traditional supply-side agents.

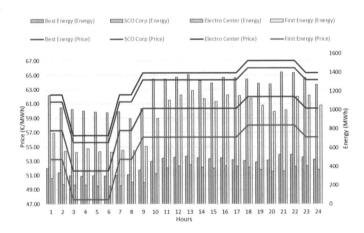

Fig. 1. Retailer offers submitted to the simulated day-ahead market

Table 2. Simulation results: simple offers

Agent	Total Energy (MWh)	Scenario 1		Scenario 2	
		Energy Traded (MWh)	Success (%)	Energy Traded (MWh)	Success (%)
Thermal 1	15072.0	15072.0	100.0	15072.0	100.0
WindPower	Traded	1511.0	100.0	20605.1	100.0
CCGT 1	19920.0	18433.4	92.5	14184.7	71.2
CCGT 2	23760.0	15813.4	66.6	9713.7	40.9
CCGT 3	28224.0	3592.8	12.7	0.0	0.0
Best Energy	10188.1	8999.5	88.3	9667.9	94.9
SCO Corp	29051.3	29051.3	100.0	29051.3	100.0
Electro Center	7658.6	7658.6	100.0	7658.6	100.0
First Energy	21784.5	8713.2	40.0	13197.7	60.6

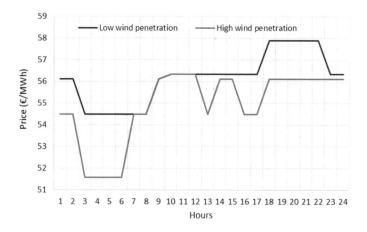

Fig. 2. Day-ahead market-clearing prices

3.2   Locational Marginal Pricing: Linear Offers

The offers of the producer agents are prepared by considering the data presented in Table 1. For the retailer agents, the simple offers are adapted by considering a slope (δ) of 5% (i.e., for every hour, the price offered decrease 5%, in the case of the minimum offer, and increase 5%, in the case of the maximum offer).

Table 3 and Fig. 3 present the simulation results. In particular, Fig. 3 depicts the market-clearing prices for both scenarios (i.e., low and high wind generation). By comparing the results with the results of the previous section, we can conclude that the market-clearing prices are higher now (as well as the traded energy). Thus, linear offers seem to bring more liquidity to the market (and more dynamism).

Table 3. Simulation results: linear offers

Agent	Total Energy (MWh)	Scenario 1		Scenario 2	
		Energy Traded (MWh)	Success (%)	Energy Traded (MWh)	Success (%)
Thermal 1	15072.0	15072.0	100.0	15072.0	100.0
WindPower	Traded	8469.2	100.0	14947.8	100.0
CCGT 1	19920.0	13978.0	70.2	10559.7	53.0
CCGT 2	23760.0	14047.7	59.1	10059.6	42.3
CCGT 3	28224.0	13881.2	49.2	9102.7	32.3
Best Energy	10188.1	7246.1	71.1	9298.3	91.3
SCO Corp	29051.3	26427.3	91.0	30895.4	106.3
Electro Center	7658.6	9567.4	124.9	9956.3	130.0
First Energy	21784.5	15249.2	70.0	15249.2	70.0

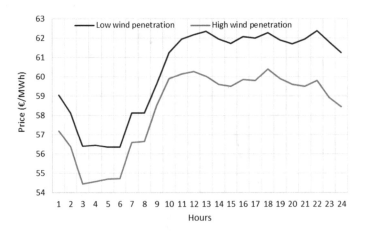

Fig. 3. Day-ahead market clearing-prices

4   Conclusion

This paper presented an overview of the MATREM system and described a case-study to analyze the impact of different levels of wind power generation on the day-ahead market prices, as well as to compare the effect of using simple and linear offers on market results. The simulation results indicated that large penetrations of variable generation lead to a reduction of market prices. Also, linear offers seem to lead to a somewhat more fair system, by bringing (to a certain extent) more dynamism to the market.

References

1. Stoft, S.: Power Systyem Economis: Designing Markets for Electricity. IEEE Press and Wiley Interscience (2002)
2. Shahidehpour, M., Yamin, H., Li, Z.: Market Operations in Electric Power Systems. Wiley, Chichester (2002)
3. Kirschen, D., Strbac, G.: Fundamentals of Power System Economics. Wiley, Chichester (2004)
4. European Union: Directive 2009/28/EC of the European Parliament and of the Council on the promotion of the use of energy from renewable sources and amending and subsequently repealing Directives 2001/77/EC and 2003/30/EC, 23 April (2009)
5. Lopes, F., Coelho, H. (eds.): Electricity Markets with Increasing Levels of Renewable Generation: Structure. Operation. Agent-based Simulation and Emerging Designs. Springer, Cham (2018)
6. Lopes, F.: MATREM: an agent-based simulation tool for electricity markets. In: Electricity Markets with Increasing Levels of Renewable Generation: Structure, Operation, Agent-based Simulation and Emerging Designs. SSDC, vol. 144, pp. 189–225. Springer, Cham (2018)
7. Lopes, F.: Coelho, H: Electricity markets and intelligent agents. part ii: agent architectures and capabilities. In: Electricity Markets with Increasing Levels of Renewable Generation: Structure. Operation, Agent-based Simulation and Emerging Designs. SSDC, vol. 144, pp. 49–77. Springer, Cham (2018)
8. Bellifemine, F., Caire, G., Greenwood, D.: Developing Multi-Agent Systems with JADE. Wiley, Chichester (2007)
9. Lopes, F., Sá, J., Santana, J.: Renewable generation, support policies and the merit order effect: a comprehensive overview and the case of wind power in Portugal. IN: Electricity Markets with Increasing Levels of Renewable Generation: Structure. Operation, Agent-based Simulation and Emerging Designs. SSDC, vol. 144, pp. 227–263. Springer, Cham (2018)
10. Algarvio, H.F., Lopes, F., Santana, J.: Multi-agent retail energy markets: bilateral contracting and coalitions of end-use customers. In: 12th International Conference on the European Energy Market (EEM 2015), pp. 1–5. IEEE (2015)

Fault Tolerance in DisCSPs: Several Failures Case

Fadoua Chakchouk[1,2(✉)], Sylvain Piechowiak[1], René Mandiau[1], Julien Vion[1], Makram Soui[2], and Khaled Ghedira[2]

[1] LAMIH UMR CNRS 8201, University of Valenciennes, Valenciennes, France
fadoua.chakchouk@univ-valenciennes.fr
[2] ENSI, University of Manouba, Manouba, Tunisia

Abstract. To solve a distributed problem in presence of a failed entity, we have to find a way to accomplish the failed entity tasks. In this paper, we present an approach which guarantees the resolution of DisCSPs in presence of failed agents. This approach is based on local CSPs replication principle: each failed agent local CSP is replicated in another agent which will support it. Obtained results confirm that our approach can solve a DisCSP in presence of failed agents by giving a solution when it exists.

Keywords: DisCSP · Robustness · Fault tolerance · Replication
Agent failure

1 Introduction

Distributed systems as Multi-Agent Systems (MAS) are defined as a set of entities that appear to users as a single system. This kind of system is characterized by a partial failure notion: if a component fails, it can affect *the proper operation of other components, while at the same time leaving yet other components totally unaffected* [8]. A failed system is defined in the literature as a system which cannot achieve its goals. In this paper, we are interested to take account agents' failures in a Multi-Agent System to solve Distributed Constraint Satisfaction Problems (DisCSP). In centralized systems, the failure of an agent causes the abandon of the solution search. However, in distributed systems, we defend the idea that the failure of an agent can be covered by replicating the failed agent tasks to another agent.

This paper presents an approach to solve DisCSP that can support the failure of more than one agent. It is based on replication principle: each failed agent local CSP is replicated in another active agent. The organization of this paper is as follows: Sect. 2 presents DisCSP definition and the fault tolerance in multi-agent systems. Section 3 describes different steps of our proposed approach. Section 4 presents results obtained with the proposed method to solve DisCSPs. Finally, Sect. 5 concludes and gives our perspectives.

© Springer International Publishing AG, part of Springer Nature 2019
F. De La Prieta et al. (Eds.): DCAI 2018, AISC 800, pp. 204–212, 2019.
https://doi.org/10.1007/978-3-319-94649-8_25

2   DisCSP and Fault Tolerance

To solve a DisCSP, each agent has a set of variables to which it assigns values from a predefined domains. These agents communicate together in order to satisfy a set of constraints that connect variables. This paper is intended to solve DisCSPs where each agent has more than one variable. We focus on Multi-ABT algorithm [6]. In fact, this algorithm is well-known in this field.

Results provided by this algorithm in presence of failures can be wrong. According to Tanenbaum and Steen [8], the distribution concept of a system aims to *recover from partial failures without seriously affecting the overall performance*. A failure can affect agents, such as crash failures (sudden stop of an agent), or byzantine failures (if an agent provides wrong results which are considered correct). It can also affect communication between agents, such as omission failures (if an agent does not respond to requests), or Timing failures (if an agent exceeds a response time interval).

To cover these failures, several methods are proposed in the literature, based on agents replication, or the utilization of sentinels. An approach based on replication agent is proposed by Fedoruck et al. [3]. This approach principle introduces proxies which reveal replicates group as one entity. Some proxies choose and manage replication modes, while other proxies manage all internal and external group communications. Other approaches based on critical agents replication are proposed by Ductor et al. [2] or Guessoum et al. [4]. A critical agent is the agent having the higher probability to fail. Sentinels were proposed for the first time by [5] as entities that monitor agents features, and protect them from undesirable events.

3   Proposed Approach

In this paper, we propose an approach that handles the crash failure in a DisCSP. To solve DisCSP in presence of failed agent, its local CSP should be solved. Our approach aims to find a global solution by solving failed agent CSP. The approach detects the failed agent, and assigns its CSP to another agent by replicating the CSPs of failed agents in other ones. This assignment is done by an agent called Dispatcher Agent (A_{Dis}). In this section, we present the solving process details: the failed agent detection, and the handling failure details.

3.1   Failure Detection

To detect a failed agent, additional messages are exchanged between agents.

Check (state) Message. This message is sent by an agent A_i to its neighbor A_j. Tanenbaum [8] introduced this kind of message within distributed systems: if A_i does not receive any solving message from one neighbor A_j for a time interval, it sends to it this message to check its state. If A_i does not receive a response for this message from A_j, it considers A_j as a failed agent. The Check Message has the highest priority; i.e. if an agent receives this message, it suspends its behavior to reply to it.

Active ( ) Message. This message is sent by A_j to A_i as a response to a Check message received from A_i.

isFailed (A_j) Message. It is sent by an agent A_i to the agent A_{Dis} to inform it that an agent A_j is failed. Receiving this message, Agent A_{Dis} replicates each failed agent CSP to another agent.

A failed agent is an agent which can neither receive nor send messages. The failure is detected during the global DisCSP solving.

3.2 Failure Handling Algorithm

The proposed approach is composed from the solver agents A_i of the DisCSP, and the Dispatcher Agent A_{Dis}. Each agent A_i has a list called CSP_{add}, where it will store the replicas sent by Agent A_{Dis} later. A replicas is a copy of a failed agent local CSP.

The agents start the CSP solving as follows (Algorithm 1): Each solver agent *self* solves its local CSP and interacts with its neighbors according to Multi-ABT algorithm (Line 1). During the CSP solving, if the agent *self* does not receive solving messages from one neighbor A_j after a time interval (*TimeOut*) (Line 2), *self* sends to A_j a *Check(state)* message (Line 3). If *self* does not receive an *Active* message from A_j for an interval time (*TimeOutState*) (Line 4), A_i considers it as a failed agent (Line 5), and informs its Agent A_{Dis} by sending to it *isFailed(A_j)* message (Line 6).

After identifying failed agents (Algorithm 2) and informing Agent A_{Dis}, Agent A_{Dis} sends the failed agents' CSPs replicas to the solver agents. Each agent A_i records the received replicas in its CSP_{add} list. Then, it merges the CSP_{add} with its local CSP by executing *MergingCSP()* algorithm (Line 2). Each agent informs its neighbors that it supports A_j local CSP by sending *MergeCSP(A_i, A_j)* message (Line 3), which contains the ID of a failed agent (A_j), and its delegate (A_k). Receiving this message, each agent deletes failed agents from its neighbors list, replace them by the delegate ones, and transmits the *MergeCSP(A_i, A_j)* message to its neighbors.

Algorithm 1. Failure detection

Input: CSP : local CSP to solve
1 Multi-ABT:local CSP solving;
2 **if** $TimeOut(A_j), A_j \in Neighbors(self)$ **then**
3   *self* sends $Check(state) to(A_j)$;
4   **if** *self does not receive Active() from (A_j) after TimeOutState(A_j)* **then**
5   A_j is considered failed;
6   *self* sends $isFailed(A_j)$ to A_{Dis};

Algorithm 2. After failure detection

Input: A_j : the failed agent

1 **foreach** $(CSP(A_j) \in CSP_{add}(self))$ **do**

2   MergingCSP();

3     $self$ sends $MergeCSP(A_i, A_j)$ to its neighbors;

3.3 Replication Process

This process aims to replicate the local CSP of failed agents in other ones. It ensures that an agent can have more than one copy of different CSPs, but a CSP is replicated within only one agent. According to this process, a CSP of a failed agent is supported by only one agent. This process is executed only and only if the Dispatcher Agent A_{Dis} receives information of failed agents. We suppose that Agent A_{Dis} knows already the failed agents number (before starting DisCSP solving).

Algorithm 3. ReplicateCSP: Dispatcher Agent

Input: CSPs : list of failed agents local CSPs

1 **foreach** $A_i, i \in \{1, .., m\}$ **do**

2   **if** A_i *is not failed* **then**

3        $List(A_i) \leftarrow \{ \ \}$;

4   **repeat**

5   **foreach** $A_j \in neighbors(A_i)$ **do**

6   **if** $\neg replicated(CSP_j)$ **then**

7                 $List(A_i) \leftarrow List(A_i) \cup \{CSP_j\}$;

8                 $CSPs \leftarrow CSPs - \{CSP_j\}$;

       until $CSPs = \{ \ \}$;

To replicate local CSPs, the Agent A_{Dis} has the list of failed agents' CSPs as an input (Algorithm 3). Firstly, it proceeds by creating an empty list $List(A_i)$ for each active agent A_i (Lines 1–3) where it records the replicas of CSPs assigned to each agent. Then, it sends the lists to agents in the end of replication process. After that, for each agent (A_i)'s neighbor (Line 5), *Dispatcher Agent* checks if its CSP (CSP_j) is replicated into another agent or not. If not, it assigns CSP_j to Agent A_i by adding it to $List(A_i)$ (Line 6–7). If all neighbors CSPs are replicated into agents other than A_i, Agent A_i will not have any additional CSP (its CSP_{add} remains empty during the DisCSP resolution). After replicating CSP_j into Agent A_i, *Dispatcher Agent* deletes CSP_j from the list of unassigned local CSPs (Line 8). This process is repeated until replicating all the local CSPs of failed agents (obtaining an empty list of unassigned CSPs).

3.4   Merging Process

The Merging algorithm (Algorithm 4) is executed by Agent A_i having at least one replicas of a failed agent CSP. The goal is to merge CSPs into a single one.

Algorithm 4. MergingCSPs

   Input: $CSP(A_j)$: CSP of failed agent
 1 $Variables(A_i) \leftarrow Variables(A_i) \cup Variables(A_j)$;
 2 $IntraC(A_i) \leftarrow IntraC(A_i) \cup IntraC(A_j) \cup InterC(A_i, A_j)$;
 3 $InterC(A_i, A_k) \leftarrow InterC(A_i, A_k) \cup \{InterC(A_j, A_{k \neq i})\}$;

Agent A_i starts this process by merging its variables with those of A_j by adding them to its variables list (Line 1), as well as its list of constraints: it updates its intra-agent constraints $(IntraC(A_i))$ by adding to it $intraC(A_j)$ list and $interC(A_j, A_i)$ list that connect A_i to A_j (Line 2). The rest of inter-agent constraints of A_j $(interC(A_j, A_k))$ are added to the inter-agent constraints list of A_i (Line 3).

After merging CSPs, Agent A_i updates its neighbors list by adding Agent A_j neighbors to its neighbors list, and informs them that it supports Agent A_j CSP. To transmit this information, Agent A_i sends $MergeCSP(A_i, A_j)$ message to Agent A_j neighbors. Receiving this message, agents update their neighbors list by removing Agent A_j and replacing it by Agent A_i.

4   Hypothesis and Experiments

This section presents several hypothesis used to realize experiments (Sect. 4.1) and obtained results (Sect. 4.2).

4.1   Hypothesis

During the experiments, DisCSPs are randomly generated having as parameters $\langle m, n, d, p \rangle$ such as: (i) m is the number of agents, (ii) n is the number of variables for each agent with d as a domain size, (iii) p is the hardness of each DisCSP constraint. The generated DisCSPs are presented as connected graph. During Multi-ABT execution, failures are simulated either before receiving a first message, or just after sending a first solution. An agent has 10 s to reply to a *Check* message, otherwise, it will be considered as a failed one. This interval is sufficient to give an agent time to interrupt its behavior and respond to *Check* message.

To evaluate our proposed approach, results are compared by increasing the number of failed agents, and the number of agents. The comparison is done according to: the number of exchanged messages to solve the DisCSP and to detect and handle the failures, and the CPU time calculated, from the beginning of the DisCSP solving to the end of the slowest agent behavior, and from the detection of a failure until the resumption of the DisCSP resolution. Results presented in the next section concern DisCSP generated with parameters $\langle m, 4, 4, 0.5 \rangle$.

4.2   Experiments and Results

This section presents results obtained by varying the number of failures. Experiments are done with JADE multi-agent platform. Results of simulations were obtained on a computer equipped with 2.4 GHz Intel Core i7 and 8GB of RAM. The number of instances is 50 for each experiment. In theory, by increasing the number of failed agents, the number of messages exchanged decreases, since the number of communicating agents decreases. Also, the total CPU time increases. In fact, the time spent to detect and handle the failures is an extra time added to the initial one. Through these experiments, we aim to validate these hypothesis, and to improve results obtained by Chakchouk et al. [1].

Table 1. DisCSP solving number of messages

Failed agents number		Agents number				
		4	6	8	10	12
1 failure	Multi-ABT	64.35	428.82	1394.75	2506.2	4237.6
	Additional	40.7	26.7	41.2	68.4	118.9
2 failures	Multi-ABT	52.3	414.15	1262.15	2770.71	5358.4
	Additional	27.5	21.15	38.45	64.75	105.42
3 failures	Multi-ABT	50.07	366.25	1163.2	2305.7	4140.17
	Additional	16	18.25	34.05	51.7	88.64
4 failures	Multi-ABT	-	203.47	868.95	2243.3	5200.55
	Additional	-	14.73	28.35	51.45	89.1
6 failures	Multi-ABT	-	-	523.05	1661.75	4348.9
	Additional	-	-	23	35.2	53.4
8 failures	Multi-ABT	-	-	-	-	3252.44
	Additional	-	-	-	-	39.45

Table 1 presents the impact of the failed agent's number variation on the exchanged messages number. It contains the number of solving messages (Multi-ABT messages) in presence of failures, and the number of exchanged messages to detect and handle failures (additional messages). We can observe that, in terms of the number of exchanged solving messages, the number decreases by increasing the failed agents number. In fact, merged CSPs number increases by increasing the number of failed agents. Then, the number of inter-agent constraints decreases, and each delegated agent has fewer neighbors with whom it communicates.

Note also that the number of additional messages decreases by increasing the number of failed agents. In fact, the most exchanged message is the checking message *Check*. Since the number of failures increases, the number of replies to it, by sending *Active* message, decreases. In addition, if an agent is active, it can receive a *Check* message several times from the same agent. However, once declared failed, no more *Check* messages are sent to it. As after merging of CSPs, the diffusion of the *NewCSP* message decreases with the increase of failed agents number.

Table 2. DisCSP solving CPU time

Failed agents number		Agents number				
		4	6	8	10	12
1 failure	Total CPU (s)	21.44	23.53	27.24	32.65	41.2
	Additional CPU (s)	1.2	1.32	2.21	4.14	5.35
	Dispatching CPU (10^{-3} s)	2.75	2.64	3	3.05	3.55
2 failures	Total CPU (s)	29.35	27.75	36.37	30.01	34.69
	Additional CPU (s)	1.32	1.87	2.78	2.91	5.81
	Dispatching CPU (10^{-3} s)	4	5.26	7.06	6.88	9.44
3 failures	Total CPU (s)	31.53	35.14	37.94	39.07	47.85
	Additional CPU (s)	2.72	2.82	2.93	3.51	4.71
	Dispatching CPU (10^{-3} s)	4.56	6.42	7.95	10.2	11.5
4 failures	Total CPU (s)	-	42.24	38.97	43.29	53.45
	Additional CPU (s)	-	3.12	3.4	3.56	6.29
	Dispatching CPU (10^{-3} s)	-	8.89	9.9	11.84	13.6
6 failures	Total CPU (s)	-	-	59.97	56.88	73.59
	Additional CPU (s)	-	-	4.34	3.93	8.05
	Dispatching CPU (10^{-3} s)	-	-	12.9	16.21	25.82
8 failures	Total CPU (s)	-	-	-	-	91.18
	Additional CPU (s)	-	-	-	-	11.49
	Dispatching CPU (10^{-3} s)	-	-	-	-	32.8

Table 2 shows the variation of the CPU time calculated from the beginning of the DisCSP solving to the end of the slowest agent behavior (total CPU), by increasing the number of failed agents. Also, it contains the CPU time spent to detect and handle failures (additional CPU), and that spent by Dispatcher Agent to replicate failed agents' CSPs (dispatching CPU). We observe that by increasing the number of failed agents, the additional CPU time increases. In fact, the detection of one more failure requires an additional exchange and waiting for messages. During the replication process, the Dispatcher Agent browses all the CSPs of failed agents and all their neighbors, which explains the increase of the dispatching CPU by increasing the number of failures.

The increase of these CPU times, as well as the resumption of the resolution after failures detection, generate the increase of the total CPU time. In fact, after merging CSPs, the delegated agents reproduce all their local solutions, and all the agents resume the DisCSP resolution from the beginning.

The obtained results show that additional processes produce an increase in terms of CPU time, but the additional CPU time is almost negligible compared to the total CPU time spent to solve the DisCSP. These processes, also, produce a decreasing of the exchanged messages number. The most important is that at the end of DisCSP solving, we obtain a solution if it exists.

5   Conclusion

This paper describes an approach, which is applied on Multi-ABT, to solve a DisCSP if more than one agent fail. This method is based on the replication of local CSP of each failed agent: each local CSP of failed agents has a replicas in another active agent. If an agent fails, its neighbor, which owns its CSP replicas, supports it by merging the replicas with its own CSP. This leads to obtain a new local CSP. The changes are done only in a local CSP, i.e. the global DisCSP is still the same one.

Experiments results show that this approach give expected results (a solution if it exists, otherwise no solution). Also, they show that our method increases in term of CPU time, but decreases in term of exchanged messages. This approach can be adapted to be applied with other algorithms than Multi-ABT. The next step of this work will be try other kind of failures, such as the presence of malicious or liar agent. These failures introduce the trust and reputation notion between agents [7].

References

1. Chakchouk, F., Vion, J., Piechowiak, S., Mandiau, R., Soui, M., Ghedira, K.: Replication in fault-tolerant distributed CSP. In: IEA/AIE 2017. Springer, Heidelberg (2017)
2. Ductor, S., Guessoum, Z., Ziane, M.: Adaptive replication in fault-tolerant multi-agent systems. In: Proceedings of International Conference on Intelligent Agent Technology, pp. 304–307 (2011)
3. Fedoruk, A., Deters, R.: Improving fault-tolerance by replicating agents. Proc. First Int. Joint Conf. Auton. Agents Multiagent Syst. Part **2**, 737–744 (2002)
4. Guessoum, Z., Briot, J.P., Faci, N.: Towards fault-tolerant massively multiagent systems. In: International Workshop on Massively Multiagent Systems, pp. 55–69. Springer, Heidelberg (2004)
5. Hägg, S.: A sentinel approach to fault handling in multi-agent systems, pp. 181–195. Springer, Heidelberg (1997)

6. Hirayama, K., Yokoo, M., Sycara, K.: An easy-hard-easy cost profile in distributed constraint satisfaction. Inf. Process. Soc. Jpn. J. **24**, 2217–2225 (2004)
7. Huynh, T.D., Jennings, N.R., Shadbolt, N.R.: An integrated trust and reputation model for open multi-agent systems. Auton. Agents Multi-Agent Syst. **13**(2), 119–154 (2006)
8. Tanenbaum, A.S., Steen, M.V.: Distributed Systems: Principles and Paradigms, 2nd edn. Prentice-Hall Inc., Upper Saddle River (2006)

Computer Vision and the Internet of Things Ecosystem in the Connected Home

Carlos Lopez-Castaño[1]([✉]) [iD], Carlos Ferrin-Bolaños[2] [iD],
and Luis Castillo-Ossa[1,3] [iD]

[1] Facultad de Ingeniería Y Arquitectura, Universidad Nacional de Colombia,
Manizales, Caldas, Colombia
caralopezcas@unal.edu.co
[2] Universidad del Valle, Cali, Valle del Cauca, Colombia
[3] GITIR Grupo Investigación Tecnologías Información y Redes,
Universidad de Caldas, Manizales, Colombia

Abstract. An automatic food replenishment system for fridges may help people with cognitive and motor impairments to have a constant food supply at home. More even, sane people may benefit from this system because it is difficult to know accurately and precisely which goods are present in the fridge every day. This system has been a wish and a major challenge for both white good companies and food distributors for decades. It is known that this system requires two things: a sensing module for food stock tracking and another actuating module for food replenishment. The last module can be easily addressed since nowadays there exist many smartphone applications for food delivering, in fact, many food distributors allow their end-users to schedule food replenishment. On the contrary, food stock tracking is not that easy since this requires artificial intelligence to determine not only the different type of goods present in the fridge but also their quantity and quality. In this work, we address the problem of food detection in the fridge by a supervised computer vision algorithm based on Fast Region-based Convolutional Network and an internet of things ecosystem architecture in the connected home for getting high performance on training and deployment of the proposed method. We have tested our method on a data set of images containing sixteen types of goods in the fridge, built with the aid of a fridge-cam. Preliminary results suggest that it is possible to detect different goods in the fridge with good accuracy and that our method may rapidly scale.

Keywords: Computer vision · Connected home · Ecosystem
Internet of things · Machine learning

1 Introduction

One of the most challenging independent tasks for cognitive and motor impaired people is refilling the fridge with food. On the one hand, it is not easy for them to determine precisely which type of food is currently present in the fridge. In addition, they may not remember the quantity and quality of goods. On the other hand, this task may be very difficult for motor impaired people because their food providers may be at

© Springer International Publishing AG, part of Springer Nature 2019
F. De La Prieta et al. (Eds.): DCAI 2018, AISC 800, pp. 213–220, 2019.
https://doi.org/10.1007/978-3-319-94649-8_26

long distances from their home. This considerably reduces its independence. However, the task of food replenishment is not only a problem of impaired people but also it is a generalized problem. For instance, a sane person may eventually not remember accurately which goods need to be replenished in the fridge while buying at the supermarket. As a consequence, people do not often refill their fridge efficiently: people refill fridge of unnecessary goods or do not refill necessary food at all.

In order to replenish food efficiently, an automatic food replenishment system is needed. However, this has been an aim not only for end-users but also for food providers and white goods companies. Such a system requires two things: a sensing module for food stock tracking and another actuating module for food replenishment. The last module can be easily addressed since nowadays there exist many smartphone applications for food delivering which are connected with 24/7 food distributors, in fact, many food distributors allow their end-users to schedule food replenishment. On the contrary, food stock tracking in the fridge is not easy since it requires not only a suitable image acquisition system but also an accurate and efficient artificial intelligence to determine different types of food present in the fridge, their quantity, and more important, their quality.

Food detection in the fridge has not been widely investigated because white goods companies are recently incorporating cameras for observing into the fridges. Some white goods companies such as MABE, LG, Whirlpool, Samsung and others have recently developed fridges with built-in cameras named fridge-cams. More even, in [1] it is reported a commercial wireless camera that fits inside any fridge. Built-in and portable fridge-cams have attracted the attention of end-users and its adoption will increase naturally in a near future, supplying the need of knowing which foods are present in the fridge and consequently avoiding food wastage.

Notwithstanding lack of research related to food detection in the fridge, some important references deserves to be mentioned. For instance, authors of [2] propose a computer vision method for determining fridge contents by analyzing images of goods acquired with a camera mounted below the shelf, which allows to see the footprints of the objects on the shelf. Although the method is very fast and simple it fails for scaling toward many type of foods, because some foods may eventually have the same footprint on the shelf, leading naturally to ambiguity troubles. One solution for the limitation of the aforementioned system were proposed in [3, 4], where the use of Radio Frequency Identification (RFID) allows the univocally determination of foods, nevertheless this solution needs to be in contact with the food products, in consequence it will not be suitable for all type of foods. Recently, some news on the web [5] show that Microsoft and Liebherr are collaborating with the development of Liebherr's new generation of the *SmartDeviceBox*. They are planning to use deep learning algorithms to determine automatically fridge content with the aid of a built-in camera. However, in our review, no patent or paper was found related to this development. It is worth to say that deep learning algorithms [6, 7] have outperformed many image classification challenges and it is natural to think of using this algorithms for the food detection problem, which is addressed in this work.

2   Methodology

In Fig. 1 we propose an IoT architecture with computer vision module for object detection and recognition, cloud computing, smartphone for visualization, and Wi-Fi to support intercommunication.

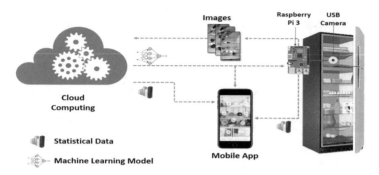

Fig. 1. IoT architecture with computer vision sensor for object detection and recognition in the fridge.

For computer vision it is necessary to strategically locate a camera that allows to see the food. This is not a new idea; in fact, as it was stated in the introduction section, many white good companies have developed built-in and portable fridge-cams. In our case, we propose the use of a Raspberry pi 3 with an USB camera for acquiring images and send them to a remote server. Then, the remote server trains a computer vision algorithm, using high performance computing capabilities for rapid training.

Once the model has been trained, it is sent back to Raspberry pi, and it now can detect and recognize objects in the fridge. An application allows defining regions for training before sending images to the remote server. Likely, statistics over time related to feeding habits can be stored in the remote server and visualized through the smartphone. Wi-Fi protocol supports the intercommunication among smart objects. This architecture answers the question related to the possibility of using computer vision in the Internet of Things ecosystem in the connected home for using cameras as sensors. Clearly, cloud computing will be a key tool for managing big data related to huge amount of images. This characteristic of the architecture will allow that our method scales to increase the number of food types.

The question now is: which type of computer vision is needed for food detection and recognition in the fridge? Well, as we cannot establish a set of feature for unknown objects (thousands of types of food), we conceive a computer vision which automatically extracts those features and build a model for classification. One of the states of the art algorithm which can handle the problem of finding automatically features for training is the Fast Region-Based Convolutional Neural Network (Fast R-CNN) [8].

Fast R-CNN is a rapid version of R-CNN [9]. R-CNN was presented in 2014 by Ross Girshick, and it outperformed previous state-of-the-art approaches on one of the major object recognition challenges in the field: Pascal VOC [10]. The basic idea of

R-CNN is to take a deep neural network which is originally trained for image classi-fication using millions of annotated images and modify it for the purpose of object detection. The basic idea from the first R-CNN paper is illustrated in the Fig. 2: (1) given an input image, (2) in a first step, a large number of region proposals [11] are generated. (3) These region proposals, or Regions-of-Interests (ROIs), are then sent independently through the network which outputs a vector of e.g. 4096 floating point values for each ROI. Finally, (4) a classifier is learned, which takes the 4096 float ROI representation as input and outputs a label and confidence to each ROI.

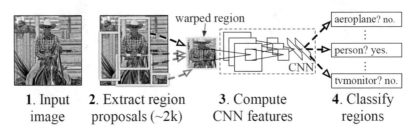

1. Input   2. Extract region   3. Compute   4. Classify
   image   proposals (~2k)   CNN features   regions

Fig. 2. R-CNN Region with Convolutional Neural Networks features: Architecture. Taken from [9].

While this approach works well in terms of accuracy, it is very costly to compute since the Neural Network has to be evaluated for each ROI. Fast R-CNN addresses this drawback by only evaluating most of the network (to be specific: the convolution layers) a single time per image. According to [8], this leads to a $213\times$ speed-up during testing and a $9\times$ speed-up during training without loss of accuracy. This is achieved by using a ROI pooling layer which projects the ROI onto the convolutional feature map and performs max pooling to generate the desired output size that the following layer is expecting. This idea defines the architecture of Fast R-CNN depicted in Fig. 3. In this case, the ROI pooling layer is put between the last convolutional layer and the first fully connected layer just as in the *AlexNet* [12].

A Fast R-CNN network has two sibling output layers. The first outputs a discrete probability distribution per ROI, $p = (p_0,.. .., p_K)$, over $K + 1$ categories. Often, p is computed by a softmax over the $K + 1$ outputs of a fully connected layer. The second sibling layer outputs bounding-box regression offsets, $t^k = \left(t_x^k, t_y^k, t_w^k, t_h^k \right)$, for each of the K object classes, indexed by k. We use the parameterization for t^k given in [9], in which t^k specifies a scale-invariant translation and log-space height/width shift relative to an object proposal. Each training ROI is labeled with a ground-truth class u and a ground-truth bounding-box regression target v. A multi-task loss L on each labeled ROI jointly trains for classification and bounding-box regression:

$$L(p, u, t^u, v) = L_{class}(p, u) + \gamma \, [u \geq 1] L_{location}(t^u, v) \tag{1}$$

Where L_{class} is log loss for true class u. The second task loss, $L_{location}$, is defined over a tuple of true bounding-box regression targets for class u, $v = (v_x; v_y; v_w; v_h)$,

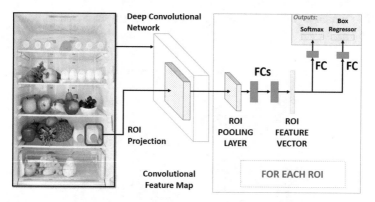

Fig. 3. Fast R-CNN architecture. An input image and multiple regions of interest (ROIs) are input into a fully convolutional network. Each ROI is pooled into a fixed-size feature map and then mapped to a feature vector by fully connected layers (FCs). The network has two output vectors per ROI: softmax probabilities and per-class bounding-box regression offsets. The architecture is trained end-to-end with a multi-task loss.

and a predicted tuple $t^u = \left(t_x^u, t_y^u, t_w^u, t_h^u\right)$, again for class u. The Iverson bracket indicator function [u $\geq$ 1] evaluates to 1 when u $\geq$ 1 and 0 otherwise. By convention the catch-all background class is labeled u = 0. For background ROIs there is no notion of a ground-truth bounding box and hence $L_{location}$ is ignored. For bounding-box regression, the loss can be obtained from:

$$L_{location}(t^u, v) = \sum_{i \in \{x,y,w,h\}} \text{smooth}_{L_1}\left(t_i^u - v_i\right) \tag{2}$$

In this case:

$$\text{smooth}_{L_1}(x) = \begin{cases} 0.5x^2 & if\,|x| < 1 \\ |x| - 0.5 & otherwise \end{cases} \tag{3}$$

is a robust L_1 loss that is less sensitive to outliers than the L_2 loss used in R-CNN [9] and SPPnet [13]. When the regression targets are unbounded, training with L_2 loss can require careful tuning of learning rates in order to prevent exploding gradients. Equation 3 eliminates this sensitivity. The hyper-parameter γ in Eq. 1 controls the balance between the two task losses. Ground-truth regression targets v_i are normalized to have zero mean and unit variance. By now we will use $\gamma = 1$ as in [8].

Since object detection methods as Fast R-CNN often output multiple detections which fully or partly cover the same object in an image. These ROIs need to be merged to be able to count objects and obtain their exact locations in the image. This is traditionally done using a technique called Non Maximum Suppression (NMS) [14]. The version of NMS we used does not merge ROIs but instead tries to identify which ROIs best covers the real locations of an object and discards all other ROIs. This is implemented by iteratively selecting the ROI with highest confidence and removing all other ROIs which significantly overlap this ROI and are classified to be of the same class. See Fig. 4.

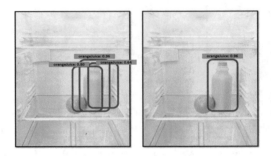

Fig. 4. Region outputs with confidence for every class (left) and final result (right) after Non Maximum Suppression algorithm [14].

3 Results

We built a dataset of 24 positive images with 16 types of food ('avocado', 'orange', 'butter', 'champagne', 'eggBox', 'gherkin', 'yoghurt', 'ketchup', 'orangeJuice', 'onion', 'pepper', 'tomato', 'water', 'milk', 'tabasco', and 'mustard') and 10 images for testing. With the help of Microsoft Cognitive Toolkit [15], we built a Fast-RCNN model with selective search [16] for generating possible object locations in positive images. As results, on the one hand, after training our Fast R-CNN (re-training *AlexNet* model [12]) we get 94% of accuracy and AUC of 0.8 for object detection from the set of 16 classes trained.

In Fig. 5 we can see ROC curve for this experiment. On the other hand, Fig. 6 depicts average precision for each class and mean average precision for all the classes. As we can see from Fig. 6, eight classes (half of the total classes conceived in this experiment) were above the mean average precision for recognition. It is important to say that 'tomato', 'yoghurt' and 'gherkin' reached the highest average precision. This demonstrates that we can improve the result for other classes.

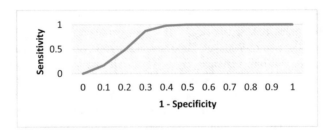

Fig. 5. ROC curve for object detection. AUC = 0.8.

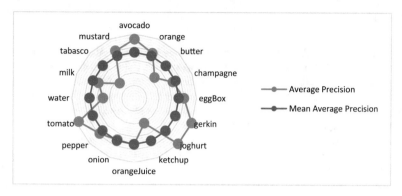

Fig. 6. Single precision in recognition of each class. Average precision for all classes = 0.71.

These results show that in fact we can have a computer vision with good accuracy to detect and recognize objects in the fridge. Nevertheless, a bigger dataset will be necessary for recognizing robustly more objects, with higher average precision. Fortunately, for every fridge, volumetric space will limit amount of variety of food present. This means that it is possible to create a computer vision specialized for a single fridge for every end-user which will easily scale due to its internet of things-based architecture and cloud computing support.

4   Conclusion

A computer vision method based on deep learning was proposed for addressing the problem of food detection in the fridge. This supervised system needs to be supported by an IoT architecture with cloud computing capabilities for easily scaling to more food types and more training images. Our architecture and experiments show that it is possible to have good accuracy for detecting goods in the fridge surpassing drawbacks of proposed methods in [2, 4]. It is worth to say that our method can be easily adapted for the detection of other type of elements in the connected home, and for other tasks, for instance, for food detection in the kitchen or even for cooking activities recognition from sequences of images.

As future work, our proposal will be evaluated with a large dataset and different conditions such as illumination and field of view. We also consider that for having a fully automatic food stock tracking module, this technology should be completely unsupervised in the future; however a suitable IoT architecture will be necessary to avoid human supervision when training.

Acknowledgments. This is a working progress being part of a research project (Computational Prototype for IoT Environments) among Colombia Institutions: Universidad Nacional, Universidad de Caldas and MABE, with code 36715.

References

1. S. A. LTD: FridgeCAM (2016). https://smarter.am/fridgecam/. Accessed 30 Dec 2018
2. Färnström, F., Johansson, B., Åström, K.: Computer vision for determination of fridge contents. In: Proceedings SSAB 2002, Symposium on Image Analysis, pp. 45–48 (2002)
3. Floarea, A.D., Sgârciu, V.: Smart refrigerator : a next generation refrigerator connected to the IoT. In: Proceedings of the 2016 8th International Conference Electronics Computers and Artificial Intelligence (ECAI) (2017)
4. Hachani, A., Barouni, I., Ben Said, Z., Amamou, L.: RFID based smart fridge. In: 2016 8th IFIP International Conference on New Technologies, Mobility and Security (NTMS), pp. 1–4 (2016)
5. Hazen, T.J.: Microsoft and Liebherr Collaborating on New Generation of Smart Refrigerators (2016). https://blogs.technet.microsoft.com/machinelearning/2016/09/02/microsoft-and-lieb herr-collaborating-on-new-generation-of-smart-refrigerators/ Accessed 15 Dec 2017
6. LeCun, Y., Bengio, Y., Hinton, G.: Deep learning. Nature **521**(7553), 436–444 (2015)
7. Goodfellow, I., Bengio, Y., Courville, A.: Deep Learning. MIT Press, Cambridge (2017)
8. Girshick, R.: Fast R-CNN. In: IEEE International Conference on Computer Vision (ICCV) 2015, pp. 1440–1448 (2015)
9. Girshick, R., Donahue, J., Darrell, T., Malik, J.: Rich feature hierarchies for accurate object detection and semantic segmentation. In: IEEE Conference on Computer Vision and Pattern Recognition 2014, pp. 580–587 (2014)
10. Everingham, M., Eslami, S.M.A., Van Gool, L., Williams, C.K.I., Winn, J., Zisserman, A.: The Pascal Visual Object Classes Challenge: A Retrospective. Int. J. Comput. Vis. **111**(1), 98–136 (2015)
11. Hosang, J., Benenson, R., Schiele, B.: How good are detection proposals, really? In: Proceedings of the British Machine Vision Conference 2014, pp. 24.1–24.12 (2014)
12. Krizhevsky, A., Sutskever, I., Hinton, G.E.: Imagenet classification with deep convolutional neural networks.In: Advances in Neural Information Processing System, pp. 1–9 (2012)
13. He, Kaiming, Zhang, Xiangyu, Ren, Shaoqing, Sun, Jian: Spatial Pyramid Pooling in Deep Convolutional Networks for Visual Recognition. In: Fleet, David, Pajdla, Tomas, Schiele, Bernt, Tuytelaars, Tinne (eds.) ECCV 2014. LNCS, vol. 8691, pp. 346–361. Springer, Cham (2014). https://doi.org/10.1007/978-3-319-10578-9_23
14. Hosang, J., Benenson, R., Schiele, B.: Learning non-maximum suppression. In: IEEE Conference on Computer Vision and Pattern Recognition (CVPR) 2017, pp. 6469–6477 (2017)
15. Microsoft: Microsoft Cognitive Toolkit. Microsoft (2016)
16. Uijlings, J.R.R., van de Sande, K.E.A., Gevers, T., Smeulders, A.W.M.: Selective search for object recognition. Int. J. Comput. Vis. **104**(2), 154–171 (2013)

Distributed System Integration Driven by Tests

Jose-Luis Poza-Lujan[1]([⊠]), Juan-Luis Posadas-Yagüe[1],
and Stephan Kröner[2]

[1] University Institute of Control Systems and Industrial Computing (ai2),
Seattle, USA
{jopolu, jposadas}@ai2.upv.com
[2] Preservation and Restoration of Cultural Property Department (DCRBC),
Universitat Politècnica de València (UPV), Camino de vera, s/n.,
46022 Valencia, Spain
ustephan@upvnet.upv.es

Abstract. In complex distributed systems, the integration phase implies a lot of actions due to it is necessary to know how a component interacts with others. Usually, in the system design phase, modules are defined in a hierarchy in order to be easily integrated based on direct dependencies between the modules. That implies a sequential process of integration. In order to accelerate the integration process, agile-inspired integration method has been designed. The method is based in the moment that a unitary test of a component is passed, the dependencies can be started to be tested. The method has been applied in an intelligent system implemented in an indoor drone. First results show that the integration process based on this method is really accelerated, but the coordination between partners and the communication channels have a lot of influence to achieve the process with some minimum quality.

Keywords: Distributed system · System integration
Distributed agile methodologies · Quality measurement

1 Introduction

Engineering projects have well-defined phases necessary to achieve the aims of the project [1]. Usually, a project starts with the classical structure: requirements, design, implementation, evaluation and deployment. In distributed systems, the complexity of these phases grown exponentially [2]. This because each component of the distributed system, have its own particularities to be connected with the others (specific technology, communication protocols, message format, and so on).

Classical methods that arises in a sequential process in the development process considers the system as a hierarchy in which the dependencies between components are clearly defined. In complex systems or unpredicted systems, agile methodologies can improve the results of the project [3].

The testing phase usually is the previous phase to deploy the system in the operational environment. That implies that when components of the distributed system are tested, the project is near to the end [4]. Consequently, test is oriented to tune the integration.

F. De La Prieta et al. (Eds.): DCAI 2018, AISC 800, pp. 221–229, 2019.
https://doi.org/10.1007/978-3-319-94649-8_27

From the point of view of the distributed systems, there are two different type of test to do: unitary test and integration test [5]. Unitary tests are oriented to prove the correct operation of a specific software function or an isolated component. These types of tests can be done easily due to the component to be tested, usually, has not strong dependencies from other components. Even if the component to be tested has some dependencies from other components, these other components can be simulated by means of "dummy" components. In the case of the integration tests, the process is more complex. It is necessary to consider some new challenges [6]:

- Complex "systems of systems" are characterized by a controlled and sometimes limited integration of individual autonomous systems. Often, there are conflicts between requirements of integration and autonomy.
- Causes for heterogeneity are different database management and operating systems utilized, as well as the design autonomy among component systems.
- Much of the distribution is due to the existence of individual systems before overall systems are built (integration of legacy systems).

The management of the tests to be performed becomes difficult when the components have a lot of interdependencies [7]. To manage the integration process, an agile method [8] has been developed. The method is based on Test Driven Development (TDD) [9] and is called Test Driven Integration (TDI), which is an approach of similar methodologies [10]. The aim of the method is to use unitary tests to create the integrations tests simultaneously. The method is based on the connection of functionalities of each component of the distributed system. When the connection of functionalities of two components has been tested (with successful results) other different components functionalities can be tested. This connection is considered a unitary test. As a result, the connection between components of the distributed system is based on the tests that components pass and not in a sequential and planned process.

This paper is organized as follows: Sect. 2 presents the theoretical concepts of the proposed method. To give an example of the method, Sect. 3 presents the system where the method has been tested. Section 4, provides the analysis of the distributed integration made with the system presented and, finally, Sect. 5 summarizes the article.

2   Integration Method

In distributed systems that have certain homogeneity in their components, as the use of the same communication channel in every component or similar interfaces, the integration system phase is easy to plan [11]. However, as system heterogeneity grows, the integration complexity increases exponentially. Consequently, create an integration plan is a difficult task to be done and to be managed. In the TDI method, the number of different components and the heterogeneity of them is not a problem, because each component will be integrated only with the related components.

2.1   Overall Perspective

Initially, a system component (columns in Fig. 1) has a lot of functionalities. Each functionality can be tested isolated (Unitary Test). When a functionality test of one component exceeds a predefined threshold, this component "unlocks" the possibility to be tested in conjunction with the related components. The steps of the integration tests process (rows in Fig. 1) are completed when a functionality of one component has been tested with the integration with the other components.

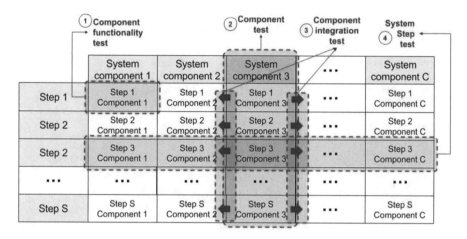

Fig. 1. General schema of the method proposed

In the Fig. 1, the combination of a row (step in the component testing process) and the component is considered a functionality. The Fig. 1 represents a system in which the maximum functionality of each component is S, but this aspect it is not necessary for the whole system.

2.2   Quality Triggers

In the TDI, each component must be tested and must have a quality measurement of the tests results. For example, in the case study presented in the next section, an intelligent autonomous indoor drone system to be used in creative industries, the position of the drone in the indoor environment can have different accuracy. To test the drone navigation is necessary an accuracy in the position in the environment over a few centimeters (less of five centimeters). But, to test the drone recording maneuvers, the position accuracy must be less of two centimeters. The different needs of different components that depends on the position, implies that it is not necessary to wait for a position accuracy needed to the recording system, to test the navigation module.

Consequently, when starting a module test, it is not necessary to wait that unitary test of a component has finished. When a unitary test passes a specific threshold, the component that depends on this threshold can be tested.

3   Case Study

3.1   System Used to be Integrated

The system to be integrated is part of the H2020 European Project Arts indoor RPAS Technology (AiRT). AiRT is a distributed system composed of several devices connected to provide the whole service of recording video or photographs in in-door scenarios using Remotely Piloted Aircraft System (RPAS). Principally, there are two main working areas: Drone and Environment Infrastructure. Each area has common components in the same three systems: Intelligent Flight Control System (IFCS), RPAS system, and Indoor Positioning System (IPS). These components are shown in the Fig. 2.

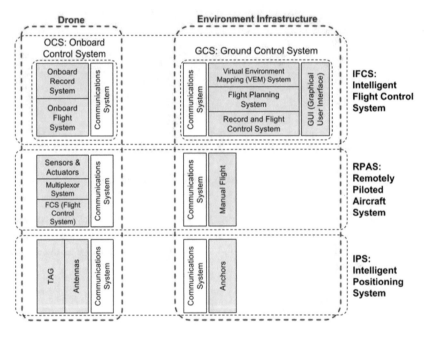

Fig. 2. AiRT system components

The environment infrastructure is the ground system in charge of supporting the drone operability. It is com-posed of several subsystems: the IPS anchors (environment antennas) that give support to Drone indoor positioning, the remote control via radio to control the flight manually, and the Ground Control System (GCS). The GCS is in charge of create and update the Virtual Environment Map (VEM), to generate the flight plan by means of the Flight Planning System. Finally, the Record and Flight Control System control and monitor the Drone flight.

In the same way that the Environment Infrastructure, Drone has components, corresponding with the previous presented three subsystems. Concerning the IPS, Drone incorporates four antennas to receive the anchors signal, and one board (caller "tag") that processes the signals and generates the position. Related with the RPAS, Drone include all sensors like distance or the Inertial Measurement Unit (IMU), a multiplexor system that provides the source that control the Drone (manually flight or automatic flight), and the Flight Control System (FCS) that it is in charge of controlling all the parameters of the flight: drone position and orientation, camera parameters, gimbal parameters. Finally, the third system is the Onboard Control System (OCS) it includes also the VEM manager (synchronized with the GCS). It is in charge of detecting the cloud of points in front of the drone, sending them back to the GCS, receive the flight plan and transfer it to the FCS, control the flight plan according to the operator requirements.

3.2   Component Communication

As explained in the previous section, in the distributed system, components have dependencies between them. In the case of the AiRT system, the dependencies increase the difficulty because of the communication channels to be used are different.

The communications channels implemented in the AiRT system depend on the needs of each components. For example, drone communicates via Wi-Fi (IEEE 802.11) [12] to the pilot modules (manual and automatic). In an emergency, if the IEEE 802.11 channel cannot be used, the 433 MHz channel is ready to be used as a secure communication channel. Internally, in the Drone, the I2C [13] channel is used to connect all components. The dependency between components depends on the communication channel used. For example, the most part of the Sensors are connected with the multiplexor or the FCS (depends if the sensor must to be shared) via the same I2C bus.

3.3   TDI Schema to be Used

Due to the quality of the integration is measured in the position error, the method to integrate, and test, the different elements must consider the increase of the accuracy when a component is added. In the AiRT project, due to the participation of three technical partners, the integrations tests have a lot of interdependencies (Fig. 3). For example, the IPS must be used by the drone since the antennas and tag are placed inside the vehicle. Likewise, the IPS anchors must send the signal to the IFCS. Subsequently, the integration does not imply one connection between two specific components but the connections of a component with more than one.

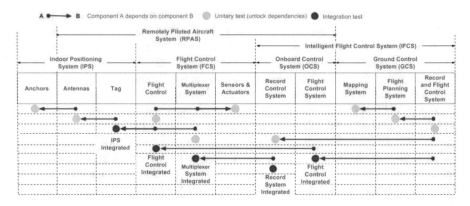

Fig. 3. General plan to integrate different components using the TDI method.

4 Quality of the Integration

4.1 Variables to Measure

The TDI method establishes a measurement of the quality to determine if a component has been passed the test and to consider if this component is integrated with the others. The quality parameters must be directly linked with the functionality of the component, or the functionality of the system. In the case of the AiRT project, it is convenient to use both: the error of the data provided for the component, and the error of the Drone navigation. This last measurement deserves a detailed explanation.

Drone must to perform a mission. This mission can be decomposed into a sequence of different positions in the space, also called "waypoint", To obtain a representative measurement about the quality of the drone mission, it is necessary to review all the waypoints involved in the complete path measuring the deviations between two consecutives waypoints (Fig. 4). In all cases, when a component is integrated, the accuracy of the path changes. Measuring the position error, it is possible to detect the contribution of the last component integrated.

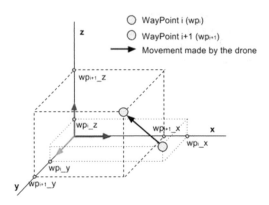

Fig. 4. Quality of the mission based on the path followed by the drone.

Drone navigation has three different variables to measure: position, pose and time. From these variables, the corresponding errors can be calculated:

- Ept(x, y, z, wpi): Position Error: defines if the Drone has achieved the expected position x, y, and z, in the waypoint i. This parameter is measured in centimeters (cm).
- Eps(Ψ, θ,Φ, wpi): Pose Error: defines if the drone has achieved the expected pose in the yaw angle (Ψ), pitch angle (θ), and the roll angle (Φ) in the waypoint i. This parameter is measured in degrees (°).
- Et(t, wpi): Time error: defines if the drone has arrived at the expected time t at the waypoint i. This parameter is measured in milliseconds (ms).

Errors defined in the previous parameters are expressed as the absolute value of the difference between the expected parameter and the measured parameter in each way-point (position, pose and time).

4.2   Quality Parameters

By means the measurement of the variables previously presented, it is possible to verify the quality values of the boundaries defined above: Ept, Eps and Et. The parameters Ept, Eps and Et are measured in an absolute value. Since Ept is measured in cm, Eps is measured in degrees, and Et is measured in milliseconds. This allows to use a comparative scale and all of these parameters can then be normalized in percentage. To normalize the parameters in percentages, it is necessary to define a maximum Error value (for each parameter). Consequently, the normalized value is the relation between the Error measured and the Maximum Error Deviation (Eq. 1).

$$Error(normalized) = Error(measured)/Max\_error\_allowed \qquad (1)$$

For example, if the maximum position error allowed is 10 cm. An error measured of 1 cm provides an error normalized to 0.1 (or 10% if percentage scale is used). Please keep in mind, that normalized errors do not have magnitudes. Since different quality levels are considered, a scale with a normalized error has been introduced (Table 1).

Table 1. Normalized error value and quality test result

Normalized error value	Quality test result
<10%	Excellent
>10%, $\leq$ 33%	Good
>33%, $\leq$ 66%	Acceptable
>66%, $\leq$ 100%	Below acceptable, needs improvement
>100%	Fail

Obviously, the critical decision is to establish the value of the maximum error allowed in each measured variable. This value will change depending on the test group to be performed.

5 Conclusions

The system integration is based fully on the agile concept "Test Driven". This concept allows testing both every single component and different components together (provided that they have passed the individual test). This is especially important when the different elements of the system contain components to be integrated from all partners, as in the case of the AiRT system. By this method, it is not necessary to develop a time-based plan, since the successful test of a single component allows to continue with the next tests.

This method can be used in heterogeneous systems when different technologies must be used. The main advantage of this method is the possibility to have a product with a specific quality (acceptable) and the continuous integration of components can increase its quality. There are systems when this method is not possible to be used. For example, when all the components are necessary to operate; for example, in hardware based projects. In the case of the system presented, Drone flight system (engines and flight control) are necessary to be tested before to be integrated with the IPS or the safety subsystem.

Acknowledgments. The work described in this paper has received funding from the European Union's Horizon 2020 research and innovation programme under grant agreement no. 732433 (reference: H2020-ICT-2016-2017, www.airt.eu). This paper reflects the views of the authors and not necessary the position of the Commission.

References

1. Varajão, J., Colomo-Palacios, R., Silva, H.: ISO 21500: 2012 and PMBoK 5 processes in information systems project management. Comput. Stand. Interfaces **50**, 216–222 (2017)
2. Nidiffer, K.E., Dolan, D.: Evolving distributed project management. IEEE Softw. **22**(5), 63–72 (2005)
3. Sutherland, J., Viktorov, A., Blount, J., Puntikov, N.: Distributed scrum: agile project management with outsourced development teams. In: 40th Annual Hawaii International Conference on System Sciences, HICSS 2007, p. 274a. IEEE, January 2007
4. Royce, W.W.: Managing the development of large software systems: concepts and techniques. In: Proceedings of the 9th International Conference on Software Engineering, pp. 328–338. IEEE Computer Society Press, March 1987
5. Bertolino, A.: Software testing research: achievements, challenges, dreams. In: 2007 Future of Software Engineering, pp. 85–103. IEEE Computer Society, May 2007
6. Hasselbring, W.: Information system integration. Commun. ACM **43**(6), 32–38 (2000)
7. Black, R.: Managing the Testing Process. Wiley, Hoboken (2002)
8. Beck, K., Beedle, M., Van Bennekum, A., Cockburn, A., Cunningham, W., Fowler, M., Kern, J.: The agile manifesto (2001)
9. Beck, K.: Test-Driven Development: By Example. Addison-Wesley Professional, Boston (2003)
10. Tahvili, S., Saadatmand, M., Larsson, S., Afzal, W., Bohlin, M., Sundmark, D.: Dynamic integration test selection based on test case dependencies. In: 2016 IEEE Ninth International Conference on Software Testing, Verification and Validation Workshops (ICSTW), pp. 277–286. IEEE, April 2016

11. White, B., Lepreau, J., Stoller, L., Ricci, R., Guruprasad, S., Newbold, M., Joglekar, A.: An integrated experimental environment for distributed systems and networks. ACM SIGOPS Oper. Syst. Rev. **36**(SI), 255–270 (2002)
12. Crow, B.P., Widjaja, I., Kim, J.G., Sakai, P.T.: IEEE 802.11 wireless local area networks. IEEE Commun. Mag. **35**(9), 116–126 (1997)
13. Semiconductors, P.: The I2C-bus specification. Philips Semicond. **9397**(750), 00954 (2000)

Coordination Platform for a Swarm of Mobile Robots

John Chavez[(✉)], Jonatan Gómez, and Ernesto Córdoba

Universidad Nacional de Colombia, Bogotá, Colombia
{jjchavez,jgomezpe,ecordoban}@unal.edu.co

Abstract. In this paper the automatic design of behaviors for a swarm of robots is explored. In order to build behaviors for robots automatically a computational platform is proposed. The proposed platform is composed by three major components. The first component is a description format which allows to specify robot properties, basic behaviors and tasks. The second component is a genetic programming implementation along with a physics-based simulator, this component builds in an automatic way expression trees which represent robot behaviors. The final component is a behaviors allocation module to assign expression trees to real robots. The proposed computational platform is deployed in a experimental manufacturing cell.

Keywords: Swarm robotics · Computational platform
Automatic design

1 Introduction

The research in coordination of multi-robot systems has been an active area since 1980's [1], research whose main objective is the development of coordination mechanisms that allow the deployment of groups of robots to solve real world problems such as transport of materials and products in manufacturing cells and many others. The final result of a design process for coordinating a group of robots are the rules that indicate to every member of the set how to behave in order to solve a specific task in a context or application environment [2]. In general, the application or deployment environments of multi-robot systems are unstructured and highly dynamic which implies the behaviors generated for every robot during the design process need to be robust, flexible and scalable to the whole of robots [7].

Swarm robotics is a bio-inspired technique that addresses the problem of robot coordination by extrapolating the concept of self-organization observed in natural swarms [3]. Here, and in the same way as it has been observed in insect societies, it is expected that the order in the group of robots arise as a result of the interactions between the elements of the set and that the group of robots exhibit the properties of adaptability, robustness, and scalability observed in natural swarms. However, from the engineering point of view the design of a

© Springer International Publishing AG, part of Springer Nature 2019
F. De La Prieta et al. (Eds.): DCAI 2018, AISC 800, pp. 230–237, 2019.
https://doi.org/10.1007/978-3-319-94649-8_28

self-organized multi-robot system is not a trivial task [5], the behavior of the system must be modeled in terms of the dynamics of the individual-individual-environment interactions. Interactions that must be encoded in the robot controllers. These controllers establish relationships between the perception system and the effector system of the robots, which define the interactions between the robots and their environment to solve the objective problem. Determining what the interaction rules are at the individual level to produce a desired pattern of self-organization is known as the design problem [3].

Designing the behaviors for each of the members that lead a desired emergent result is difficult (design problem), among other things, due to the inherent complexity of the system. In general, it is about breaking down the global pattern of the robot set into more simple interactions between the members of the robot set taking into account the dynamics of the environment and then such interactions must be coded in the robot controllers. In order to address these difficulties, some research projects have proposed the use of optimization algorithms along with representations of controllers (state machines, neural networks, etc.) construct and tune the behaviors for a set of robots [4]. However, the lack of design methodologies, environments and integrated development tools hinders the automatic development of self-organizing robot swarms [6]. This paper presents a computational platform for the automatic design of behaviors for a heterogeneous group of robots in the field of evolutionary swarm robotics. This platform allows to describe robots, tasks and behaviors, simulate robots and build control software for such robots automatically using artificial evolution.

2   Platform Architecture

The computational platform (CP) design process is guided by the next three requirements. The CP must include a mechanism to specify the different elements involved into the automatic behaviors design process, that is: a group of robots (swarm), tasks and their restrictions and basic robot capabilities. The CP must integrate a module for automatic behaviors design, this module should work with the element descriptions (robots, tasks, etc.). The CP must allow to deploy the built behaviors on real robots. These requirements are achieved as follows. In order to describe the robots and the tasks that the swarm solves, a description format is attached to the computational environment. The description format allows defining the most relevant aspects of swarm individuals related to their mechanical properties, elements of perception of the environment and effectors. In the same way, it is necessary to specify the task that the robot set solves in a self-organized way. The description format is a markup format description based on XML.

The computational platform, following the approximation of the area of evolutionary robotics to the design problem [3], integrates a mechanism based on artificial evolution [10] for the automatic construction of high-level behaviors based on primitive behaviors. This is a bottom-up approach [2] in which complex control programs, in the form of expression trees [11], are induced from primitive behaviors (basic robot capabilities) using genetic programming [11].

Every expression tree represent a program that is executed by an interpreter (controller) on the computer on board the robots. Primitive behaviors are defined in the computational platform, in the same way as the swarm and the tasks, through the description format. In each of the cycles of the evolutionary process the generated behaviors are evaluated in a virtual environment. The virtual environment contains the model of the physical environment where the robots are deployed, as well as the models of each member of the swarm. In the simulation system, the global dynamics of the swarm that arises from the behaviors represented in the expression trees are evaluated, which in turn lead to the robot-robot and robot-environment interactions that produce the desired self-organizing pattern, as described by Trianni [3]. The fitness function is defined as part of the task to be performed in the description format. This function quantifies the performance of the swarm in each evolutionary step.

Finally, in order to facilitate the behaviors allocation and validate their effect on real robots, communication modules have been included in the computational platform to assign the constructed behaviors to robots at the end of the evolutionary process. The proposed conceptual scheme for the computational platform is illustrated in Fig. 1.

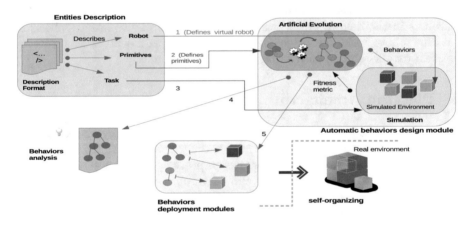

Fig. 1. Coordination platform architecture.

The robots, the primitives and the task that the swarm must perform are described using the description format (upper part of the figure) by mean of a set of XML marks. The primitive behaviors and the task must be defined into the robot controllers and the simulator module, respectively. From the description of the robots a virtual representation of them is created in the simulator (arrow 1), as well as a virtual representation of the physical environment in which the robots interact. In a similar way, from primitive behaviors description (arrow 2) self-organizing behaviors for the robots are constructed. In the center of the diagram the interaction between the mechanism of automatic construction of self-organizing behaviors and the virtual environment is illustrated.

An implementation of genetic programming builds and sends high-level programs (complex behaviors) to the simulator in which the virtual robots execute them and the swarm performance is evaluated. The task description defines how the performance is evaluated (arrow 3), the defined functions using the description format are executed in order to evaluate the performance of the individuals. Two possible actions to be carried out are illustrated, at the lower part of the diagram, once the behavior generation cycle has been completed. The behaviors can be directly assigned to the physical robots or be available for analyzing the rules of generated self-organizing behavior. A system for communication, allocation and execution of the generated programs has been included in the computational platform.

2.1   Description Format

In order to provide a mechanism for describing the elements required to design behaviors for a swarm of robots, a description format has been developed. This description format allows to specify the following three elements: members of the swarm, tasks that the swarm makes and a set of primitive behaviors that each swarm robot has. The description format is a markup language composed by elements that are used to specify the aforementioned elements. The proposed description format is named Swarm Description Language (SDL).

SDL is composed by a set of marks for describing hierarchical entities. SDL contains marks to define three atomic entities and two composite entities. Atomic entities are descriptions of physical objects (robots) or abstract objects (tasks and primitives). The atomic entities in SDL are: a robot description, a task description and the group of primitives description. A composite entity is a logical grouping of atomic entities. Composite entities in SDL are: a swarm description, which associates a robot set with a primitives set, a process description that associates a swarm with a task that is required to be solve by the swarm. Figure 2 shows the relationships that can be established between entities in SDL. A robot entity has a primitive set (basic robot capabilities), a swarm is composed by a group of robots and the process entity groups a swarm and a task. The structure of the SDL language is illustrated in Fig. 2. Each entity in SDL is defined by a set of XML marks. SDF (Simulation Description Format) [9] is integrated to SDL description format for describing the robots of the swarm.

Robot Description. The SDL allows to define three aspects of a robot: physical properties, perception system and effectors. The marks for physical properties allow to describe the robot as a physical entity in a virtual environment (embodiment) or the robot physical structure. The marks for perception and effectors allow to describe the elements the robot has to perceive and interact with the environment (situatedness), that is, it allows to specify the sensors and effectors that the robot has. A code snippet representing a robot description using SDL is shown in Fig. 3 (left).

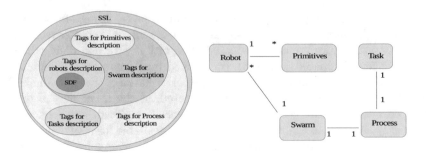

Fig. 2. SDL structure (left) and relationships between their entities (right).

Task Description. In the CP a task is defined by a set of functions (objectives) that are called during the evaluation phase of the behavior construction process. These functions are executed in a specific order, the order defines the task constraints. In SDL a task is specified by declaring the names, parameters and return data types for all the functions that make up the task. Along with the functions declaration a restriction set can be defined: sequence and activation time. With a sequence restriction an order in the functions execution can be imposed, with an activation time restriction is possible to specify when the functions are executed. The functions that make up a task are defined in the simulator module and get called during behaviors validation phase. In the Fig. 3 (center) a task description in SDL is shown.

Primitive Behaviors Description. In order to describe primitive behaviors that can be used by the CP for the automatic construction of complex behaviors for a group of robots, a set of marks has been defined in the description format which allows the user to specify primitives in the form of functions. A primitive behavior is modeled as a function with parameters and return value. The parameters and return values have a specific data type. The functions that represent the primitive behaviors are declared in SDL and these must be implemented in every robot controller as well. The functions declaration are used by CP to specify the primitive set which in turn is used for building tree programs (complex behaviors). In Fig. 3 (right) a primitive behavior declaration using SDL is shown.

2.2   Swarm Simulation

In order to approximate the conditions found in physical environments in which a real-world swarm is planned to be deployed and to validate the behaviors built during the automatic design process, a physics-based simulator has been adapted. From the robot descriptions made in SDL virtual instances of the robots are created and deployed in a simulated environment. As is illustrated in Fig. 4 the simulation system is made up of several elements. A virtual world is created, this world is composed by the virtual robots and the environment where the

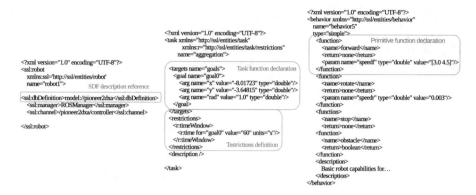

Fig. 3. Robot (left), task (center) and primitive behaviors (right) desclaration using SDL.

robots interact. Every virtual robot has a controller that implements the functions of the primitive behaviors declared in SDL. Every robot controller interprets the expression trees generated by the genetic programming implementation and sends commands to the virtual robots. The commands are executed by the robots and robot-robot-enviroment interactions are expected to be generated in a similar way as the real environment.

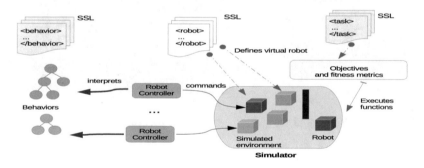

Fig. 4. Simulation system structure.

The simulator contains an implementation of the functions that represent the task objectives defined by the user using SDL. These functions are called periodically during the automatic building process of behaviors and they can access to the simulation state and generate information that influences the robots state. The function that quantifies the performance of the swarm is specified as part of the task in its declaration in SDL. Gazebo [8] is used as the simulation environment in the CP and is extended, according to model in Fig. 3, to support: the robot controllers, the functions that represent the tasks, automatic loading of the SDL descriptions, and the information exchange modules with the implementation of genetic programming to quantify the performance of the behaviors generated in the automatic behaviors process.

3   Computational Platform Deployment

The CP is deployed in a real world environment as a tool for automatic behaviors design for robots, such environment is the experimental manufacturing cell named Laboratorio Fábrica Experimental (LabFabEx) at Universidad Nacional de Colombia. At LabFabEx the CP is integrated as a tool that allows to investigate how emergent behaviors can be designed for robots in real world applications: robots for product making in flexible manufacturing cells. In Fig. 5 the process flow for behaviors design using the CP at LabFabEx is depicted. The CP functionality is exposed through a WEB interface where the user describes: the robots, their primitive behaviors, the enviroment and the task using SDL format, along with other configuration parameters such the population size, generations number and stop criterion. Then the CP builds possible solutions for the task at hand. Finally, when the stop criterion is met the resulting behaviors are available to be analyzed or assigned to robots in the manufacturing plant. Videos demonstrating the CP functionality are available online at [12].

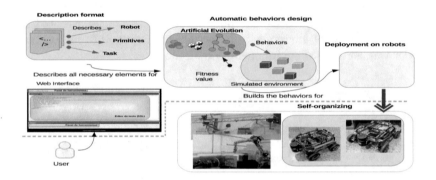

Fig. 5. The CP as a tool for evolutionary swarm robotics research at LabFabEx.

4   Conclusions and Future Work

A computational platform for automatic behaviors design and allocation for a group of robots was presented. The proposed platform has three main parts to facilitate the design process of behaviors for a group of mobile robots using artificial evolution. The SDL format allows to describe the different elements required for the automatic design process. SDL includes SDF format providing a whole range of possibilities for describing different robot types such as differential and omnidirectional mobile robots and serial and parallel architecture robots, as well as a primitive set for every type of robot and tasks to be executed by the group of robots. A genetic programming implementation in the CP builds behaviors from the primitive set for every robot in the group, this behaviors could be the building blocks for the group of robots can resolve more complex tasks. The proposed platform can simulate a wide range of environments and robots providing a virtual world where the generated behaviors can be tested. Finally, an allocation module allows to deploy generated behaviors on real robots. As is described, the proposed platform groups the main components for behaviors design in the evolutionary swarm robots field. Future work includes extending the SDL format for complex tasks description such as material transportation on manufacturing cells and other tasks in different real world scenarios. Future work also includes adapting the CP communication modules for allocating behaviors to a wide range of robots as well as to extend the platform simulation capabilities integrating other physics-based simulators.

References

1. Parker, L.: Distributed intelligence: overview of the field and its application in multi-robot systems. J. Phys. Agents (JoPha) **2**(1), 5–14 (2008). https://doi.org/10.14198/jopha.2008.2.1.02
2. Brambilla, M., Ferrante, E., Birattari, M., Dorigo, M.: Swarm robotics: a review from the swarm engineering perspective. Swarm Intell. **7**(1), 1–41 (2013). https://doi.org/10.1007/s11721-012-0075-2
3. Trianni, V.: Evolutionary swarm robotics. In: Studies in Computational Intelligence, vol. 108(1) (2008). https://doi.org/10.1007/978-3-540-77612-3
4. Francesca, G., Brambilla, M., Brutschy, A., Trianni, V., Birattari, M.: AutoMoDe: a novel approach to the automatic design of control software for robot swarms. Swarm Intell. **8**(2), 89–112 (2014). https://doi.org/10.1007/s11721-014-0092-4
5. Ohkura, K., Yasuda, T., Matsumura, Y.: Coordinating the collective behavior of swarm robotics systems based on incremental evolution. In: Proceedings of 2013 IEEE International Conference on Systems, Man, and Cybernetics (2013). https://doi.org/10.1109/smc.2013.687
6. Dorigo, M., Floreano, D., Gambardella, L., et al.: Swarmanoid: a novel concept for the study of heterogeneous robotic swarms. IEEE Robot. Autom. Mag. **20**(4), 60–71 (2013). https://doi.org/10.1109/mra.2013.2252996
7. Hecker, J., Moses, M.: Beyond pheromones: evolving error-tolerant, flexible, and scalable ant-inspired robot swarms. Swarm Intell. **9**(1), 43–70 (2015). https://doi.org/10.1007/s11721-015-0104-z
8. Koenig, N., Howard, A.: Design and use paradigms for gazebo, an open-source multi-robot simulator. In: 2004 IEEE/RSJ International Conference on Intelligent Robots and Systems (IROS) (IEEE Cat No 04CH37566). https://doi.org/10.1109/iros.2004.1389727
9. Simulation Definition Format. http://www.ncbi.nlm.nih.gov
10. Lee, J., Ahn, C., An, J.: An approach to self-assembling swarm robots using multitree genetic programming. Sci. World J. **2013**, 1–10 (2013). https://doi.org/10.1155/2013/593848
11. Koza, J.: Genetic programming: a paradigm for genetically breeding populations of computer programs to solve problems (1990)
12. Computational platform demonstrative videos (2018). https://github.com/labfabexun/cp-sdl

Distributed Group Analytical Hierarchical Process by Consensus

M. Rebollo[✉], A. Palomares, and C. Carrascosa

Universitat Politècnica de València, Camino de Vera s/n, 46022 Valencia, Spain
{mrebollo,apalomares,carrasco}@dsic.upv.es

Abstract. The analytical hierarchical process (AHP) is a multi-criteria, decision-making process. This work presents a method to be applied in group decisions (GAHP) using a combination of consensus process and gradient ascent to reach a joint agreement. The GAHP problem is modeled through a multilayer network, where each one of the criteria is negotiated by consensus with the direct neighbors on each layer of the network. Furthermore, each node performs a transversal gradient ascent and corrects the deviations from the personal decision locally. The process locates the optimal global decision, taking into account that this global function is never calculated nor known by any of the agents. If there is not an optimal global decision, but a set of suboptimal choices, agents are automatically divided into different groups that converges into these suboptimal decisions.

Keywords: Complex networks · Consensus · Gradient · AHP
GAHP · Agreement

1 Introduction

The Analytic Hierarchical Process (AHP) is a multi-objective optimization method. The decision makers provide subjective evaluations regarding the relative importance of the different criteria and the preference of each alternative for each criterion [6]. The result is a ranking of the considered choices that includes the relative score assigned to each one of these options. The AHP can be used for a single user to take a decision, but also for a group to achieve a joint agreement. However, these approaches assume that all the actors know each other and they can exchange information. This work proposes a method for group decision making based on AHP (GAHP), where the participants are connected through a network, and they interact exclusively with their direct neighbors.

The AHP begins with the definition of the criteria used to evaluate the alternatives, organized as a hierarchy. The importance of each criteria is defined through its weight $w^\alpha \in [0, 1]$. For example, let's assume that a new leader has to

---

This work is supported by the PROMETEOII/2013/019 and TIN2015-65515-C4-1-R projects of the Spanish government.

F. De La Prieta et al. (Eds.): DCAI 2018, AISC 800, pp. 238–246, 2019.
https://doi.org/10.1007/978-3-319-94649-8_29

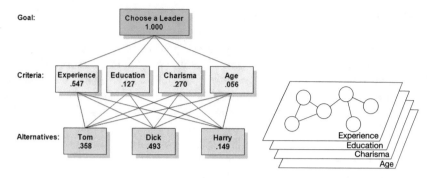

Fig. 1. Example of criteria hierarchy for a AHP and correspondence in a multilayer network

Table 1. (Left) Local priority matrix with the relative importance of each candidate regarding to their experience. (Right) Final priorities for the candidates. Dick is the selected one.

	Tom	Dick	Harry	Priority (l_i^α)	Exp	Edu	Char	Age	Goal
Tom	1	1/4	4	**0.217**	0.119	0.024	0.201	0.015	*0.358*
Dick	4	1	9	**0.717**	0.392	0.010	0.052	0.038	**0.492**
Harry	1/4	1/9	1	**0.066**	0.036	0.093	0.017	0.004	*0.149*

be chosen among three candidates: Tom, Dick, and Harry. The age, experience, education, and charisma of these candidates are considered to evaluate them (see Fig. 1). Once the criteria are defined, a pairwise matrix is created, assigning a relative judgment or preference value to each pair of alternatives which has priorities. The value a_{ij} represents the preference of the alternative i over the alternative j for the considered criteria, and $a_{ij} = 1/a_{ji}$. From this pairwise matrix, the local priority l_i^α is calculated, which defines the preference of the alternative i for the criterium α (Table 1).

This paper proposes a method to solve GAHP combining consensus and gradient algorithms through a multilayer network. A consensus process is performed in each layer, trying to achieve a joint decision for the corresponding criteria for all the agents. Simultaneously, a gradient ascent is executed across the layers, trying to compensate the concession an agent does in one criteria to keep its utility value as high as possible. This process converges to the optimal decision if some conditions are fulfilled.

2   Related Technologies

Group Analytical Hierarchical Process (GAHP) is the generalization of AHP for a group of decision makers. Each agent has its judgement matrix and its final priorities and they negotiate to achieve a unique, agreed solution. There are mainly two methods to aggregate individual preferences: aggregation of individual judgments (AIJ) and aggregation of individual preferences (AIP) [10]. The former aggregates the complete decision matrix, where as the latter aggregates the priorities only. Ossadnik et al. [9] make a comparative analysis of different GAHP approaches. Huang et al. [4] combines judgements and preferences, which leads to a more satisfactory group decision. Other techniques, such as simulated annealing have been also used, trying to solve the GAHP when there are significative differences among the options [1]. Finally, the word *consensus* appears in other works related with group decision making and GAHP, but with a general meaning, and not referred to the method used in this paper [2,5]. There are other approaches applied successfully to multi-objective optimization that can be adapted [7], but they are out of the scope of this paper.

Consensus means reaching an agreement on the value of a variable. Agents are connected through an acquaintances network whose topology constraints the possible interaction between them. The theoretical framework for solving consensus problems in agent networks was formally introduced by Olfati–Saber and Murray [8]. It has been demonstrated that a convergent and distributed consensus algorithm in discrete-time exists and it converges to the average of their initial values.

$$x_i(t+1) = x_i(t) + \varepsilon \sum_{j \in N_i} (x_j(t) - x_i(t)) \tag{1}$$

where N_i denotes the neighbors of i and ε is the step size, $0 < \varepsilon < \min_i 1/d_i$, being d_i the degree of node i.

Agreement processes frequently involve the optimization of some global utility function. Decentralized approaches take advantage of scalability, adaptation to dynamic network topologies and can handle data privacy. Coupled optimization problems can be solved using a variety of distributed algorithms. Particularly, the combination of consensus and *gradient* models can be expressed as a two-step process [11].

$$x_i(t+1) = \sum_j w_{ij} x_j(t) - \alpha \nabla f_i(x_i(t)) \tag{2}$$

where $W = [w_{ij}]$ is a double stochastic, symmetric matrix and $\nabla f_i(x_i(t))$ performs a gradient descent to minimize a cost function.

To generalize this method to a multidimensional case, *multilayer networks* are introduced. In multilayer networks, links of different type exist among the nodes and they are separated in layers. The interdependence among layers is defined through cross-links between the nodes that represent the same entity in each network. These cross-links models the transference of information that passes from one layer to the others. A multilayer network is formally defined as a pair $M = (G, C)$ where $G = \{G^1, \ldots, G^p\}$ is a family of graphs $G^\alpha = (E^\alpha, L^\alpha), \forall \alpha \in [1, p]$ called layers, and $C = \{L^{\alpha\beta} \subseteq E^\alpha \times E^\beta, \forall \alpha, \beta \in [1, p], \alpha \neq \beta\}$ is the set of connections between two different layers G^α and G^β [3]. The elements of each L^α are called *intralayer* connections and the elements of C are the *interlayer* ones or *crossed layers*.

3 Decentralized AHP Using Consensus in Multiplex Networks

3.1 Consensus and Gradient Combination

Let's consider the agents connected in an undirected network. A multilayer network is defined, where each layer represents one of the final criteria. For example, in the example shown in Fig. 1, four layers are created: experience, education, charisma and age. Each layer is weighted using the weight defined for the criteria. A utility function is used by the participant to perform the gradient ascent, trying to keep as near as possible to its preferred distribution.

Each participant has its own criteria (see Fig. 2, left), and the goal of the system is to agree the best candidate according to all the agents involves in the decision. Therefore, a consensus process is executed in each layer to find the weighted average. However, this process considers the criteria as independent, and it does not converge in the value that optimizes the decision. The combination of the consensus process with a gradient ascent, as it is defined in Eq. 3, corrects the deviation produced by the consensus and each participant tries to maintain the decision that maximizes its own local utility. This decentralized process leads to a consensus value near to the global optimum, considered as the sum of the local utility functions. Observe that this global utility function is never calculated and the participants reach this value exchanging information with their direct neighbors.

$$x_i^\alpha(t+1) = \overbrace{x_i^\alpha + \frac{\varepsilon}{w_i^\alpha} \sum_{j \in N_i^\alpha} (x_j^\alpha(t) - x_i^\alpha(t))}^{g(x_i^\alpha)} + \underbrace{\varphi \nabla u_i(x_i^1(t), \ldots, x_i^p(t))}_{h(x_i^\alpha)} \qquad (3)$$

Individual agents are not conscious of a final, global solution, but of the convergence to an agreed compromise among its near neighbors. Furthermore, the system is scalable since new nodes can be added without additional notifications to the rest of the network.

3.2 Utility Function

Utility functions have some common properties in any optimization problem: independence, completeness, transitivity, and continuity. As we propose a model with cooperative agents, we shall assume that each local utility functions have a maximum and this maximum is the starting point for all the agents. Further-more, the function must be a decreasing one. The normal distribution fulfills all these properties. Therefore, it has been the selected one for the utility function u_i of the agents. We can assume that agents are initially situated in its maximum value, which corresponds to the mean value of the utility function. The weight assigned to the term can be used in the dispersion measure. These functions are combined in one unique utility function for the agent.

$$u_i^\alpha(x_i^\alpha) = e^{-\frac{1}{2}\left(\frac{x_i^\alpha - l_i^\alpha}{1 - w_i^\alpha}\right)^2} \qquad u_i(x_i) = \prod_\alpha u_i^\alpha(x_i^\alpha) \tag{4}$$

This definition corresponds to a renormalized, multi-dimensional, gaussian dis-tribution such that the maximum utility for the agent i is $u_i(x_i(0)) = 1$. The global utility of the system is the sum of the individual utilities of the agents. This value is never calculated, and the function is known by none of the partic-ipants in the GAHP.

$$U = \sum_i u_i(x_i) \tag{5}$$

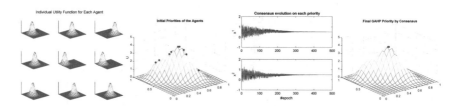

Fig. 2. Experiment 1. From left to right: local utility function of the 9 agents, initial situation over the global utility, evolution of the consensus values and final situation after convergence

4   Maximizing Utility of GAHP by Consensus

This section presents 3 synthetic examples with 9 agents with different initial configurations that are going to take a decision using GAHP. These examples illustrate that, in occasions, the best solution is reached by all agents agreeing (experiment 1). However, in other occasions is better to divide the whole group in smaller coalitions, so that each one of the small groups leads to different consensus and the whole system reaches a higher utility (experiment 2). A bi-dimensional example has been chosen to be able to represent it graphically.

4.1   Experiment 1: Joint Solution

This experiment presents a set of 9 agents that reach a consensus solution where all of them agree. Figure 2 left shows the utility function calculated from the initial preferences of each participant according to Eq. 5, and the initial and final status of the process (center and right images of the figure, respectively). When the combined process stops, all the participants have reached the same point, which corresponds to the common decision agreed by the agents. For this solution to exist, the only condition is that all the participants have a positive utility $u_i > 0$ along the complete solution space. Figure 2 shows the evolution of the value for each criterion for each one of the participants along the process. It converges to the final decision. If these values are considered as the x and y coordinates, it matches with the point that corresponds to the optimal solution.

4.2   Experiment 2: Breaking in Separated Agreements

Our proposal is to allow break links among the agents when the local utility of the agent in the solution is (near) zero. When an agent detects that the solution guides towards a point with zero-utility, the agent can decide to break the link to those neighbors who are pulling from the preferences. As Fig. 3 shows, in this case, the network is broken into groups, each one of them converges to a different agreement. The optimal decision is located by the group formed by those participants whose utility function is positive in the best solution. This solution is reached if the agents with zero-utility are just removed from the system. Despite doing so, we allow these participants to reach another decision forming a separate group. Figure 3 shows the evolution of the criteria in such a case. It can be observed how more than one decision is taken. The network is divided into 2 groups, each one in a local maximum value.

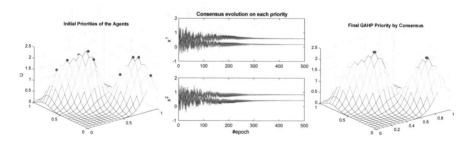

Fig. 3. Experiment 2. GAHP process allowing to break links and reconnect to near neighbors. This solutions guarantees the convergence of a subgroup to the best possible decision.

5   Comparing with Other Approaches

In this section, we show the results obtained with the proposed method compared to other GAHP approaches [9]. The experimental set consists of 4 agents who have to agree in the evaluation criteria for a decision. Two versions of consensus process are included: consensus over judgments (COJ) and over priorities (COP).

Table 2. Case example: results of the evaluation regarding Pareto optimality

Agent	Scenario	A_1	A_2	A_3	A_4	Pareto Opt.
Individual Preferences						
DM_1	1	0.236	0.418	0.164	0.181	$A_1 > A_3$
	2	0.161	0.421	0.159	0.259	$A_1 > A_3$
DM_2	1	0.490	0.127	0.173	0.210	$A_1 > A_3$
	2	0.485	0.140	0.204	0.174	$A_1 > A_3$
DM_3	1	0.238	0.262	0.063	0.437	$A_1 > A_3$
	2	0.190	0.248	0.104	0.458	$A_1 > A_3$
Agent	Scenario	A_1	A_2	A_3	A_4	Pareto Opt.
Group Preferences						…satisfied?
AIJ(WAMM)	1	0.290	0.321	0.133	0.254	$A_1 > A_3$ Yes
	2	0.181	0.353	0.184	0.281	$A_1 < A_3$ No
AIJ(WGMM)	1	0.279	0.344	0.130	0.246	$A_1 > A_3$ Yes
	2	0.164	0.370	0.190	0.276	$A_1 < A_3$ No
AIP(WAMM)	1	0.312	0.300	0.146	0.241	$A_1 > A_3$ Yes
	2	0.264	0.302	0.161	0.272	$A_1 > A_3$ Yes
AIP(WGMM)	1	0.318	0.288	0.149	0.244	$A_1 > A_3$ Yes
	2	0.252	0.297	0.172	0.279	$A_1 > A_3$ Yes
LFA(WAMM)	1	0.253	0.343	0.128	0.275	$A_1 > A_3$ Yes
	2	0.177	0.370	0.165	0.288	$A_1 > A_3$ Yes
LFA(WGMM)	1	0.289	0.343	0.133	0.234	$A_1 > A_3$ Yes
	2	0.165	0.375	0.190	0.269	$A_1 < A_3$ No
PDR	1	0.370	0.264	0.041	0.325	$A_1 > A_3$ Yes
	2	0.329	0.276	0.063	0.331	$A_1 > A_3$ Yes
COJ	1	0.238	0.262	0.063	0.436	$A_1 > A_3$ Yes
	2	0.288	0.3651	0.152	0.1934	$A_1 > A_3$ Yes
COP	1	0.313	0.299	0.146	0.241	$A_1 > A_3$ Yes
	2	0.505	0.048	0.312	0.174	$A_1 > A_3$ Yes

Experiments compare our method with the methods examined in the reference mentioned above: (i) aggregation of individual judgments (AIJ), (ii) aggregation of individual priorities (AIP), (iii) loss function approach (LFA), and (iv) GAHP with preferential differences and ranking (PDR). Arithmetic and geometric means are considered in all cases (except GAHP).

Scenarios 1 and 2 evaluates the Pareto Optimality which says that if all group members prefer alternative A_1 to alternative A_3, then the group should prefer A_3 too. The difference between both scenarios is that, in scenario 2, A_1 and A_3 are closer to each other (see the values for DM_1 in Table 2)

Another measure is the homogeneity condition, that says that if each individual judges an alternative A_1 μ -times as large as another alternative A_2, then the synthesized judgement A_1 should be μ -times as large as another alternative A_2 also. Scenarios 3 and 4 evaluates this property. In scenario 3, all agents prefer A_1 to A_3 with a factor $\mu_{1,3} = 2$. In scenario 4, all agents judge alternative A_4 as 1.6 times as preferable as A_3, so $\mu_{4,3} = 1.6$ (Table 3)

The results show that just AIP and our proposal (COP and COJ) perform properly in all cases, ensuring the Pareto optimality and the homogeneity of the results. COJ fails in the homogeneity condition for scenario 4, with a value of 1.57 (instead of 1.6). Nevertheless, is the best of the results that fails this test and is close enough to the correct solution.

Table 3. Case example: results of the evaluation regarding homogeneity condition

Agent	Scenario	A_1	A_2	A_3	A_4	Homogeneity
Individual Preferences						
DM_1	3	0.332	0.338	0.166	0.164	$\mu_{1,2} = 2.00$
	4	0.264	0.364	0.143	0.229	$\mu_{4,3} = 1.60$
DM_2	3	0.458	0.119	0.229	0.194	$\mu_{1,2} = 2.00$
	4	0.445	0.139	0.160	0.257	$\mu_{4,3} = 1.60$
DM_3	3	0.205	0.257	0.102	0.436	$\mu_{1,2} = 2.00$
	4	0.232	0.290	0.183	0.295	$\mu_{4,3} = 1.60$
Method	Scenario	A_1	A_2	A_3	A_4	Homogeneity
Group Preferences						... satisfied?
AIJ(WAMM)	3	0.290	0.271	0.212	0.228	$\mu_{1,3} = 1.37$ No
	4	0.268	0.304	0.149	0.279	$\mu_{4,3} = 1.88$ No
AIJ(WGMM)	3	0.302	0.311	0.171	0.216	$\mu_{1,3} = 1.77$ No
	4	0.268	0.320	0.145	0.267	$\mu_{4,3} = 1.84$ No
AIP(WAMM)	3	0.344	0.256	0.172	0.228	$\mu_{1,3} = 2.00$ Yes
	4	0.312	0.282	0.156	0.250	$\mu_{4,3} = 1.60$ Yes
AIP(WGMM)	3	0.352	0.248	0.176	0.223	$\mu_{1,3} = 2.00$ Yes
	4	0.312	0.270	0.161	0.258	$\mu_{4,3} = 1.60$ Yes
LFA(WAMM)	3	0.273	0.293	0.163	0.271	$\mu_{1,3} = 1.68$ No
	4	0.243	0.331	0.139	0.286	$\mu_{4,3} = 2.05$ No
LFA(WGMM)	3	0.287	0.322	0.171	0.220	$\mu_{1,3} = 1.68$ No
	4	0.266	0.330	0.146	0.259	$\mu_{4,3} = 1.77$ No
PDR	3	0.352	0.238	0.073	0.337	$\mu_{1,3} = 4.83$ No
	4	0.371	0.267	0.074	0.288	$\mu_{4,3} = 3.91$ No
COJ	3	0.331	0.338	0.165	0.166	$\mu_{1,3} = 2.00$ Yes
	4	0.199	0.330	0.183	0.287	$\mu_{4,3} = 1.57$ (*)
COP	3	0.362	0.423	0.181	0.032	$\mu_{1,3} = 2.00$ Yes
	4	0.208	0.260	0.203	0.328	$\mu_{4,3} = 1.60$ Yes

6   Conclusions

This work presents a method based on a combination of consensus and gradient ascent to solve GAHP in a decentralized environment. Participants in the decision making process exchange their preferences with their direct neighbors to reach an agreement that allows the team to select the alternative with the global highest utility. The model presented makes a consensus between priorities and also judgments, fulfilling the Pareto optimality and homogeneity condition. Moreover, the comparison with other approaches gives similar solutions that other approaches but in a distributed fashion, where all the participating agents keep their preferences as private and work with their values and the ones obtained by their direct neighbors. Usually, the GAHP methods converge to the geometric mean, while this proposal converges to the maximal global utility keeping the consistency ratio below 0.1. Our method is fully decentralized, avoiding the use of a moderator to guide the group decision process, as other GAHP methods do. Moreover, in the cases where the global utility function presents local maximum points, the method suggests the division of the group in different coalitions.

References

1. Blagojevic, B., et al.: Heuristic aggregation of individual judgments in ahp group decision making using simulated annealing algorithm. Inf. Sci. **330**, 260–273 (2016)
2. Wu, J., et al.: A visual interaction consensus model for social network group decision making with trust propagation. Knowl.-Based Syst. **122**, 39–50 (2017)
3. Boccaletti, S., et al.: The structure and dynamics of multilayer networks. Phys. Rep. **544**(1), 1–122 (2014)
4. Huang, Y.S., et al.: Aggregation of utility-based individual preferences for group decision-making. Eur. J. Oper. Res. **229**(2), 462–469 (2013)
5. Dong, Q., Saaty, T.L.: An analytic hierarchy process model of group consensus. J. Syst. Sci. Syst. Eng. **23**(3), 362–374 (2014)
6. Ishizaka, A., Labib, A.: Review of the main developments in the analytic hierarchy process. Expert Syst. Appl. **38**(11), 14336–14345 (2011)
7. Kłosowski, G., Gola, A., Świć, A.: Application of fuzzy logic in assigning workers to production tasks. In: Proceedings of 13th International Conference on DCAI, pp. 505–513. Springer, Cham (2016)
8. Olfati-Saber, R., Fax, J.A., Murray, R.M.: Consensus and cooperation in networked multi-agent systems. Proc. IEEE **95**(1), 215–233 (2007)
9. Ossadnik, W., Schinke, S., Kaspar, R.H.: Group aggregation techniques for analytic hierarchy process and analytic network process: a comparative analysis. Group Decis. Negot. **25**(2), 421–457 (2016)
10. Ramanathan, R., Ganesh, L.: Group preference aggregation methods employed in ahp: an evaluation and an intrinsic process for deriving members' weightages. Eur. J. Oper. Res. **79**(2), 249–265 (1994)
11. Yuan, K., Ling, Q., Yin, W.: On the convergence of decentralized gradient descent. Tech. Rep. Report, pp. 13–61, UCLA CAM (2014)

Intelligent Flight in Indoor Drones

Giovanny-Javier Tipantuña-Topanta, Francisco Abad, Ramón Mollá,
Jose-Luis Poza-Lujan$^{(\boxtimes)}$, and Juan-Luis Posadas-Yagüe

University Institute of Control Systems and Industrial Computing (ai2),
Universitat Politècnica de València (UPV), Camino de Vera, s/n.,
46022 Valencia, Spain
{giotitoa,fjabad,rmolla,jopolu,jposadas}@ai2.upv.es

Abstract. Currently, drones are one of the most complex control systems. This control covers from the control of the stability of the drone, to the automatic control of the navigation in complex environments. In the case of indoor drones, technological challenges are specific. This paper presents an intelligent control architecture for indoor drones where security is the main axis of the system design. So, a definition of different navigation modes based on security is proposed. The drone must have different navigation modes: manual, reactive, deliberative and intelligent. For indoor navigation it is necessary to know the position of the drone, therefore the system must have a location mode similar to GPS, but that provides better accuracy. For deliberative and intelligent modes, the system must have a map of the environment, as well as a control system that sends the navigation orders to the drone.

Keywords: Distributed system · Indoor drones · Intelligent flight modes

1 Introduction

Creative industries require to control cameras many times. Many devices such as cranes, rails or portable frames are used in order to obtain interesting shots. These devices often have many drawbacks:

- They are complex to install, handle and remove.
- These devices have limited movement space.
- They are invasive on the scene they are recording. Devices like steadicams may avoid these drawbacks, but they cannot be used on any situation and they cannot be moved up from the ground.

On the other hand, unmanned aerial vehicles, known as UAVs, RPAS (remotely piloted aircraft system) or drones, obviate the drawbacks. When recording takes place indoors, let's say a television or movie set, drones can provide shots not available to current auxiliary devices because of their stability and precision [1].

Drone navigation requires to know the position of the drone at all times. In outdoor flights, drones can use GPS location systems. When working indoors, GPS has not the accuracy to allow a safe flight. So, an Indoor Positioning System (IPS) is needed. Furthermore, due to smaller spaces and increased risk of damages to property and

© Springer International Publishing AG, part of Springer Nature 2019
F. De La Prieta et al. (Eds.): DCAI 2018, AISC 800, pp. 247–254, 2019.
https://doi.org/10.1007/978-3-319-94649-8_30

people in case of an accident, a much higher accuracy is required. Typically, the necessary accuracy is in the order of tens of centimeters (that is, two orders of magnitude more precise than GPS).

Currently drone indoor navigation is mostly performed using commercial, off-the-shelf solutions, both for drone control [2] and for trajectory tracking [3]. This latter aspect is one of the most interesting as far as research is concerned [4].

The drone real-time navigation and control features are especially relevant to the creative industry. Safety is a key factor for both outdoor and indoor flight environments. In most countries, outdoor drone flights on populated areas are very restricted. Common outdoor flight environments are on non-populated areas with little elements of value around. However, even if indoor flights are not regulated, it is not strange to find elements of value like paintings, sculptures, lamps, furniture, ... Combined with smaller spaces and the presence of people, indoor drone flights should have higher levels of security than outdoor flights. This aspect therefore determines, to a large extent, all questions relating to the design of both the control architecture and the drone.

Next section describes the system drone architecture, then section three proposes different flight modes and four and five sections describe the relationship between the flight modes of the drone and the security level required. Finally, section six summarize the article.

2   Distributed System Architecture

The system is part of the H2020 European Project Arts indoor RPAS Technology (AiRT). AiRT is a distributed system composed of several devices connected to provide the whole service of recording video or photographs in in-door scenarios using Remotely Piloted Aircraft System (RPAS) [5]. Principally, there are two main working areas: Drone and Environment Infrastructure. Each area has common components in the same three systems: Intelligent Flight Control System (IFCS), RPAS system, and Indoor Positioning System (IPS) [6]. These components are shown in the Fig. 1.

The environment infrastructure is the ground system in charge of supporting the drone operability. It is com-posed of several subsystems: the IPS anchors (environment antennas) that give support to Drone indoor positioning, the remote control via radio to control the flight manually, and the Ground Control System (GCS). The GCS is in charge of create and update the Virtual Environment Map (VEM), to generate the flight plan by means of the Flight Planning System. Finally, the Record and Flight Control System control and monitor the Drone flight.

In the same way that the Environment Infrastructure, Drone has components, corresponding with the previous presented three subsystems. Concerning the IPS, Drone incorporates four antennas to receive the anchors signal, and one board (caller "tag") that processes the signals and generates the position. Related with the RPAS, Drone include all sensors like distance or the Inertial Measurement Unit (IMU), a multiplexor system that provides the source that control the Drone (manually flight or automatic flight), and the Flight Control System (FCS) that it is in charge of controlling all the parameters of the flight: drone position and orientation, camera parameters,

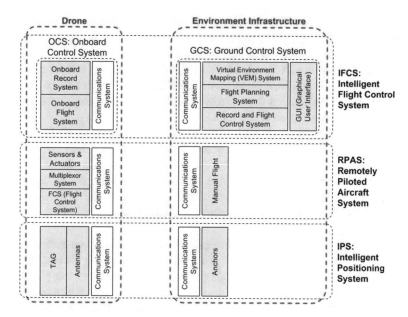

Fig. 1. AiRT system components

gimbal parameters. Finally, the third system is the Onboard Control System (OCS) it includes also the VEM manager (synchronized with the GCS). It is in charge of detecting the cloud of points in front of the drone, sending them back to the GCS, receive the flight plan and transfer it to the FCS, control the flight plan according to the operator requirements.

3   Flight Modes

Safety is the main axis of system design. There are several layers of safety throughout the whole architecture. The lowest level of safety is at the hardware level. It concerns all the devices selected for the hardware, connectors for transmitting the data from one module to another one, power cables, motors, batteries, and so on.

There is no control over the basic software layer composed of drivers, O.S. and the low-level FCS. This is not the aim of this work. On this layer, there is another level in charge of the way the drone behaves when it is flying. There are several ways of flying depending on the degree of autonomy of the drone and the level of safety of the mission to accomplish:

1. Manual Flight Mode (M.F.M.). The pilot has complete control of the drone. The drone movements have no restrictions, and the pilot can instruct it to go anywhere, regardless of any sensor reading or map configuration. The pilot has canceled the reactive mode (see later). The human pilot has full control of the drone.
2. Reactive Flight Mode (R.F.M.). It is a defensive flight mode. It is performed by the Flight Control System (FCS), taking into account the reading of the proximity

sensors. It is a priority flight mode always active unless the human pilot expressly cancels it when flying in Manual Flight Mode. It is active by default for all the other kinds of flight modes: Assisted, Mixed and Smart.

3. Assisted Flight Mode (A.F.M.). The pilot controls the drone, and he can take the drone out of the established flight plan. The difference with respect to the Manual Flight Mode (MFM) is that the reactive mode is engaged, so the pilot cannot crash into the environment even if she tries.

4. Deliberative Flight Mode (D.F.M.). It is an A.F.M. where the drone is not allowed to move into no-flight zones like populated zones, hanging cables areas,...

5. Mixed Flight Mode (Mi.F.M.). The user can explicitly stop, move forward or backward at different speeds along the flight plan. The metaphor is like having a virtual rail along the trajectory of the flight plan. The drone behaves like a 3D virtual dolly. It requires having captured the virtual map and having defined the flight plan.

6. Guided Flight Mode (G.F.M.). Automatic flight considering obstacles, restricted areas and surrounding architecture. This mode is completely autonomous. It is supervised by humans, but humans do not control the drone. Human pilot can pass to any other kind of flight mode from this one. The drone moves automatically along the trajectory of the flight plan. The drone behaves like a 3D virtual dolly as in Mi.F.M. but completely autonomous. It has the same requirements as Mi.F.M.

7. Smart Flight Mode (S.F.M.). This mode can be engaged when the drone has left the flight plan, and the pilot wants it to return to the predefined flight plan selecting the shortest itinerary and considering obstacles, restricted areas and surrounding architecture in real time. This mode is completely autonomous. It is supervised by humans, but humans do not control the drone. The implementation of this mode is beyond the goals of this work. It requires the virtual map, the flight plan and 3D sensors.

8. Emergency Flight Mode (E.F.M.). Moreover, in the event of a loss of IPS datalink, radio contact, engine failures or battery level below a safety level, a defensive failsafe behavior will be executed. Depending on the type of failure and position of the RPAS, it will start a slow landing or return automatically to the starting point (return to launch - RTL Mode). It requires Environment Scanning.

4 Security vs Flight Modes

There is a relationship between the flight modes of the drone and the security level required. See Fig. 2. Notice that R.F.M. is not properly an automatic flight mode selectable by the user but a cross-cutting safety feature to all flight modes except for the M.F.M., where the human pilot has full control of the drone. So, it is not included as a flight mode in the horizontal axis.

Notice that the Emergency Flight Mode:

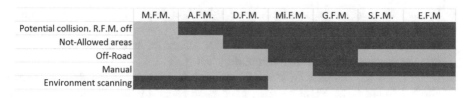

Fig. 2. Relationship between flight modes and security levels

1. Can only activate the RTL mode without entering in S.F.M. if the drone is in the path.
2. Can perform an emergency landing in any situation if it has not enough battery to come back home through the shortest path available. This is an exceptional mode that can be reached from any other state of the drone.
3. If the drone is not following the flight plan, a return straight line trajectory could be dangerous since the drone could collide with some obstacle or fly into no-flight zones. In case of detecting an obstacle, the drone cannot decide where to move to. It cannot recalculate in real time an alternative trajectory to RTL.
4. If the battery is really low, the drone can land on any safe landing point or area specified in the map.
5. Requires an Environment Scanning since this mode requires to know the path and potential obstacles (walls, environment cloud of points,…) to determine the return path to a landing point or area (typically the take off point).

Notice that M.F.M. and A.F.M.:

1. May be used in whatever situation.
2. Do not require any kind of scanning of the environment
3. Are the flight modes used currently when flying a drone in indoor scenarios using any current on-the-shelf commercial drone.
4. Do not require the use of an accurate IPS since the flight is completely manual.

Notice that D.F.M.:

1. Does not require any kind of scanning surroundings since this flight mode has to avoid restricted areas.
2. Is a flight mode that improves security for drone indoor navigation over a completely free M.F.M. since it takes into account both the R.F.M. and the not allowed areas.
3. Cannot use any current on-the-shelf commercial drone since the restriction has to be edited by the pilot in a GUI on a PC/Tablet and later, transferred to the drone FCS in order to avoid those areas when flying.
4. Requires to use an accurate IPS since current GPS has a minimum resolution of ten meters when flying out-door. On the other hand, there are many situations where GPS reception is bad or even non-existent when working indoor.

Notice that (Mi, G, S, E).F.M.:

1. Require to scan the environment since this flight mode has to avoid not only restricted areas but walls, columns, furniture,…
2. Improve security for drone navigation in indoor scenarios avoiding the pilot to crash into the surrounding walls and furniture
3. Cannot use any current on-the-shelf commercial drone since forbidden areas, allowed flying paths, flight plans,… have to be edited by the pilot in a GUI on a PC/Tablet and later, transferred to the drone FCS in order to avoid those areas when flying.
4. Require to use an accurate IPS since current GPS has a minimum resolution of ten meters when flying outdoor. On the other hand, there are many situations where GPS reception is bad or non-existent when working indoor.

5   Flight Mode State Machine

A drone may be understood as a state machine. This machine works on different flight modes. Every kind of flight mode may be understood as a state of this state machine. On Fig. 3, there is a state-transition diagram that shows how an indoor drone should work.

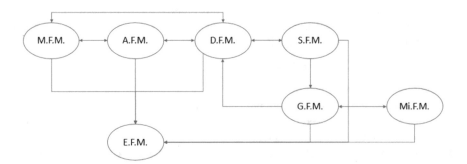

Fig. 3. State-transition diagram of all the flight modes

Notice that there is a group formed by the (M, A, D).F.M. on the left side. This is the Manual group while the group formed by the (Mi, G, S).F.M. on the right side is the automatic one.

The Manual group has three characteristics:

1. It is controlled directly by the human pilot. The drone does not move if the human pilot does not order anything (it flies in loiter mode).
2. The human pilot can change from one mode to any other simply activating the defensive/reactive flight mode or de/activating the no-flight zones. It depends on the safety risks the pilot wants to assume.
3. D.F.M. has the highest security level of all the manual flight modes. It does not allow the pilot to get the drone into no-flight zones and do not allow the drone to collide with walls. So, it is the flight mode reached as soon as the user enters in manual mode while the drone is in G.F.M.

Notice that S.F.M.:

1. Is a transition state that allows to return back to the original flight plan path trying to avoid obstacles. There are several choices for returning to the original path: going to the nearest path point, to the point where it is supposed to be at that moment according the flight plan, going to the point where the drone left the flight plan, or to a given checkpoint.
2. It is the only way to return from a manual flight mode to an automatic flight plan.
3. This mode is beyond the goals of this project and it will not be implemented by now.
4. While the drone is turning back to the original path (S.F.M.), the user can take the control of the drone again, passing to D.F.M., the immediately lower security level. This is why there is a bidirectional arrow between both flight modes. It is the responsibility of the pilot to reduce later the security passing to (M,A).F.M.
5. Once the drone has reached the flight plan trajectory, it enters into the G.F.M. and it starts to follow it as if nothing would have happened.

Notice that the Automatic group:

1. It is not controlled directly by the human pilot except for the Mi.F.M. which is a semiautomatic mode.
2. Cannot move outside the path in the flight plan. So there is no need to go to S.F.M. from G.F.M. to correct anything. The arrow is one way.
3. The human pilot can:

   - Move forward and backward the defined trajectory of the flight plan changing from G.F.M. to Mi.F.M.
   - Move out of the trajectory changing from G.F.M. to D.F.M.

Note that for current implementation, the S.F.M. is not available. So, when moving manually out of the path, the drone is always on D.F.M. and it cannot come back to G.F.M. until the drone is landed and reset.

The diagram shows that manual flight is always mandatory over any other flight mode. Manual flights are always available. They may be seen as an escape mode when the drone is in risky situations.

Some final remarks:

1. All the transitions from any state to another one are controlled from the Flight Path Manager at the land base.
2. If the manual radio control base is touched explicitly or accidentally, the drone abandon any automatic flight mode and gets into D.F.M. for security.
3. Once the drone is in any manual flight mode, the pilot can change to any other manual flight mode by selecting the new mode in the Flight Plan Manager device.

6 Conclusions

This paper has presented an architecture oriented to achieve the autonomous navigation of indoor drones whose mission is to record videos and pictures with very high definition cameras. In order to achieve this type of missions the system requires different flight modes that have been proposed: manual, reactive, deliberative and intelligent. Safety is the main axis of system design and a relationship between the flight modes of the drone and the security level required has been presented. Finally, the paper shows how an indoor drone should work by using these flight modes.

Acknowledgements. This work has received funding from the European Union's Horizon 2020 research and innovation programme under grant agreement no. 732433 (reference: H2020-ICT-2016-2017, www.airt.eu).

References

1. Castillo, P., García, P., Lozano, R., Albertos, P.: Modelado y estabilización de un helicóptero con cuatro rotores. Revista Iberoamericana de Automática e Informática Industrial RIAI **4**(1), 41–57 (2007)
2. Hussein, A., Al-Kaff, A., de la Escalera, A., Armingol, J.M.: Autonomous indoor navigation of low-cost quadcopters. In: 2015 IEEE International Conference on Service Operations and Logistics, and Informatics (SOLI), pp. 133–138. IEEE, November 2015
3. Santana, L.V., Brandao, A.S., Sarcinelli-Filho, M., Carelli, R.: A trajectory tracking and 3rd positioning controller for the AR.Drone quadrotor. In: 2014 International Conference on Unmanned Aircraft Systems (ICUAS), pp. 756–767. IEEE, May 2014
4. Martínez, S.E., Tomas-Rodriguez, M.: Three-dimensional trajectory tracking of a quadrotor through PVA control. Revista Iberoamericana de Automática e Informática Industrial RIAI **11**(1), 54–67 (2014)
5. AeroTools-UAV, julio 2017. www.aerotools-school.es
6. Pozyx - Accurate Positioning, julio 2017. http://www.pozyx.io

Privacy Preserving Expectation Maximization (EM) Clustering Construction

Mona Hamidi[1], Mina Sheikhalishahi[2(✉)], and Fabio Martinelli[2]

[1] Dipartimento Ingegneria dell Informazione e Scienze,
Universita di Siena, Siena, Italy
`mona.hamidi@student.unisi.it`
[2] Istituto di Informatica e Telematica, Consiglio Nazionale delle Ricerche, Pisa, Italy
`{mina.sheikhalishahi,fabio.martinelli}@iit.cnr.it`

Abstract. This paper presents a framework for secure *Expectation Maximization (EM)* clustering construction over partitioned data. It is assumed that data is distributed among several (more than two) parties either *horizontally* or *vertically*, such that for mutual benefits all the parties are willing to identify clusters on their data as a whole, but for privacy restrictions, they avoid to share their datasets. To this end, in this study general algorithms based on *secure sum* is proposed to securely compute the desired criteria in constructing clusters' scheme.

Keywords: Privacy · *Expectation Maximization* (EM) clustering
Data sharing

1 Introduction

The collection of digital data brought by governments, companies, and individuals, has created fabulous opportunities for knowledge-based decision making. For mutual benefits, or by regulations that require certain data to be published, there is a demand for the exchange and publication of data among various parties [9]. Such an intelligence can be exploited from shared data to improve revenues, prevent loss coming from brand-new potential cyber-threats, analyze statistical records of medical data. For example, hospitals and health centers to shape the causes and symptoms related to a new pathology are interested to publish their own patient records. Independently from the final goal, unfortunately data sharing brings the issues and drawbacks related to individual privacy. Hence, the most desirable strategy is the one which enables data sharing in secure environment, such that it preserves the individual privacy requirements while at the same time the data are still practically useful for data analysis.

Clustering is a very well-known tool in unsupervised data analysis, which has been the focus of significant researches in different studies, spanning from information retrieval, text mining, data exploration, to medical diagnosis [1].

© Springer International Publishing AG, part of Springer Nature 2019
F. De La Prieta et al. (Eds.): DCAI 2018, AISC 800, pp. 255–263, 2019.
https://doi.org/10.1007/978-3-319-94649-8_31

Clustering refers to the process of partitioning a set of data points into groups, in a way that the elements in the same group are more similar to each other rather than to the ones in other groups.

The problem of data clustering becomes challenging when data is distributed between two (or more) parties and for privacy concerns the data holders refuse to publish their own dataset, but still they are interested to shape more accurate clusters, identified on richer set of data. To this end, we propose a framework which serves as a tool for several agents to detect the cluster structures on the whole of their data, in terms of EM clustering, without needing to reveal the original data. The *Expectation Maximization* (EM) clustering is an iterative method based on maximum likelihood principle [8], where each cluster is represented as a probability distribution. To find *securely* the two criteria which represents a cluster on whole data, based on *Guassian* distribution, secure sum protocol is exploited among participated parties.

The rest of this paper is structured as follows. Section 3 presents some preliminary notations exploited in this study, including *EM clustering* and *secure sum protocol*. Section 4 presents the system model in four subsections: privacy preserving EM clustering, stopping criterion, privacy preserving EM clustering over horizontally partitioned data and privacy preserving EM clustering over vertically partitioned data. Related work on private data clustering is presented in Sect. 2. Finally Sect. 5 briefly concludes and propose the possible future directions.

2   Related Work

The problem of privacy preserving data clustering is generally addressed for the specific case of k-means clustering, either when data is distributed between two parties [4] or more than two parties [7]. In [14] a method for k-means clustering when data is vertically partitioned among multi parties has been addressed. The *secure permutation algorithm* has been utilized to keep the privacy of the distributed data. In the studies presented in [2,3], *agglomerative hierarchical clustering algorithm* is constructed when data is distributed between two (or more) parties, either horizontally or vertically. In [11,12], the problem of privacy preserving data clustering is addressed for a divisive hierarchical clustering algorithm when data is distributed between two and among several parties, respectively. Differently from above mentioned works, in present work, we studied the problem of secure *EM clustering* with the main focus on the estimation of the unknown parameters and find the maximum likelihood, either when data is distributed either *vertically* or *horizontally* among several parties.

The authors in [8] present a technique which uses EM algorithm to cluster the distributed data which is *horizontally* partitioned among several parties such that no individual data leaks. *Secure summation* [6] is used to preserve the privacy of *horizontally* partitioned data. However, we used *secure sum protocol* introduced in [10] which is simple and effective for preserving the individual data privacy, on both *vertically* and *horizontally* partitioned data among several parties [13].

3   Preliminary Notations

In this section, we present some background knowledge which are exploited in our proposed framework.

3.1   Expectation Maximization Clustering

Clustering is a data mining algorithm which groups similar samples into clusters so that objects within any cluster are similar or close [5]. In EM clustering, each cluster is represented by a probability distribution which typically is *Gaussian*. The parameters of the distribution, i.e. *mean* and *covariance*, are required to be estimated to compute the likelihood of a point belonging to a cluster. The process starts by choosing and placing the *Gaussians* randomly, then for each data point the probability of being in a specific cluster is computed. Unlike k-means clustering, in EM algorithm, the belongness of a point to a cluster is represented by a value between zero and one. Similar to k-means clustering, the process iterates until convergence, and in each iteration the parameters need to be recomputed in order to adjust better. Formally, suppose D be a dataset with n records $(x_1, \ldots, x_n)$ where each record is described by m attributes. In the beginning k clusters are arbitrary chosen with the Gaussian distribution function $f(x, \Psi)$ where Ψ is a vector of the unknown parameters $[\Psi_i = (\mu_i, \Sigma_i, \Pi_i)$ $\forall i = 1, \ldots, k]$. The likelihood of the distribution i is presented as the following:

$$f(x; \Psi_i) = \frac{1}{(2\pi)^{\frac{m}{2}} |\Sigma_i|^{\frac{1}{2}}} \exp(-\frac{(x - \mu_i)^T (x - \mu_i)}{2\Sigma_i}) \tag{1}$$

where Σ_i is the covariance matrix of the distribution, $|\Sigma_i|$ is the determinant of the covariance matrix, μ_i is the mean vector of the Gaussian distribution, and Π_i is the prior probability of the distribution i. The *Gaussian* mixture model over k clusters is presented as $f(x; \Psi) = \sum_{i=1}^{k} \Pi_i f(x; \Psi_i)$. It is often convenient to work with the logarithm of the likelihood function, i.e. $\log L(\Psi) = \log f(x; \Psi)$.

Given a dataset, EM algorithm is an iterative procedure to estimate the parameters Ψ which maximizes the logarithm of the likelihood $\log L(\Psi)$. Let j be the index for the data and i be the index for the clusters, the posterior probability $F(\Psi_i; x_j)$ denotes the j'th data point belongs to i'th cluster, which for simplicity we denote it as $Z_{ij} = \frac{\Pi_i^{t+1} f_i(x_j; \Psi_i^{t+1})}{\sum_{i=1}^{k} \Pi_i^{t+1} f_i(x_j; \Psi_i^{t+1})}$.

Thus, considering the whole dataset D the complete log likelihood would be:

$$\log L(\Psi) = \sum_{i=1}^{k} \sum_{j=1}^{n} Z_{ij} \log f(x_j; \Psi_i)$$

$$= -\frac{m}{2} \log(2\pi) - \frac{1}{2} \sum_{i=1}^{k} \sum_{j=1}^{n} Z_{ij} [\frac{(x_j - \mu_i)^T (x_j - \mu_i)}{\Sigma_i} + \log |\Sigma_i|] \tag{2}$$

3.2   Secure Sum Protocol

In the following, we present *secure sum protocol* proposed in [10] proven to be resistant against colluding. Assume that P parties $l_1, \ldots, l_P$ involve in a cooperative secure sum computation, where each party is able to break their private number into a fixed number of segments, such that the addition of segments is equal to their private number. In the proposed protocol the number of segments is equal to the number of parties (P). The values of each segment is randomly selected by the associated party and it is secret from other parties. Then, each party holds one segment of their data and sends $P-1$ other segments to the other $P-1$ parties. In this way, at the end each party holds P segments, where only one belongs to the party and the others are collected from remaining parties, one from each. Now, the *secure sum protocol* can be applied to obtain the sum of all the segments. According to this protocol, one of the parties is required to be selected as the protocol initiator party that starts the computation by sending the data segment to the next party in the ring. The receiving party adds their data segment and to the received partial sum and then sends the result to the next party in the ring. This process is repeated till all the segments of all the parties are added and the sum is announced by the initiator party. In this scenario, even if two adjacent parties maliciously cooperate in order to discover the data of middle party, they just get some segments of the real data. For the sake of simplicity, in the proposed algorithms, *secure sum protocol* will be represented as $\mathcal{SSP}$.

4   System Model

Suppose that several (more than two) data holders are interested in detecting the structure of clusters (through EM clustering algorithm) on the whole of their datasets. However, for privacy concerns, they are not willing to publish or share their main datasets. As mentioned before, it is assumed that clustering on the whole data (as in general cases) produces better results comparing to clustering on individual dataset. To this end, we propose algorithms, based on *secure computation protocols*, which verifies the criteria for constructing EM clustering. Thence, the structure of EM algorithm is shaped on the whole data, without revealing data to other parties. It is assumed that participated parties are *honest* (but curious) and they completely follow the protocols. Based on the framework defined in Sect. 3.1, a privacy preserving EM algorithm over *horizontally* and *vertically* partitioned data is presented.

4.1   Privacy Preserving EM Clustering

Suppose that data D is partitioned among $P\ (>2)$ parties. Assume parties want to cluster their joint dataset without revealing their private data except for the final result. In the first iteration each party P_l randomly initialize z_{ijl} to be 0 or 1, afterwards party P_l computes its parameter values on each iteration t as the following:

$$A_{il} = \sum_{j=1}^{n_l} Z_{ijl}^t x_j \tag{3}$$

$$B_{il} = \sum_{j=1}^{n_l} Z_{ijl}^t \tag{4}$$

$$C_{il} = \sum_{j=1}^{n_l} Z_{ijl}^t (x_j - \mu_i^{t+1})(x_j - \mu_i^{t+1})^T \tag{5}$$

With the use of local values in Formulas 3, 4, and 5, and *secure sum protocol*, the parameters *mean, covariance*, and *prior probability*, are respectively computed (securely) as the following:

$$\mu_i^{t+1} = \frac{\sum_{j=1}^n Z_{ij}^t x_j}{\sum_{j=1}^n Z_{ij}^t} = \frac{\sum_{l=1}^p A_{il}}{\sum_{l=1}^p B_{il}} \tag{6}$$

$$\Sigma_i^{t+1} = \frac{\sum_{j=1}^n Z_{ij}^t (x_j - \mu_i^{t+1})(x_j - \mu_i^{t+1})^T}{\sum_{j=1}^n Z_{ij}^t} = \frac{\sum_{l=1}^p C_{il}}{\sum_{l=1}^p B_{il}} \tag{7}$$

$$\Pi_i^{t+1} = \frac{\sum_{j=1}^n Z_{ij}^t}{n} = \frac{\sum_{l=1}^p B_{il}}{n} \tag{8}$$

At the end of each iteration, the participated parties need to update the Z_{ijl} as the following:

$$Z_{ijl} = \frac{\Pi_i^{t+1} f_i(x_{jl}; \mu_i^{t+1}, \Sigma_i^{t+1})}{\sum_{i=1}^k \Pi_i^{t+1} f_i(x_{jl}; \mu_i^{t+1}, \Sigma_i^{t+1})} \tag{9}$$

The process will be iterated until stopping criterion meets, which will be discussed in the next section.

4.2 Stopping Criterion

EM algorithm stops when $\log L(\Psi^{t+1}) - \log L(\Psi^t)$ is less than a preselected threshold. In distributed architecture, all parties require to set together this threshold, say ϵ. Then, at the end of each iteration, each party *locally* computes on her own dataset the amount of $\log L(\Psi^t) = \sum_{i=1}^k \sum_{j=1}^{n_l} [\log \Pi_i f(x_j; \Psi_i^t)]$. Through *secure sum protocol* all parties find *globally* whether $[\log L(\Psi^{t+1}) - \log L(\Psi^t)] \leq \epsilon$ or not. Once the stopping criterion is met at the iteration $(t+1)$, each party makes her own data clusters, meaning that $x_j \in$ cluster g $(1 \leq g \leq k)$ if $Z_{gj}^{t+1} = \max_{1 \leq i \leq k} Z_{ij}^{t+1}$ [8].

In what follows, we address the problem of secure EM clustering in two scenarios of data being distributed either *horizontally* or *vertically* among several parties in Sects. 4.3 and 4.4, respectively.

4.3   Privacy Preserving EM Clustering over Horizontally Partitioned Data

Let's assume that data is distributed *horizontally* among several (more than two) parties. This means that each data holder has information about all the features but for different collection of objects. The participated parties are interested in clustering their joint datasets without revealing their private data except for the final result.

Each party P_l owns n_l records described by m attributes. Suppose that each party P_l has $X_l = (x_{l1}, \dots, x_{ln_l})$ records, where each record is an m dimensional vector $x_j = (x_{j,1}, \dots, x_{j,m})$. The total number of data among P parties equals to $n = \sum_{l=1}^{P} n_l$ which can securely be obtained through *secure sum protcol*. With the use of Eqs. 3, 4 and 5, in each iteration the values of A, B and C are computed locally by each party, then the global values of μ_i^{t+1}, Σ_i^{t+1} and Π_i^{t+1} are securely obtained through *secure sum protocol*. When all the participated parties obtain the global parameters, they update Z_{ijl}^{t+1}, using Eq. 9 and repeat the process for the next iteration until stopping criterion meets. Algorithm 1 details the process of secure EM clustering over horizontally partitioned data.

Algorithm 1. Secure EM Algorithm Over Horizontally Partitioned Data

---

Data: Each party $P_l (1 \leq l \leq P)$ owns n_l records as $X_l = (x_{l1}, \dots, x_{ln_l})$; the preselected stop criterion threshold ϵ.

Result: All parties obtain securely EM clustering structure on whole data

1 *initialize*
2 Each party P_l randomly initialize Z_{ijl} as 0 or 1.
3 $n = \sum_{l=1}^{p} n_l$ is computed securely through *secure sum protocol*.
4 t=0
5 **while** $|\log L(\Psi^{t+1}) - \log L(\Psi^t)| \leq \epsilon$ **do**
6   **for** $i = 1, \dots, k$ **do**
7   Each party P_l calculates A_{il}^{t+1} (Eq. 3) and B_{il}^{t+1} (Eq. 4) locally.
8          $A_i^{t+1} \leftarrow SSP_{l=1}^{P}(A_{il}^{t+1})$
9          $B_i^{t+1} \leftarrow SSP_{l=1}^{P}(B_{il}^{t+1})$
10   The first party computes μ_i^{t+1} (Eq. 6) and sends it to the other parties.
11   Each party calculates C_{il}^{t+1} (Eq. 5).
12         $C_i^{t+1} \leftarrow SSP_{l=1}^{P}(C_{il}^{t+1})$
13   The first party computes Σ_i^{t+1} (Eq. 7), Π_i^{t+1} (Eq. 8) and sends the results to the other parties.
14   Each party updates Z_{ijl}^{t+1} (Eq. 9).
15   **end**
16     $t = t + 1$.
17 **end**

4.4 Privacy Preserving EM Clustering over Vertically Partitioned Data

Lets us consider that P parties $P_1, P_2, \ldots, P_P$ are interested in detecting the clusters on the whole of their datasets, when data is partitioned *vertically* among $P > 2$ parties. This means that each party holds the same set of objects, but described with different set of attributes. More precisely, let $\mathcal{A} = \{A_1, A_2, \ldots, A_m\}$ be the set of attributes used to describe each record of data, where P_1 has the description of data based on $\mathcal{A}^1 = \{A_1, \ldots, A_{m_1}\}$, P_2 owns the set $\mathcal{A}^2 = \{A_{m_1+1}, \ldots, A_{m_1+m_2}\}$, $\ldots$, and P_P has the information of attributes $\mathcal{A}^P = \{A_{m(P-1)+1}, \ldots, A_{m(P-1)+m_P}\}$ describing the same set of objects. This means that party P_l has some part of information about record x described on her own set of attributes, which we denote it as x_r (r varies from $m(l-1) + 1$ to $m(l-1) + m_l$ ranging on the number of attributes owned by party P_l). Now the participated parties are willing to cluster their joint datasets through EM algorithm without revealing their data. All the parties randomly initialize Z_{ij} to 0 or 1 in the same way, meaning if the first record held by the first party randomly initialize Z_{ij} to zero, all other parties should initialize Z_{ij} of their first record to zero. In this way, all the participated parties know about their initialization.

Algorithm 2. Secure EM Algorithm over Vertically Partitioned Data

---

Data: P parties; each party $P_l (1 \leq l \leq P)$ owns n records with m_l attributes, preselected stop criterion threshold ϵ.

Result: All parties obtain securely EM clustering structure on whole data

1 *initialize*
2 all the parties together decide to randomly initialize Z_{ij} to 0 or 1.
3 t=0
4 **while** $|\log L(\Psi^{t+1}) - \log L(\Psi^t)| \leq \epsilon$ **do**
5 $\quad$ **for** $i = 1, \ldots, k$ **do**
6 $\quad\quad$ $Z_i \leftarrow \sum_{j=1}^n Z_{ij}$.
7 $\quad\quad$ P_l : party l locally computes $A_{il}^{t+1} = \sum_{r=m(P-1)+1}^{r=m(P-1)+m_P} Z_i^t x_r$.
8 $\quad\quad$ $A_i^{t+1} \leftarrow SSP_{l=1}^P (A_{il}^{t+1})$.
9 $\quad\quad$ The first party computes $\mu_i^{t+1} = \frac{A_i^{t+1}}{Z_i^t}$ and sends it to other parties.
10 $\quad\quad$ P_l : party P_l locally computes
$\quad\quad$ $C_{il}^{t+1} = \sum_{r=m(l-1)+1}^{r=m(l-1)+m_l} Z_i^t (x_r - \mu_i^{t+1})(x_r - \mu_i^{t+1})^T$.
11 $\quad\quad$ $C_i^{t+1} \leftarrow SSP_{l=1}^P (C_{il}^{t+1})$.
12 $\quad\quad$ The first party calculates $\Sigma_i^{t+1} = \frac{C_i^{t+1}}{Z_i^t}$ and sends it to other parties.
13 $\quad\quad$ All parties update Z_i^{t+1} (Eq. 9).
14 $\quad$ **end**
15 $\quad$ $t = t + 1$.
16 **end**

Values of A and C can be computed locally by each party, then they will share these values using *secure sum protocol* in order to obtain the global parameter μ_i^{t+1} and Σ_i^{t+1}. Unlike *horizontally* partitioned data, in *vertically* partitioned data, the parameters Z_{ij} and Π_i are mutual among all the parties. When all parties obtain the global parameters, they will update Z_{ij}^{t+1}, and repeat the process for the next iteration until stopping criterion meets. Algorithm 2 details the process of secure EM clustering over vertically partitioned data.

5   Conclusion and Future Directions

In this work, a framework was proposed which can be exploited for more than two parties to construct EM clustering on whole of their data without revealing the original datasets. To this end, secure computation algorithms are proposed to obtain the required criteria for detecting the clusters on the whole data. Two scenarios of data being distributed either horizontally or vertically have been considered. While the method is simple and efficient, the privacy of the individual datasets have been preserved.

In the future works we would like to address the problem of secure construction of EM algorithm when data is distributed between two parties either *vertically* and *horizontally*. Moreover, we plan to calculate the communication cost among parties for constructing EM clustering securely.

Acknowledgment. This work was supported by the H2020 EU funded project C3ISP [GA #700294].

References

1. Berkhin, P.: A survey of clustering data mining techniques. In: Kogan, J., Nicholas, C., Teboulle, M. (eds.) Grouping Multidimensional Data, pp. 25–71. Springer, Heidelberg (2006)
2. Hamidi, M., Sheikhalishahi, M., Martinelli, F.: A secure distributed framework for agglomerative hierarchical clustering construction. In: Proceedings of 26th PDP 2018 Parallel, Distributed, and Network-Based Processing (2018)
3. Hamidi, M., Sheikhalishahi, M., Martinelli, F.: Secure two-party agglomerative hierarchical clustering construction. In: Proceedings of 4th ICISSP 2018 International Conference on Information Systems Security and Privacy (2018)
4. Jagannathan, G., Pillaipakkamnatt, K., Wright, R.N.: A new privacy-preserving distributed k-clustering algorithm. In: SDM, pp. 494–498. SIAM (2006)
5. Jagannathan, G., Wright, R.N.: Privacy-preserving distributed k-means clustering over arbitrarily partitioned data. In: Proceedings of the Eleventh ACM SIGKDD International Conference on Knowledge Discovery in Data Mining KDD 2005, pp. 593–599. ACM, New York (2005)
6. Benaloh, J.C.: Secret sharing homomorphisms: keeping shares of a secret. In: Proceedings of the 14th ACM Conference on Computer and Communications Security CCS 2007 (1987)
7. Jha, S., Kruger, L., McDaniel, P.: Privacy Preserving Clustering, pp. 397–417. Springer, Heidelberg (2005)

8. Lin, X., Clifton, C., Zhu, M.Y.: Privacy-preserving clustering with distributed EM mixture modeling. Knowl. Inf. Syst. **8**(1), 68–81 (2005)
9. Martinelli, F., Saracino, A., Sheikhalishahi, M.: Modeling privacy aware information sharing systems: a formal and general approach. In: Proceedings of 2016 IEEE Trustcom/BigDataSE/ISPA, Tianjin, China, pp. 767–774, 23–26 August 2016
10. Sheikh, R., Kumar, B., Mishra, D.K.: A distributed k-secure sum protocol for secure multi-party computations. CoRR abs/1003.4071 (2010)
11. Sheikhalishahi, M., Martinelli, F.: Privacy preserving clustering over horizontal and vertical partitioned data. In: Proceedings of 2017 IEEE Symposium on Computers and Communications, ISCC 2017, Heraklion, Greece, pp. 1237–1244, 3–6 July 2017
12. Sheikhalishahi, M., Martinelli, F.: Privacy preserving hierarchical clustering over multi-party data distribution. In: Proceedings of 10th International Conference on Security, Privacy, and Anonymity in Computation, Communication, and Storage SpaCCS 2017, pp. 530–544 (2017)
13. Sheikhalishahi, M., Martinelli, F.: Privacy-utility feature selection as a privacy mechanism in collaborative data classification. In: Proceedings of 26th IEEE International Conference on Enabling Technologies: Infrastructure for Collaborative Enterprises, WETICE 2017, pp. 244–249 (2017)
14. Vaidya, J., Clifton, C.: Privacy-preserving k-means clustering over vertically partitioned data. In: Proceedings of the Ninth ACM SIGKDD International Conference on Knowledge Discovery and Data Mining KDD 2003, pp. 206–215. ACM, New York (2003)

A Framework for Group Decision-Making: Including Cognitive and Affective Aspects in a MCDA Method for Alternatives Rejection

João Carneiro[1]($\boxtimes$), Luís Conceição[1], Diogo Martinho[1],
Goreti Marreiros[1], and Paulo Novais[2]

[1] GECAD – Research Group on Intelligent Engineering and Computing
for Advanced Innovation and Development, Institute of Engineering,
Polytechnic of Porto, Porto, Portugal
{jomrc,lmdsc,diepm,mgt}@isep.ipp.pt
[2] ALGORITMI Centre, University of Minho, Braga, Portugal
pjon@di.uminho.pt

Abstract. With the evolution of the organizations and technology, Group Decision Support Systems have changed to support decision-makers that cannot be together at the same place and time to make a decision. However, these systems must now be able to support the interaction between decision-makers and provide all the relevant information at the most adequate times. Failing to do so may compromise the success and the acceptance of the system. In this work it is proposed a framework for group decision using a Multiple Criteria Decision Analysis method capable of identify inconsistent assessments done by the decision-maker and identify alternatives that should be rejected by the group of decision-makers. The proposed framework allows to present more relevant information throughout the decision-making process and this way guide decision-makers in the achievement of more consensual and satisfactory decisions.

Keywords: MCDA · Consensus-based approach
Group decision support systems · Group decision-making
Cognitive decision-making

1 Introduction

Decision-making has always been a core process of any organization [1, 2]. Nowadays, most of the decisions taken inside organizations are made in group [1]. There are many advantages associated to group decision-making which allow better decisions to be made [3]. For instance, Dennis [4] stated some advantages, such as: to share workloads, to build social networks, to gain support among stakeholders, to train less experienced group members and most importantly to improve the quality of the decision. Other advantages include more knowledge being exchanged between decision-makers, better evaluation of the alternatives (compared to individual decision-making), increased acceptance of a decision and a better comprehension of the problem and the decision [5–7]. Group Decision Support Systems (GDSS) have been studied throughout the last

© Springer International Publishing AG, part of Springer Nature 2019
F. De La Prieta et al. (Eds.): DCAI 2018, AISC 800, pp. 264–275, 2019.
https://doi.org/10.1007/978-3-319-94649-8_32

decades with the objective of supporting decision-makers in group decision-making processes. With the appearance of global markets, the growth of multinational enterprises and a more global vision of the planet, we easily find chief executive officers and top managers (decision-makers) spread around the world, in different countries and with different time zones. Because of this, time and location pose as two major constraints to support group decision-making [5, 8].

To provide an answer and operate correctly in this type of scenarios, traditional GDSS have evolved to what we identify today as Web-based Group Decision Support Systems (Web-based GDSS). Web-based GDSS support the decision-making process by using main characteristics of ubiquity ("anytime" and "anywhere") [9, 10]. This evolution follows the evolution of general technology, as well as to the need of enhancing the efficiency of the group decision-making processes. There are some works in the literature that address the term of Web-based GDSS [9]. The Web-based GDSS may present different complexity levels. They can provide information about decision-maker preferences and other simple statistical information [11, 12]. The current challenge is to develop systems that can properly support the group decision-making process when decision-makers are dispersed [13–16]. For this, it is essential that each decision-maker can correctly define his preferences and intentions for each problem. After that, the system should support and guide that decision-maker throughout the entire decision-making while providing all the relevant information at the most adequate times. Otherwise, the decision-maker may never trust the information provided by system which is critical for its overall success. Therefore, it makes sense to think in multiple criteria decision analysis (MCDA) [17] which allows decision-makers to share information through problem configurations [18] and is appropriate to deal with both complex decision problems that involve the interaction of decision-makers with conflicting opinions and also deal with multiple qualitative and quantitative objectives [19, 20].

In this paper, we proceed with our ongoing research in the context of Web-based GDSS by examining one of the many aspects that can affect group decision-making which is the definition of negative preferences while modelling user preferences. We take advantage of the model proposed in [21] and instead of only suggesting the most adequate alternatives to each decision-maker based on each decision-maker criteria preferences' configurations we formulated a model that can suggest the rejection of alternatives whenever certain situations are verified. With this model it is possible to present more relevant information throughout the decision-making process and this way the system can guide decision-makers in the achievement of more consensual and satisfactory decisions.

The rest of the paper is organized as follows: Sect. 2 describes the proposed model as well as all the considered steps, and in Sect. 3 we present some conclusions and some guidelines regarding future work that we aim to carry on.

2 Method

As mentioned before, the framework proposed in this work is a variant of the Cognitive Analytic Process method which was previously presented in [21] and has been formulated in order to consider the possibility of suggesting alternatives to be rejected (eliminated) by decision-makers throughout the decision-making process. The model is divided in six main steps: the first step is the definition of the multi-criteria problem; the second step is the definition of the weights associated to each alternative and criterion; the third step is the adjustment of the weights associated with the alternatives to consider credibility, expertise and styles of behavior (these notions have been previously introduced in [22, 23]); the fourth step is the classification of each criterion and the classification of each alternative based on the new adjusted values; the fifth step is the selection of the alternatives that have the worst classification; the last step is the measurement of the consistency for each selected alternative.

2.1 Step 1: Multi-criteria Problem Definition

In the first step the multi-criteria problem is defined, which includes criterion, alternative and decision-maker definitions.

Definition 1: Let D be a decision matrix, $D = A * C$ where:

- C is a set of criteria $C = \{c_1, c_2, \ldots, c_n\}$, $n > 0$;
- A is a set of alternatives $A = \{a_1, a_2, \ldots, a_m\}$, $m > 0$.

Rule 1: $\forall a_i \in A, \forall c_j \in C, c_{j_{a_i}} \in D$, each alternative $a_i \in A$ is related with each criterion $c_j \in C$. There cannot be an existing alternative with values for criteria that are not considered in the problem.

Definition 2: A criterion $c_i = \{id_{c_i}, v_{c_i}, m_{c_i}\}$ consists of:

- $\forall c_i \in C, i \in \{1, 2, \ldots, n\}$;
- id_{c_i} is the identification of a particular criterion;
- v_{c_i} is the value of a particular criterion (Numeric, Boolean or Classificatory);
- m_{c_i} is the greatness associated with the criterion (Maximization or Minimization).

Definition 3: An alternative $a_i = \left\{id_{a_i}, \left[c_{1_{a_i}}, c_{2_{a_i}}, \ldots, c_{n_{a_i}}\right]\right\}$ consists of:

- $\forall a_i \in A, i \in \{1, 2, \ldots, m\}$;
- id_{a_i} is the identification of a particular alternative;
- $\left[c_{1_{a_i}}, c_{2_{a_i}}, \ldots, c_{n_{a_i}}\right]$ is the instantiation of each criterion.

Definition 4: Let D' be a normalized decision matrix such that: $\forall a_i \in A_{D'} \wedge \sum_{n=1}^{\langle C_{D'} \rangle} c_{n_{a_i}} = 1$.

Definition 5: Let DM be a set of decision-makers where $DM = \{dm_1, dm_2, \ldots, dm_k\}$ and $k \in \{1, 2, \ldots, n\}$.

2.2   Step 2: Definition of Alternatives and Criteria Weights

For the second step it will be defined the weight given by each decision-maker towards each alternative and each criterion. This will result in two preference matrices containing all the weights for each alternative and criterion.

Definition 6: Let $w_{dm_{ia_j}}$ be the weight or preference given to a certain alternative a_j by a decision-maker dm_i and $a_j \in A$.

Rule 2: A decision-maker dm_i can define a set of alternatives weights where:

- $W_{dm_i} = \left\{ w_{dm_{ia_1}}, w_{dm_{ia_2}}, \ldots, w_{dm_{ia_n}} \right\}, n > 0, \forall j \in \{1, 2, \ldots, n\}, 0 \le w_{dm_{ia_{nj}}} \le 1;$
- $\langle W_{dm_i} \rangle = \langle A \rangle.$

Definition 6.1: Let $wc_{dm_{ic_j}}$ be the weight or preference given to a certain criterion c_j by a decision-maker dm_i and $c_j \in C$.

Rule 2.1: A decision-maker dm_i can define a set of criteria weights where:

- $WC_{dm_i} = \left\{ wc_{dm_{ic_1}}, wc_{dm_{ic_2}}, \ldots, wc_{dm_{ic_n}} \right\}, n > 0, \forall j \in \{1, 2, \ldots, n\}, 0 \le wc_{dm_{ic_{nj}}} \le 1;$
- $\langle WC_{dm_i} \rangle = \langle C \rangle.$

Definition 7: Let W_{DM} be the set of alternatives weights of a set of decision-makers DM where: $W_{DM} = \left\{ W_{dm_1}, W_{dm_2}, \ldots, W_{dm_z} \right\}, z > 1.$

Definition 7.1: Let WC_{DM} be the set of alternatives weights of a set of decision-makers DM where: $WC_{DM} = \left\{ WC_{dm_1}, WC_{dm_2}, \ldots, WC_{dm_z} \right\}, z > 1.$

Definition 8: Let AP_{DM} be an alternatives preference matrix, where:

$$AP_{DM} = A \times W_{DM} = \begin{bmatrix} w_{dm_{1a_1}} & w_{dm_{2a_1}} & \cdots & w_{dm_{ia_1}} \\ w_{dm_{1a_2}} & w_{dm_{2a_2}} & \cdots & w_{dm_{ia_2}} \\ \vdots & \vdots & \cdots & \vdots \\ w_{dm_{1a_m}} & w_{dm_{2a_m}} & \cdots & w_{dm_{ia_m}} \end{bmatrix}$$

Definition 8.1: Let CP_{DM} be a criteria preference matrix, where:

$$CP_{DM} = C \times WC_{DM} = \begin{bmatrix} wc_{dm_{1c_1}} & wc_{dm_{2c_1}} & \cdots & wc_{dm_{ic_1}} \\ wc_{dm_{1c_2}} & wc_{dm_{2c_2}} & \cdots & wc_{dm_{ic_2}} \\ \vdots & \vdots & \cdots & \vdots \\ wc_{dm_{1c_m}} & wc_{dm_{2c_m}} & \cdots & wc_{dm_{ic_m}} \end{bmatrix}$$

2.3   Step 3: Adjustment of Alternatives Weights Using Credibility and Expertise of Decision-Makers

In this step, the weight given for each alternative is readjusted with the credibility, style of behavior and expertise values of each decision-maker [22, 23]. The weight given for each alternative is used as well as the $AP_{DMcredible}$ matrix that contains all the weights given for each alternative by the decision-makers that are credible.

Formula 1 correlates the style of behavior with the credibility using the values of concern for self and concern for others that have been selected by the decision-maker. Formula 2 readjusts the value obtained in Formula 1 according to the expertise level of the decision-maker. The expertise levels considered where defined in [23] and are Expert, High, Medium, Low and Null.

Definition 9: Let $DMcredible_{dm_i}$ be the set of decision-makers that decision-maker dm_i considers as credible.

We can now define Formula 1 which correlates the style of behavior with the credibility values:

$$\forall dm_i \in DM, \forall a_j \in A, w_{dm_{i_{a_j}}} = \frac{w_{dm_{i_{a_j}}} \times CS_{dm_i} + \left(\frac{TP}{ND}\right) \times CO_{dm_i}}{CS_{dm_i} + CO_{dm_i}} \tag{1}$$

Where:

- $w_{dm_{i_{a_j}}}$ is the weight given to the alternative a_j by decision-maker dm_i;
- CS_{dm_i} is the value of the Concern for Self of decision-maker dm_i chosen style of behavior;
- TP is the sum of the given weights to alternative a_j by each one of the credible decision-maker in $DMcredible_{dm_i}$;
- ND is the number of credible decision-makers such that $ND = \langle DMcredible_{dm_i}\rangle$;
- CO_{dm_i} is the value of the Concern for Others of decision-maker dm_i chosen style of behavior.

The weight of each alternative can now be readjusted with the decision-maker expertise level using Formula 2.

$$\forall dm_i \in DM, \forall a_j \in A, w_{dm_{i_{a_j}}} = \frac{w_{dm_{i_{a_j}}} \times e_{dm_i} + \left(\frac{TP}{ND}\right) \times e'_{dm_i}}{e_{dm_i} + e'_{dm_i}} \tag{2}$$

Where:

- $w_{dm_{i_{a_j}}}$ is the readjusted weight given to alternative a_j by decision-maker dm_i using formula 1;
- e_{dm_i} is the expertise level of decision-maker dm_i;

- *TP* is the sum of the given weights to alternative a_j by each one of the credible decision-makers in $DMcredible_{dm_i}$;
- *ND* is the number of credible decision-makers such that $ND = \langle DMcredible_{dm_i} \rangle$;
- e'_{dm_i} is the inverse of expertise level of decision-maker dm_i.

2.4 Step 4: Alternatives Classification

After the weight of each alternative has been readjusted using both Formula 1 and Formula 2 we can now classify both alternatives and criterions. For this we first define a F_{Dif} function which return the difference between the maximum and minimum weights found in a W weight set.

Definition 10: Let F_{Dif} be a function that returns the difference between maximum and minimum weights given to the alternatives or criteria that belong to a set of W weights.

$$F_{Dif} : W = \begin{cases} max(W) - min(W), & if\ max(W) \neq min(W) \\ max(W) \end{cases}$$

This means that $F_{Dif}(W_{DM})$ returns the difference between the alternative with the greatest weight and the alternative with the lowest weight, while for $F_{Dif}(WC_{DM})$ returns the difference between the criterion with the greatest weight and the criterion with the lowest weight. The result of F_{Dif} can now be classified in five different levels according Table 1.

Table 1. F_{Dif} levels

Level (*l*)	F_{Dif}
5	$\geq 0, 80$
4	$\geq 0, 60$
3	$\geq 0, 40$
2	$\geq 0, 20$
1	$< 0, 20$

Since each criteria or alternative weighting is done in a scale of [0, 1] the minimum difference between two criteria or alternatives is less than 0.2 and the maximum difference is greater than 0.8. Measuring the difference (using function F_{Dif}) between the criterion or alternative with more weight and the criterion or alternative with less weight we can obtain (according to Table 1) the *l* value.

After identifying the *l* value we can then perform Algorithm 1 to measure the classification done for each criterion $imp_{c_{j_{dm_i}}}$ or alternative $imp_{a_{j_{dm_i}}}$.

```
foreach cⱼ ∈ C
    k ← 1
```

$$\text{while} \left(wc_{dm_{i_{c_j}}} \leq \max(WC_{dm_i}) - k \times \frac{F_{Dif}(WC_{dm_i})}{l} \text{ AND } k \leq 5 \right)$$

```
    k ← k + 1
    endWhile
```

$$imp_{c_{j_{dm_i}}} \leftarrow 6 - k$$

```
endfor
```

Algorithm 1. Criterion importance classification algorithm

To classify the importance of each alternative Algorithm 1 is also performed with the only difference being that the algorithm is applied for each $a_j \in A$ and the weights considered belong to the set W_{DM}. After $imp_{c_{j_{dm_i}}}$ and $imp_{a_{j_{dm_i}}}$ have been identified for each criterion and alternative we can now classify them according the assigned value as can be seen in Table 2.

Table 2. Importance classification

Value	imp	Definition
5	VI	Very Important
4	I	Important
3	M	Medium
2	NI	Not Important
1	IN	Insignificant

2.5 Step 5: Alternatives Selection

In the fifth step it will be identified the alternatives with the worst classification among all the classifications for each decision-maker. For this a set of $AIMP_{dm_i}$ must first be defined containing all the importance values for each alternative for the decision-maker dm_i.

Definition 11: Let $AIMP_{dm_i}$ be the set of importance values of all the alternatives for the decision maker dm_i, where:

- $AIMP_{dm_i} = \left\{ imp_{a_{1_{dm_i}}}, imp_{a_{2_{dm_i}}}, \ldots, imp_{a_{j_{dm_i}}} \right\}, j > 0;$
- $\langle AIMP_{dm_i} \rangle = \langle A \rangle.$

Definition 11.1: Let $CIMP_{dm_i}$ be the set of importance values of all the criteria for the decision maker dm_i, where:

- $CIMP_{dm_i} = \left\{ imp_{c_{1_{dm_i}}}, imp_{c_{2_{dm_i}}}, \ldots, imp_{c_{j_{dm_i}}} \right\}, j > 0;$
- $\langle CIMP_{dm_i} \rangle = \langle C \rangle.$

We can now define the matrix AE_{DM} which contains all the alternatives importance for all decision-makers.

Definition 12: Let AE_{DM} be an alternatives evaluation matrix, where:

$$AE_{DM} = A \times AIMP_{DM} = \begin{bmatrix} imp_{a_{1_{dm_1}}} & imp_{a_{1_{dm_2}}} & \cdots & imp_{a_{1_{dm_i}}} \\ imp_{a_{2_{dm_1}}} & imp_{a_{2_{dm_2}}} & \cdots & imp_{a_{2_{dm_i}}} \\ \vdots & \vdots & \cdots & \vdots \\ imp_{a_{j_{dm_1}}} & imp_{a_{j_{dm_2}}} & \cdots & imp_{a_{j_{dm_i}}} \end{bmatrix}$$

```
value ← 1
while (flag==false && value ≤ 5) do
   flag ← true
   foreach (aⱼ ∈ A)
      foreach (dmᵢ ∈ DM)
         if (AE_DM[aⱼ, dmᵢ] > value) then flag ← false
      endfor
      if(flag == true) then insert aⱼ into selectedAlts
   endfor
   value ← value + 1
end while
```

Algorithm 2. Alternative selection algorithm

After the evaluation matrix AE_{DM} has been defined we can now perform Algorithm 2 which selects the alternatives with the worst classification *selectedAlts* based on its importance value. Algorithm 2 iterates through the entire AE_{DM} matrix and first searches through all the alternatives importance values and attempts to find at least an alternative whose importance has the lowest possible value (which corresponds to Insignificant classification) for all the decision-makers. If an alternative is not found, the algorithm reiterates again with an increased value and the process is repeated until at least an alternative has been found. This means that in the worst case scenario all the alternatives would have been classified as Very Important by at least one of decision-makers. In this case all the alternatives will be selected for Step 6.

2.6   Step 6: Alternatives Weighting Consistency

In the last step of the proposed model the worst alternative will be identified among the alternatives that were selected in the previous step (*selectedAlts*).

To identify the worst alternative, it is necessary to measure the consistency between the weight given to the alternative by each decision-maker and their preference for each criterion. For this we use the normalized values from the D' matrix and we classify the instantiation of each criterion for each alternative using the same process applied to

Step 4. As a result, we will have a set $CIMP_{a_j}$ which contains the importance of each criterion for the alternative a_i and is defined as follows:

Definition 13: Let $CIMP_{a_j}$ be the set of importance values of all the criteria for the alternative a_i, where:

- $CIMP_{a_j} = \left\{ imp_{c_{1_{a_j}}}, imp_{c_{1_{a_j}}}, \ldots, imp_{c_{i_{a_j}}} \right\}, j > 0;$
- $\langle CIMP_{a_j} \rangle = \langle C \rangle.$

After this, we define a function $F_{Consistency}$ which returns the difference between the importance of a criterion to a decision-maker dm_i and the importance value of the same criterion for an alternative a_j.

Definition 14: Let F_{Cons} be a function that returns the difference between the importance of a criterion c_i to a decision-maker dm_k and the importance value of the same criterion for an alternative a_j.

$$F_{Cons} : dm_k, c_i, a_j = \begin{cases} imp_{c_{i_{a_j}}} - imp_{c_{i_{dm_k}}}, & \text{if } imp_{c_{i_{dm_k}}} \geq 4 \\ 0 \end{cases}$$

It is important to note that F_{Cons} only returns the difference of criteria that are weighted as Very Important or Important by the decision-maker (value must be 4 or 5). This is done to make sure that the evaluation done to each alternative by the decision-maker is consistent with the criteria that decision-maker considers to be important. We can then define the consistency matrix CM_{dm_k} of the decision-maker dm_k as:

Definition 15: Let CM_{dm_k} be a consistency matrix of the decision-maker dm_k, where:

$$CM_{dm_k} = \begin{bmatrix} F_{Cons}(dm_k, c_1, a_1) & F_{Cons}(dm_k, c_2, a_1) & \cdots & F_{Cons}(dm_k, c_i, a_1) \\ F_{Cons}(dm_k, c_1, a_2) & F_{Cons}(dm_k, c_2, a_2) & \cdots & F_{Cons}(dm_k, c_i, a_2) \\ \vdots & \vdots & \cdots & \vdots \\ F_{Cons}(dm_k, c_1, a_j) & F_{Cons}(dm_k, c_2, a_j) & \cdots & F_{Cons}(dm_k, c_i, a_j) \end{bmatrix}$$

Finally, we define a function $F_{TotalConsistency}$ which returns the sum of all consistency values for an alternative a_j of the decision-maker dm_k.

Definition 16: Let $F_{TotalConsistency}$ be a function which returns the sum of all consistency values for an alternative a_j of the decision-maker dm_k.

$$F_{TotalConsistency} : dm_k, a_j = \sum_{i=1}^{\langle C \rangle} F_{Con}(dm_k, c_i, a_j)$$

The average consistency value acv between all $F_{TotalConsistency}$ for each decision-maker is then measured according to Formula 3:

$$\forall a_j \in A, acv = \frac{\sum_{k=1}^{\langle DM \rangle} F_{TotalConsistency}(dm_k, a_j)}{DM} \qquad (3)$$

Where:

- *DM* is the list of all decision-makers.

After the average consistency value is measured for each alternative in *selectedAlts*, the alternative with the lowest average consistency value is then suggested to all decision-makers as an alternative that should be rejected.

3 Conclusions and Future Work

As organizations and technology evolve, GDSS have also changed to be able to support decision-makers in a context where trying to establish a face-to-face meeting to deal with a problem has become a rather difficult task. However, the new Web-based GDSS have not been well accepted by organizations and we know from our previous works that the interaction between decision-makers and the system is an essential point regarding the acceptance of systems by users and the organizations themselves.

The model here proposed is a variant of a work that has been previously proposed and we define how we can measure the consistency between the assessment done by the decision-maker towards each alternative and the evaluation that was expected according to the assessment the decision-maker provided towards each criterion. This measure is essential to help us understand the impact that subjectivity has in the assessment done by the decision-maker. By identifying inconsistent alternatives assessment according to the criteria that a decision-maker considers to be important or very important, it will be possible to inform the decision-maker that the assessment provided was not the correct one and that there may be situations where it makes sense to reject a certain alternative. Likewise, there may be situations where the opposite situation is verified which means that a user might have provided a lower assessment than the expected one and his preference for a certain alternative should be increased.

Another important remark of this work is the inclusion of factors such as the credibility, expertise and styles of behavior which are part of the interactions that happen in real situations and that should be considered to better support the decision-making process.

Finally, but not less important, all the information that is measured using this model, such as the identification of very important alternatives or criteria or the identification of inconsistent and consistent evaluations can be easily provided to the decision-maker at any moment of the decision-making process. This will guide him and the group to obtain more consensual and satisfactory decisions. Furthermore, the system will also be ready to suggest the best decision at any moment of the decision-making process.

As future work, the first step is to evaluate the proposed model using a real case study in order to understand the impact that inconsistencies and that the rejection of alternative have in the achievement of more consensual and satisfactory decisions (compared to models that do not consider these concepts). Another point worth of studying is to consider the possibility of suggesting the reevaluation of preferences whose assessment should have been higher by decision-maker and its impact in the consensus and satisfaction that can be obtained. We also intend to study and include the concept of restrictions which the decision-maker can define throughout the decision-making process and that will let him accept or reject a certain alternative.

Acknowledgments. This work was supported by NIS Project (ANI|P2020 21958) and has received funding from FEDER Funds through P2020 program and from National Funds through FCT - Fundação para a Ciência e a Tecnologia (Portuguese Foundation for Science and Technology) under the projects UID/EEA/00760/2013 and UID/CEC/00319/2013.

References

1. Lunenburg, F.C.: Decision making in organizations. Int. J. Manage. Bus. Adm. **15**, 1–9 (2011)
2. Wilson, D.C.: Decision-making in organizations. In: Managing Organizations: Current Issues, vol. 43 (1999)
3. Luthans, F.: Organizational Behavior, vol. 46, p. 594. McGraw-Hill, Irwin (2011)
4. Dennis, A.R.: Information exchange and use in group decision making: you can lead a group to information, but you can't make it think. MIS Q. **20**, 433–457 (1996)
5. Lunenburg, F.C.: Group decision making. Nat. Forum Teach. Educ. J. **20**(3), 1–7 (2010)
6. Proctor, T.: Creative Problem Solving: Developing Skills for Decision Making and Innovation. Routledge, New York, NY (2011)
7. Gunnarsson, M.: Group Decision Making. Verlag, Frederick (2010)
8. Huber, G.P.: Issues in the design of group decision support systems. MIS Q. Manage. Inf. Syst. **8**, 195–204 (1984)
9. Kwon, O., Yoo, K., Suh, E.: UbiDSS: a proactive intelligent decision support system as an expert system deploying ubiquitous computing technologies. Expert Syst. Appl. **28**, 149–161 (2005)
10. Morente-Molinera, J.A., Wikström, R., Herrera-Viedma, E., Carlsson, C.: A linguistic mobile decision support system based on fuzzy ontology to facilitate knowledge mobilization. Decis. Support Syst. **81**, 66–75 (2016)
11. Jay F Nunamaker, J.R.: Future research in group support systems: needs, some questions and possible directions. Int. J. Hum. Comput. Stud. **47**, 357–385 (1997)
12. Shim, J.P., Warkentin, M., Courtney, J.F., Power, D.J., Sharda, R., Carlsson, C.: Past, present, and future of decision support technology. Decis. Support Syst. **33**, 111–126 (2002)
13. Carneiro, J., Martinho, D., Marreiros, G., Novais, P.: Intelligent negotiation model for ubiquitous group decision scenarios. Front. Inf. Technol. Electron. Eng. **17**, 296–308 (2016)
14. Sanchez-Anguix, V., Julian, V., Botti, V., García-Fornes, A.: Tasks for agent-based negotiation teams: analysis, review, and challenges. Eng. Appl. Artif. Intell. **26**, 2480–2494 (2013)
15. Sánchez-Anguix, V., Botti, V., Julián, V., García-Fornes, A.: Analyzing intra-team strategies for agent-based negotiation teams. In: The 10th International Conference on Autonomous Agents and Multiagent Systems, vol. 3, pp. 929–936 (2011). International Foundation for Autonomous Agents and Multiagent Systems
16. Carneiro, J., Martinho, D., Marreiros, G., Jimenez, A., Novais, P.: Dynamic argumentation in UbiGDSS. Knowl. Inf. Syst. **55**(3), 633–669 (2018)

17. Greco, S., Figueira, J., Ehrgott, M.: Multiple Criteria Decision Analysis. Springer, Berlin (2005)
18. Dehe, B., Bamford, D.: Development, test and comparison of two multiple criteria decision analysis (MCDA) models: a case of healthcare infrastructure location. Expert Syst. Appl. **42**, 6717–6727 (2015)
19. Ram, C., Montibeller, G., Morton, A.: Extending the use of scenario planning and MCDA for the evaluation of strategic options. J. Oper. Res. Soc. **62**, 817–829 (2011)
20. Golmohammadi, D., Mellat-Parast, M.: Developing a grey-based decision-making model for supplier selection. Int. J. Prod. Econ. **137**, 191–200 (2012)
21. Carneiro, J., Conceição, L., Martinho, D., Marreiros, G., Novais, P.: Including cognitive aspects in multiple criteria decision analysis. Ann. Oper. Res. **265**(2), 269–291 (2018)
22. Carneiro, J., Saraiva, P., Martinho, D., Marreiros, G., Novais, P.: Representing decision-makers using styles of behavior: an approach designed for group decision support systems. In: Cognitive Systems Research (2017)
23. Carneiro, J., Martinho, D., Marreiros, G., Novais, P.: Including credibility and expertise in group decision-making process: an approach designed for UbiGDSS. In: World Conference on Information Systems and Technologies, pp. 416–425. Springer, Cham (2017)

Domain Identification
Through Sentiment Analysis

Ricardo Martins$^{(\boxtimes)}$, José João Almeida, Pedro Henriques, and Paulo Novais

Algoritmi Centre, University of Minho, Braga, Portugal
ricardo.martins@algoritmi.uminho.pt, {jj,prh,pjon}@di.uminho.pt

Abstract. When dealing with chatbots, domain identification is an important feature to adapt the interactions between user and computer in order to increase the reliability of the communication and, consequently, the audience and decrease its rejection avoiding misunderstandings.

In order to adapt to different domains, the writing style will be different for the same author. For example, the same person in the role of a student writes to his professor in a different style than he does for his brother.

This article presents a process that uses sentiment analysis to identify the average emotional profile of the communication scenario where the conversation is done. Using Natural Language Processing and Machine Learning techniques, it was possible to obtain an index of 96.21% of correct classifications in the identification of where these communications have occurred only analysing the emotional profile of these texts.

Keywords: Emotional profile · Sentiment analysis
Machine learning · Natural processing language

1 Introduction

Along the day, a person must represent different roles: worker, father, student, boss, ... and for each role he must interact to other people according to the place they are and the reaction or feedback he receives from others. In some cases it is necessary to be more "politically correct" during the speeches, and in other cases not so much.

According to Collins dictionary [3], "if you say that someone is politically correct, you mean that they are extremely careful not to offend or upset any group of people in society who have a disadvantage, or who have been treated differently because of their sex, race, or disability." When talking to people, in a daily interaction or via chatbots, the idea is to decrease the chances of being rejected by the target audience, increasing the chances of get his speech accepted. So, identifying the audience's communication profile is the first step to provide a better experience between users and chatbots. Knowing where the conversation takes place is essential to avoid misunderstandings. For example, like in the real life, it is unacceptable to talk to professors in classroom like we talk to best

© Springer International Publishing AG, part of Springer Nature 2019
F. De La Prieta et al. (Eds.): DCAI 2018, AISC 800, pp. 276–283, 2019.
https://doi.org/10.1007/978-3-319-94649-8_33

friends during a party. So, it is not acceptable that a chatbot responses to a user different than the pattern from where the conversation takes place.

The purpose of this article is to present a classifier based in sentiment analysis which compares speeches from public and common person in social media as LinkedIn and Twitter in order to identify the emotional communication profile for each social media and predict the domain where the conversation is being held. For "domain", we consider as the social media where the text was originally posted.

The remainder of this paper is as follows: Sect. 2, introduces the concept of emotion and presents the basic emotions theory for emotion representation and analysis. Section 3 presents some work in this area to detect emotion from social media, while Sect. 4, describes the steps followed in our emotional analysis in chatbot messages and discusses some results obtained, and finally, the paper ends in Sect. 5 with the conclusion and future work.

2   Basic Emotions

Basic emotion theorists agree that all human emotion can be contained within a small set of basic emotions, which are discrete.

Many researchers have attempted to identify a number of universal basic emotions which are common for all people and differ one from another in important ways. A popular example is a cross-cultural study of 1972 by Paul Ekman [4] and his colleagues, in which they concluded that the six basic emotions are *anger, disgust, fear, happiness, sadness,* and *surprise.*

A major part of work in emotion mining and classification from text has adopted this basic emotion set. For example, in order to model public mood and emotion, Bollen et al. [2] extracted six dimensions of mood including *tension, depression, anger, vigour, fatigue, confusion* from Twitter. Strapparava and Mihalcea [11] created a large data set with six basic emotions: *anger, disgust, fear, joy, sadness* and *surprise.* For Plutchik [9], all sentiment is composed of a set of 8 basic emotions: *anger, anticipation, disgust, fear, joy, sadness, surprise* and *trust.*

However, there is no consensus on which human emotions should be categorized as basic and be included in the basic emotion set. Moreover, the emotional disambiguation is a contested issue in emotion research. For instance, it is unclear if *surprise* should be considered an emotion since it can assume negative, neutral or positive valence.

In our tests, it was used the Plutchik model because is a well-known model implemented in some libraries and toolkits for sentiment analysis used in this work.

3   Related Work

Despite the vast amount of works using sentiment analysis, none of them considers the messages emotional profile as a dimension of the communication profile. So, each work cited below has inspired partially our work as will be mentioned.

Analysing emotions in social media was suggested by Schwartz et al. [10], whose work predicts the individual well-being, as measured by a life satisfaction scale, through the language people used on social media communication channels. This is made using randomly selected posts from Facebook and a lexicon-based approach to identify the text words polarities.

Other work who inspired our analysis was introduced by Baldoni et al. [1] who has presented a project involving lexicons and ontologies to extract emotions including sadness, happiness, surprise, fear and anger, which contributed in the emotional profile creation.

The work of Widmer [12] contributed with the idea of domain identification using machine learning techniques.

4   Data Analysis

In our analysis, all data was collected from same author's public posts and texts in LinkedIn[1] and Twitter[2]. The choice of these social media is because while the audience of LinkedIn is more professional and aimed at laboral relationships - and for these reasons more politically correct - Twitter has a different audience profile, aimed at casual relationships, i.e., people on Twitter tend to expose their opinions with more freedom.

The authors, as presented in Table 1, were selected randomly according to the following criteria:

- Must have LinkedIn and Twitter profiles;
- Must have at least 10 opinion texts published in LinkedIn pulse;
- Must have at least 500 posts in Tweeter.

So, having in mind these requirements and after a search at LinkedIn and Tweeter profiles, the authors mentioned in Table 1 were chosen to provide the texts for analysis.

Table 1. Authors analysed

Author	Profession
C. Fairchild	News editor
J. Saper	Investor
J. Battelle[a]	Entrepreneur
B. McGovan	Media Trainer
A. Mitchell	Professor
L. Profeta	Medical Doctor

[a]LinkkedIn influencer

[1] http://www.linkedin.com.
[2] http://www.twitter.com.

4.1   Data Preprocessing

Preprocessing is a data mining technique that transforms raw data into an understandable form. There are in the literature several preprocessing techniques available to extract information from text, and their usage is according to the characteristics of the information desired.

In order to analyse the emotion contained into the text, all texts have been preprocessed according to the planned pipeline described in Figure 1, where each process is denoted by an acronym as follow:

1. TK - Tokenization;
2. POS-T - Part of Speech Tagging;
3. NER - Name Entity Recognition;
4. SWR - Stopwords Removal.

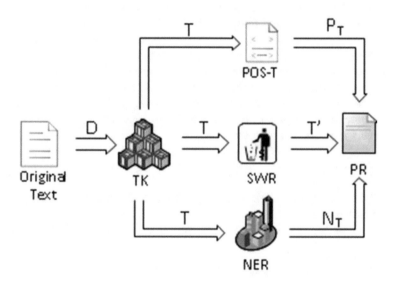

Fig. 1. Preprocessing tasks

POS-T process identifies the text grammatical structure and tags all nouns, verbs, adverbs and adjectives removing the remaining words. The reason for this text cleaning is because only these grammatical categories can bring emotional information. In a formal description, the TK process converts the original text D in a set of tokens $T = \{t_1, t_2, ..., t_n\}$ where each element contained in T is part of the original document D. Later, the POS-T labels each token with a semantic information and creates a set P, where $P_T = \{p_{(T,1)}, p_{(T,2)}, ..., p_{(T,k)}\}$ and $0 \leq k \leq n$ and $P_T \subset T$, and P is-a noun, verb, adverb, adjective.

NER process separates (nouns) names in 3 different categories: "Location", "Person" and "Organization" and removes all tokens related with these categories. As result, a set $N_T = \{n_{(T,1)}, n_{(T,2)}, ..., n_{(T,j)}\}$ is constructed based on identified word category and where $0 \leq j \leq n$ and $N_T \subset T$. This step is important to be done in parallel with POS-T because some locations can be confused with some grammatical structure (as Long Beach or Crystal Lake, for instance).

SWR process is responsible of the removal of stopwords (undesirable words in the text) from the tokens. The stopwords gathered in a predefined set $SW = \{sw_1, sw_2, ...sw_y\}$ of words, available in R through the package **tm** [5] and the SWR process result is a set $T' = T - SW$.

After the 3 preprocessing tasks finish, the result document PR must contain a set of words where $PR = T' \cap P_T \cap N_T$.

For all three tasks - POS-T, NER and TK - the Stanford Core NLP [7] toolkit was used.

4.2 Emotional Analysis

After the preprocessing, all texts were analysed against EmoLex lexicon [8] in order to detect the amount of each basic emotion for each phrase according the Plutchik model. For this step, it was used the package Syuzhet [6] available in R.

The texts' emotional values returned by the Syuzhet package is a scalar value, so it was necessary to convert them in a scale ranging from 0 to 1. This is important for determining the percentage of each basic emotion that is commonly used by the author in his texts. For this conversion, it was considered the ratio of each basic emotion in the emotion total amount. For example, the ratio (P) of the emotion (E) *fear* is calculated as:

$$P_{fear} = \frac{E_{fear}}{E_{anger} + E_{anticipation} + E_{disgust} + E_{fear} + E_{joy} + E_{sadness} + E_{surprise} + E_{trust}}$$

Finally, the average of the percentage of each basic emotion was calculated, as shown in Table 2.

These values represent the emotional profile of each author in each platform.

Table 2. Sentiment analysis from LinkedIn and Tweeter per author

Author	Source	Anger	Anticipation	Disgust	Fear	Joy	Sadness	Surprise	Trust
C. Fairchild	LinkedIn	9%	18%	5%	11%	13%	10%	7%	27%
	Twitter	7%	17%	6%	14%	12%	12%	8%	23%
J. Saper	LinkedIn	7%	20%	3%	10%	14%	8%	8%	32%
	Twitter	5%	22%	4%	7%	20%	5%	13%	24%
J. Battelle	LinkedIn	12%	17%	8%	14%	11%	10%	5%	23%
	Twitter	11%	17%	6%	12%	13%	9%	10%	22%
B. McGowan	LinkedIn	11%	18%	6%	15%	10%	11%	8%	21%
	Twitter	9%	18%	6%	11%	16%	10%	9%	20%
A. Mitchell	LinkedIn	6%	15%	3%	8%	9%	6%	5%	48%
	Twitter	3%	21%	2%	6%	17%	4%	10%	38%
L. Profeta	LinkedIn	9%	16%	8%	13%	14%	13%	8%	20%
	Twitter	9%	18%	2%	5%	25%	6%	7%	28%

4.3 Looking Closer

The first step to analyse these results aims at determine the correlation between emotions from LinkedIn and Twitter, in order to identify differences between LinkedIn emotional profile and Twitter emotional profile. These results are presented in Table 3.

Table 3. Correlation between LinkedIn emotional profile and Twitter emotional profile

Author	r^2
C. Fairchild	0.96
J. Saper	0.86
J. Battelle	0.92
B. McGowan	0.81
A. Mitchell	0.93
L. Profeta	0.79

Considering that all LinkedIn messages are politically correct, when analysing the correlations between platforms, it evidences the proximity of emotional profile between LinkedIn and Tweeter of the authors involved with communication (C. Fairchild, J. Battelle and A. Mitchell) while for the other 3 authors the correlation is not so strong, allowing to distinguish the domain under which the author is writing.

In a second step, a new analysis was made concerned with emotional profile between authors in order to identify which authors are emotionally close to others according to the social media. It is expected that the proximity remains the same in different social media. For this objective, the emotional profile of all authors was correlated with each other for both social media. Table 4 shows the results of these correlations; the strongest are overlined and the weakest are underlined.

As we delve deeply into the analysis of correlations between authors, it is possible to highlight that in general, the correlations values are lower on Twitter when compared to LinkedIn, indicating a greater emotional distance between the authors on Twitter, reinforcing the idea of a common emotional profile in LinkedIn like a "common mask for everyone" and a "emotional freedom" in Tweeter.

4.4 Machine Learning Analysis

In order to achieve the objective of identifying the domain where the text was written, it was used an approach using machine learning for classification based on emotions contained in the text as model's dimensions. For this purpose, it was created a new dataset based on the preprocessed information from the emotional

Table 4. Correlation between authors according social media

		C. Fairchild	J. Saper	J. Battelle	B. McGowan	A. Mitchell	L. Profeta
LinkedIn	C. Fairchild	1.00	**0.99**	0.92	**0.91**	0.95	0.95
	J. Saper	**0.99**	1.00	0.90	**0.89**	0.94	0.93
	J. Battelle	0.92	0.90	1.00	**0.95**	**0.86**	0.92
	B. McGowan	0.91	0.89	**0.95**	1.00	**0.81**	0.92
	A. Mitchell	**0.95**	0.94	0.86	0.81	1.00	**0.84**
	L. Profeta	**0.95**	0.93	0.92	0.92	**0.84**	1.00
		C. Fairchild	J. Saper	J. Battelle	B. McGowan	A. Mitchell	L. Profeta
Twitter	C. Fairchild	1.00	0.76	**0.92**	0.90	0.87	**0.73**
	J. Saper	**0.76**	1.00	0.85	**0.94**	0.92	0.92
	J. Battelle	0.92	**0.85**	1.00	0.94	**0.95**	0.86
	B. McGowan	**0.90**	**0.94**	0.94	1.00	0.93	0.94
	A. Mitchell	**0.87**	0.92	**0.95**	0.93	1.00	0.89
	L. Profeta	**0.73**	0.92	0.86	**0.94**	0.89	1.00

analysis. Each line of this dataset contains 11 dimensions, referring to the eight basic emotions according to the Plutchik [9] model, the polarities (positive and negative) and the source of information (social media name) regarding to the preprocessed text.

Later, this dataset was loaded and tested using several different algorithms in order to classify the social media according the emotions.

In our tests, the best classification score was obtained using a Random Forest algorithm, using a 10 fold cross-validation for training and testing, which achieved an weighted average of 96.21% of correct classified instances, which considers number of instances of each class for the weights, as presented in Table 5.

Table 5. Data classification results

TP Rate[a]	FP Rate[b]	Precision	Recall	F-Measure	Class
0.870	0.011	0.959	0.870	0.913	LinkedIn
0.989	0.130	0.963	0.989	0.976	Twitter
0.962	**0.103**	**0.962**	**0.962**	**0.962**	**Weighted Avg.**

[b]True Positive Rate
[c]False Positive Rate

5   Conclusion

This paper presents a combination of lexicon-based and machine learning approaches to explore the emotions contained in a text through practices in sentiment analysis in order to detect the emotional profile and predict the conversation environment/domain.

Based on the analysis presented, it is possible to claim that there is an emotional profile according to each domain (in this case Twitter and LinkedIn). If this emotional profile is known, it is possible to adapt the discourse according to the audience, so that there is an emotional levelling of the author's discourse to their audience. This means that it is relevant to identify that context, or discourse environment, from the analysis of the communicator emotional writing style. So systems like chatbots can adapt their emotional profile according to the interactions received from people or even other systems, interacting with the user - at least emotionally - like a human.

As future work, it is planned to expand this analysis to include the emotional intensity profile, by combining with other text analysis metrics, in order to increase the emotional profile identification.

Acknowledgements. This work has been supported by COMPETE: POCI-01-0145-FEDER-0070 43 and FCT - Fundação para a Ciência e Tecnologia within the Project Scope UID/CEC/ 00319/2013.

References

1. Baldoni, M., Baroglio, C., Patti, V., Rena, P.: From tags to emotions: ontology-driven sentiment analysis in the social semantic web. Intelligenza Artificiale **6**(1), 41–54 (2012)
2. Bollen, J., Mao, H., Pepe, A.: Modeling public mood and emotion: Twitter sentiment and socio-economic phenomena. ICWSM **11**, 450–453 (2011)
3. COBUILD Advanced English Dictionary. Politically correct definition — Collings English Dictionary
4. Ekman, P.: Basic emotions. Handb. Cogn. Emot. **98**, 45–60 (1999)
5. Feinerer, I., Hornik, K., Meyer, D.: Text mining infrastructure in R. J. Stat. Softw. **25**(5), 1–54 (2008)
6. Jockers, M.L.: Extract Sentiment and Plot Arcs from Text, Syuzhet (2015)
7. Manning, C.D., Surdeanu, M., Bauer, J., Finkel, J., Bethard, S.J., McClosky, D.: The stanford coreNLP natural language processing toolkit. In: Association for Computational Linguistics (ACL) System Demonstrations, pp. 55–60 (2014)
8. Mohammad, S.M., Turney, P.D.: Crowdsourcing a word-emotion association lexicon. Comput. Intell. **29**(3), 436–465 (2013)
9. Plutchik, R.: Emotions: a general psychoevolutionary theory. In: Approaches to Emotion, pp. 197–219 (1984)
10. Schwartz, H.A., Sap, M., Kern, M.L., Eichstaedt, J.C., Kapelner, A., Agrawal, M., Blanco, E., Dziurzynski, L., Park, G., Stillwell, D., et al.: Predicting individual well-being through the language of social media. In: Biocomputing 2016: Proceedings of the Pacific Symposium, pp. 516–527 (2016)
11. Strapparava, C., Mihalcea, R.: Learning to identify emotions in text. In: Proceedings of the 2008 ACM Symposium on Applied Computing, pp. 1556–1560. ACM (2008)
12. Widmer, G.: Tracking context changes through meta-learning. Mach. Learn. **27**(3), 259–286 (1997)

Automatic Music Generation by Deep Learning

Juan Carlos García and Emilio Serrano$^{(\boxtimes)}$ (ID)

Ontology Engineering Group, Universidad Politécnica de Madrid, Madrid, Spain
`juancarlos.garcia.torrecilla@alumnos.upm.es`, `emilioserra@fi.upm.es`

Abstract. This paper presents a model capable of generating and completing musical compositions automatically. The model is based on generative learning paradigms of machine learning and deep learning, such as recurrent neural networks. Related works consider music as a text of a natural language, requiring the network to learn the syntax of the sheet music completely and the dependencies among symbols. This involves a very intense training and may produce overfitting in many cases. This paper contributes with a data preprocessing that eliminates the most complex dependencies allowing the musical content to be abstracted from the syntax. Moreover, a web application based on the trained models is presented. The tool allows inexperienced users to generate automatic music from scratch or from a given fragment of sheet music.

Keywords: Automatic music · Deep learning
Recurrent neural networks

1 Introduction

Machines carry out a number complex tasks and processes whose automation was unthinkable years ago. Everyday there are more tasks that can be delegated to them. For some of these tasks, such as face recognition, there are not well known steps to undertake them. In these problems, the use of *machine learning* can allow computers to learn a solution from labeled examples. In the last few years, the use of *deep learning*, a subfield of machine learning, has outperformed all previous techniques in a number of problems and fields. More specifically, the use of *Recurrent Neural Networks* (RNN) is the state of the art in speech recognition and other time series problems.

The concept of art is subjective and not clearly defined. Learning creative processes is one of the most challenging problems of artificial intelligence. These processes are typical of human beings, not of living beings in general, which have a creative component that gives them greater complexity. Moreover, it is hard to define correct and incorrect cases that allow machine learning to leverage new knowledge.

This paper presents a study and comparison of several artificial neural networks architectures to model music as a time series prediction. Furthermore,

© Springer International Publishing AG, part of Springer Nature 2019
F. De La Prieta et al. (Eds.): DCAI 2018, AISC 800, pp. 284–291, 2019.
https://doi.org/10.1007/978-3-319-94649-8_34

a responsive web application based on these deep learning models is also presented. The web allows inexperienced users to explore the possibilities of artificial intelligence to enhance their music composition.

The paper revises the related works in Sect. 2. Section 3 describes the methodology followed in the research work and their different phases: data understanding, data preparation, modeling, and evaluation. Section 4 offers the experiments results that are discussed in Sect. 5. The tool implementation is described in Sect. 6. Finally, Sect. 7 concludes and gives future works.

2   Related Works

One of the best known related works in automatic music generation is BachBot [14,15,17]. This project uses *Long Short-Term Memory* (LSTM) networks to generate and harmonize musical pieces with Batch style. LSTMs are widely used in music generation. This project uses cross entropy as loss function. In a similar vein, DeepBach [11,12] presents an alternative approach for the same problem that generates musical notes by means of a pseudo-Gibbs sampling procedure.

Magenta [5] google project covers a number of models for music generation such as: Drums RNN [6], Melody RNN [7], Polyphony RNN [10] inspired by BachBot, Performance RNN [8] ane Pianoroll RNN-NADE [9]. This last model combines LSTMs with N*eural Autoregressive Distribution Estimator* (NADE). Most of these models are based on LSTM architectures and all of then employ an encoder/decoder structure to allow a sequence to sequence interface.

The LSTM-NADE combination is also employed by Johnson [13], whose recent research work employs log-likelihood as cost function. Agarwal et al. [1,2] also use LSTMs combining them with *Gated Recurrent Unit* (GRU) recurrent neural networks and achieve an accuracy of 65.5%.

Some of these works are open source [5,13,15] and have been studied for the contribution presented in this paper. However, they require a considerable programming and technical expertise to be used for music creation. Therefore, beyond the experimentation with a number of architectures and a new preprocessing method, the web application presented in this paper is an important asset for spreading the artificial intelligence assistance in music composition.

3   Methodology

This research has followed the *Cross Industry Standard Process for Data Mining methodology* (CRISP-DM) [16,18]. This methodology is composed of 6 phases: Business Understanding, Data Understanding, Data Preparation, Modeling, Evaluation, and Deployment. The following sections describe the processes undertaken in these phases over several iterations.

3.1 Data Understanding

The dataset employed in this work is the *Nottingham Music Database*[1], also employed in related works [4][2]. The data is composed of 1037 British and American folk tunes, (hornpipe, jigs, etc.) that was created by Eric Foxley and posted on Eric Foxley's Music Database.

The database was converted to ABC music notation format and was posted on abc.sourceforge.net. The ABC notation is a format of musical notation in ASCII, consisting of a header with metadata and a body with the musical content. The header is composed of several fields, some of the most relevant ones are:

- **T**: Tune title.
- **Q**: Tempo. Default: ♩ = 120.
- **M**: Meter. Default: ⁴₄.
- **L**: Unit note length. Default: ♪.
- **K**: Key signature. Default: C.

3.2 Data Preparation

The main goal in this phase has been to eliminate the existing dependencies between the notes and the header making the notes self-contained. The musical content of the body is specially conditioned by the fields K and L, see Sect. 3.1. Therefore, (1) all compositions have been transformed to the default note length; see Figure 1a. Subsequently, (2) all the compositions have been transformed to the default key signature, so that the accidents (♯, ♭) appear explicitly in each note. To avoid alterations between notes of the same bar or measure, the natural tone (♮) has been explicitly indicated in the rest of the notes; see Fig. 1b. Finally, since each tone has several representations, (3) the flat notes (♭) have been transformed to their corresponding alternative representation; see Fig. 1c. Finally, the symbols that did not provide musical content and accompanying directives were eliminated.

3.3 Modeling

The data was codified using the notes ("=A", "ˆb", "=c'"...) as temporal symbols instead of the character level approach ("=", "A", "ˆ"...). This prevents neural networks from requiring the syntax of the sheet music to be learned. Padding and masking also were employed to allow RNNs to support training situations including one-to-many, many-to-one, as also support variable length time series in the same mini-batch. Cross entropy is employed as cost function.

The TensorFlow open-source software library was used to model and evaluate different RNN architectures. The first architectures tested count with one

---

[1] https://ifdo.ca/~seymour/nottingham/nottingham.html.
[2] http://www-etud.iro.umontreal.ca/~boulanni/icml2012.

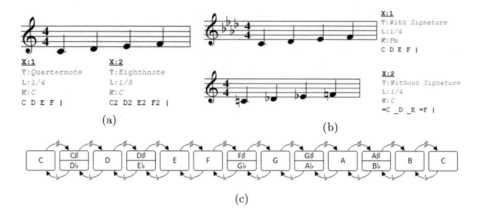

Fig. 1. Data transformations.

recurrent layer of 64 neurons and a fully connected softmax layer as output layer. Different recurrent layers have been assessed: simple RNN, LSTM, and GRU. Several time steps also have been evaluated to be considered in the truncated back propagation: 50 and 100.

– SimpleRNN with 50 timesteps.
– SimpleRNN with 100 timesteps.
– LSTM with 50 timesteps.
– LSTM with 100 timesteps.
– GRU with 50 timesteps.
– GRU with 100 timesteps.

Before these configurations, several experiments were conducted to optimize the hyperparameters. Some of these included different layer widths (16, 32, 64 and 128 neurons) in combination with different block sizes to model the tensors shape (16, 32, 64 and 128 samples).

After evaluating the performance of the different recurrent structures (simple, LSTM, GRU) with 50 and 100 timesteps, tests were performed with two GRU layers and 100 timesteps. An embedding layer was also added at the network input with dense vectors of length 5 and 10. The experiments performed are listed below:

– GRU (64 cells) - GRU (64 cells): 100 timesteps.
– GRU (256 cells) - GRU (128 cells): 100 timesteps.
– Embedding (5 length) - GRU (256 cells) - GRU (128 cells): 100 timesteps.
– Embedding (10 length) - GRU (256 cells) - GRU (128 cells): 100 timesteps.

3.4   Evaluation

Evaluating musical samples is a complex task. There is no standard procedure for evaluating musical quality. However, the standard machine learning evaluation splits data into three partitions: training, validation, and test. In this

vein, a 60%/20%/20% partition has been undertaken. *Accuracy* is reported in
the experiments, i.e. percentage of correct predictions. The test partition is not
used at all except for reporting a final accuracy of the chosen model based on
experiments over the training and validation data.

4   Results

Figure 2 (below) shows the loss values and accuracy of the experiments performed to compare different recurring layers. All of them employ 64 cells per
layer and have been trained with 64 samples per block. Executions are shown

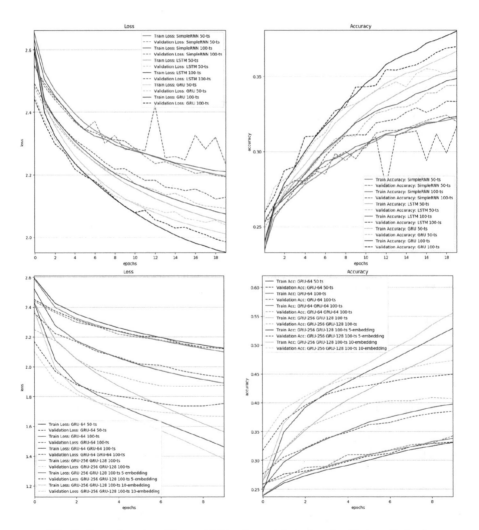

Fig. 2. Experiments with one hidden layer (above) and with two (below).

for simple recurring networks, LSTM and GRU with 50 and 100 timesteps, during 20 epochs. Figure 2 (below) details the results for the different GRU RNNs described in Sect. 3.3.

5   Discussion

The simple recurrent models shown in Fig. 2 (above) get a similar accuracy regardless of the number of timesteps. This effect may be caused because experimental results show that simple recurrent networks are not able to learn long-term dependencies [3]. In contrast, in LSTM and GRU models there is a slight variation between the test with 50 timesteps and the test with 100 timesteps. In this case, the results show that the GRU models work better than LSTM for the problem addressed, reaching an accuracy around 38% using 100 timesteps. GRU seems to benefit from the increase in timesteps, while LSTM obtains better results in the 50 timesteps test.

Experiments shown in Fig. 2 (below) present better validation results than training accuracy with few epochs. However, this tendency is reversed with more epochs presenting the classical overfitting situation. This gap is especially noticeable with the GRU(256)-GRU(128) architecture without embedding layer.

The results with a single GRU layer are quickly improved by the models with two layers. The embedding layer also provides an increase in the performance of the network, preserving similarities between input data and providing a greater number of parameters. The results also show how the models with embedding improve the models without embedding, being slightly higher the one obtained with a dense vector of dimension 10. This model obtained a 44.5% accuracy when evaluated with the test data.

Using the model to generate music and to complete fragments an acceptable musical quality is obtained (see demonstration in Sect. 6). When selecting the next note based on the most likely one, the generated sequences contain repetitions. This specially happens with unusual symbols. As in automatic text generation, the solution is to use the generated probability distribution of the softmax output layer so unlikely notes have a chance of being the next ones. This prevents the model from simply repeating known patterns and somehow providing creativity, as well as different compositions for the same musical input fragments.

6   A Deep Learning Based Web Application to Support Musical Composition

A web application based on the trained models studied in this paper has been developed. Its interface is shown in Fig. 3. The tool allows to introduce notes in the ABC format so the deep learning model can complete the musical composition. Alternatively, the music can be generated from scratch. The tool also allows to select a musical instrument to play the results in the web browser. Furthermore, the user can choose between completing by predicting the most likely next note or to use the learned probability distribution to "roll a dice" so the least likely musical notes have a chance to appear in the piece of music. Finally, the compositions can be edited, reproduced online, and saved in an MIDI file.

There is a demonstration video available online (https://youtu.be/MobbCV_SSOA/) and the source code is also available in GitHub (https://github.com/Tuxt/AutoScore/). In the next months, the web service will be deployed in one of the OEG research group servers.

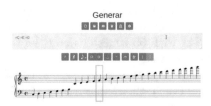

Fig. 3. Web application Interface.

7   Conclusion and Future Works

This paper describes an approach to automatic music generation through deep learning technologies based on language modeling techniques. An accuracy of 44.5% is achieved by a Recurrent Neural Network (RNN) with Gated Recurrent Units (GRUs) and embedding input layers. The main contribution regarding related works is the preprocessing and transformation of the music compositions used for training. These processes allow RNNs to reduce the necessity of learning the music sheets syntax. The web application, whose demonstration video is given together with its source code in GitHub, allows inexperienced users to use this powerful tool to enhance creativity in music composition.

Some of the future works on this research are: the extension of the training dataset, dealing with inbalanced classification, the experimentation with other neural structures such as Neural Autoregressive Distribution Estimation (NADE), the representation of samples as two-dimensional data (pitch and duration) or multidimensional time vectors, the study and elaboration of a neuronal model that contemplates the ability to produce music with accompaniment, or the implementation of a model that allows music to be created forward and backward.

Acknowledgments. This research work is supported by the Universidad Politécnica de Madrid under the education innovation project "Aprendizaje basado en retos para la Biología Computacional y la Ciencia de Datos", code IE1718.1003; and by the Spanish Ministry of Economy, Indystry and Competitiveness under the R&D project Datos 4.0: Retos y soluciones (TIN2016-78011-C4-4-R, AEI/FEDER, UE).

References

1. Agarwala, N., Inoue, Y., Sly, A.: CS224N Final Project. https://github.com/yinoue93/CS224N_proj. Accessed Dec 2017
2. Agarwala, N., Inoue, Y., Sly, A.: Music composition using recurrent neural networks
3. Bengio, Y., Simard, P., Frasconi, P.: Learning long-term dependencies with gradient descent is difficult. IEEE Trans. Neural Netw. **5**(2), 157–166 (1994)
4. Boulanger-Lewandowski, N., Bengio, Y., Vincent, P.: Modeling temporal dependencies in high-dimensional sequences: application to polyphonic music generation and transcription. In: Proceedings of the Twenty-nine International Conference on Machine Learning (ICML 2012). ACM (2012)
5. Google Brain Team. Magenta. https://github.com/tensorflow/magenta. Accessed Dec 2017
6. Google Brain Team. Magenta Drums RNN. https://github.com/tensorflow/magenta/tree/master/magenta/models/drums_rnn. Accessed Jan 2018
7. Google Brain Team. Magenta Melody RNN. https://github.com/tensorflow/magenta/tree/master/magenta/models/melody_rnn. Accessed Jan 2018
8. Google Brain Team. Magenta Performance RNN. https://github.com/tensorflow/magenta/tree/master/magenta/models/performance_rnn. Accessed Jan 2018
9. Google Brain Team. Magenta Pianoroll RNN-NADE. https://github.com/tensorflow/magenta/tree/master/magenta/models/pianoroll_rnn_nade. Accessed Jan 2018
10. Google Brain Team. Magenta Polyphony RNN. https://github.com/tensorflow/magenta/tree/master/magenta/models/polyphony_rnn. Accessed Jan 2018
11. Hadjeres, G.: DeepBach. https://github.com/Ghadjeres/DeepBach. Accessed Dec 2017
12. Hadjeres, G., Pachet, F., Nielsen, F.: Deepbach: a steerable model for bach chorales generation. arXiv preprint arXiv:1612.01010 (2016)
13. Johnson, D.D.: Generating polyphonic music using tied parallel networks. In: International Conference on Evolutionary and Biologically Inspired Music and Art, pp. 128–143. Springer, Heidelberg (2017)
14. Liang, F.: BachBot: automatic composition in the style of Bach chorales. Ph.D. thesis, Masters thesis, University of Cambridge (2016)
15. Liang, F., Gotham, M., Tomczak, M., Johnson, M., Shotton, J.: BachBot. https://github.com/feynmanliang/bachbot. Accessed Dec 2017
16. Serrano, E., Rovatsos, M., Botía, J.A.: Data mining agent conversations: A qualitative approach to multiagent systems analysis. Inf. Sci. **230**, 132–146 (2013)
17. Tomczak, M.: Bachbot. Ph.D. thesis, Masters thesis, University of Cambridge (2016)
18. Wirth, R., Hipp, J.: CRISP-DM: towards a standard process model for data mining. In: Proceedings of the 4th International Conference on the Practical Applications of Knowledge Discovery and Data Mining. Citeseer, pp. 29–39 (2000)

A Novel Hybrid Multi-criteria Decision-Making Model to Solve UA-FLP

Laura García-Hernández[1](✉) ⓘ, L. Salas-Morera[1] ⓘ, H. Pierreval[2],
and Antonio Arauzo-Azofra[1] ⓘ

[1] Area of Project Management, University of Córdoba, Córdoba, Spain
irlgahel@uco.es
[2] LIMOS UMR CNRS 6158 IFMA, Clermont-Ferrand, France

Abstract. The unequal area facility layout problem (UA-FLP) has been addressed by many approaches. Most of them only take quantitative aspects into consideration. In this paper, we will solve UA-FLP using a novel hybrid methodology that joins interactive evolutionary optimization and multi-criteria decision making. I particular, a combination of an interactive genetic algorithm and the analytic hierarchy process (AHP), is proposed. By means of this new approach, it is possible to consider both quantitative and qualitative (using the expert knowledge) criteria in order to reach an acceptable design. Our approach allows the decision maker (DM) to interact with the algorithm, guiding the search process and ranking the criteria that are more relevant in each design solution. In this way, the algorithm is adjusted to the DM's preferences through his/her subjective evaluations of the representative solutions obtained by a clustering method, and also, to the quantitative criteria. A interesting real-world data set is analysed to empirically probe the robustness of this model. Relevant results are obtained, and interesting conclusions are drawn from the application of this novel intelligent framework.

Keywords: Facility layout · Interactive genetic algorithm · AHP

1 Introduction

Facility Layout Design (FLD) determines the placement of facilities in a manufacturing plant with the aim of determining the most effective arrangement in accordance with some criteria or objectives, under certain constraints. In this respect, Kouvelis et al. (1992) provided that FLD is known to be very important for production efficiency because it directly affects manufacturing costs, lead times, work in process and productivity. According to Tompkins et al. (2010), well laid out facilities contribute to the overall efficiency of operations and can reduce between 20% and 50% of the total operating costs. There are many kinds of layout problems; A classification of them is given by Drira et al. (2007). We will focus on the Unequal Area Facility Layout Problem (UA-FLP) as formulated by Armour and Buffa (1963). In short, UA-FLP considers a rectangular plant layout that is made up of unequal rectangular facilities that have to be placed effectively in the plant layout. Aiello et al. (2012) stated that generally speaking, the problem of designing a physical layout involves the minimization

© Springer International Publishing AG, part of Springer Nature 2019
F. De La Prieta et al. (Eds.): DCAI 2018, AISC 800, pp. 292–299, 2019.
https://doi.org/10.1007/978-3-319-94649-8_35

of the material handling cost as the main objective. But, there are other authors that consider additional quantitative performance, as for example, Aiello et al. (2006), who have addressed this problem taking into account criteria that can be quantified (e.g., material handling cost, closeness or distance relationships, adjacency requirements and aspect ratio), which are used in an optimization approach.

However, Babbar-Sebens and Minsker (2012) established that these approaches may not adequately represent all of the relevant qualitative information that affect a human expert involved in design (e.g. engineers). In this way, qualitative features sometimes also have to be taken into consideration. Brintup et al. (2007) stipulated that such qualitative features are complicated to include with a classical heuristic or meta-heuristic optimization. Besides, according to García-Hernández et al. (2013) these qualitative features can be subjective, not known at the beginning and can be changed during the process. As a consequence, the participation of the decision maker (DM) is essential to include qualitative considerations in the design. Moreover, involving the DM in the process provides additional advantages which have been detailed in its work. Although, García-Hernández et al. (2013) were able to obtain UA-FLP designs that satisfy the qualitative features wanted by the DM, unfortunately, they have not considered the quantitative performance in the reached solutions. As a consequence of the reasons stated previously, taking into account both quantitative and qualitative criteria in the design, seems to be very important issue for obtaining solutions that satisfy all the criteria that exist in a real design. For that matter, García-Hernández et al. (2015) proposed an approach based on a multi-objective interactive algorithm which allowed the consideration both quantitative and qualitative criteria in the design. Unfortunately, in their approach, the DM knowledge was considered exclusively as an objective in the fitness function, so that, the approach reached solutions as 'good solutions' even when the preferences of the DM were not satisfied. This way, it seems to be a promising line of research to work with a technique that will be able to take into account both quantitative and qualitative criteria in the design.

In this work, we will address this problem using a new approach that combines interactive optimization and multi-criteria decision making. By means of this approach, a combination both quantitative and qualitative (using the expert knowledge) criteria are taken into account to determine an acceptable solution. Our approach allows the DM to interact with the algorithm, guiding the search process and ranking the criteria that are more relevant in each design case. In this way, the algorithm is adjusted to the DM's preferences through his/her subjective evaluations of representative solutions, and also, to the quantitative criteria that are described in following sections.

2 Suggested Approach

For addressing this issue, we have proposed a new approach that takes into account the important quantitative factors. From the state of the art survey, it can be extracted that it should be considered the following aspects: material handling cost, adjacency, distance requests and desired aspect ratio. Therefore, a good solution also implies the DM satisfaction about the preferences that he/she could want in the final design. So that, the proposed approach must consider the expert knowledge too for obtaining adequate solutions. Both the quantitative criteria (such as: material handling cost, adjacency, distance request, aspect ratio) and the qualitative aspects.

2.1 Proposed Methodology

A modification of the Interactive Genetic Algorithm (IGA) proposed by García-Hernández et al. (2013) is used, which allows that a DM (or several) interacts with the algorithm, guiding the search process. In this way, the algorithm is adjusted to the DM's preferences through his/her subjective evaluations. In order to not overburden the DM, this evaluations will be performed over a set of representative solutions, which are made sufficiently different and are chosen using the c-Means clustering method.

On the other hand, the quantitative factors explained previously should be also taken into consideration. For that matter, we have included the Analytic Hierarchy Process (proposed by Saaty 1980) into the approach because of it allows both qualitative and quantitative data, at the same time (Steiguer et al. 2003).

Layout Representation. In order to represent the plant layout as a chromosome, we use the Flexible Bay Structure (FBS) proposed by Tong (1991), which is currently receiving widespread attention from researchers (Wong and Komarudin (2010); Ulutas and Kulturel-Konak (2012)). This rectangular area is divided in one direction into bays of varying width. Then, each bay is subdivided to allocate the facilities that make up the layout.

Genotype. To encode a plant layout, the chromosome used is inspired from that proposed by Gomez et al. (2003), and it made up of 2 segments. The first segment represents the facility sequence that is read bay by bay, from top to bottom and from left to right. In order to interpret it, a permutation of the integers from 1 through n, is used, where n is the total number of facilities in the plant layout.

Analytic Hierarchy Process. The procedure to apply analytic hierarchy process (AHP) can be summarized by the following steps:

1. Modeling the problem as a hierarchy, obtaining the decision hierarchy tree as result. Which details: the main objective to reach, the factors or criteria to be considered, and the different alternatives for obtaining the pursued aim.
2. Determining the criteria relative importance. Once the decision hierarchy tree has been built, the DM analyzes the criteria using a series of pairwise comparisons, which are made through the scale given by Saaty (1980).
3. Establishing the criteria priorities by means of the pairwise comparisons assigned in the previous step. Priorities are relative weights associated with the criteria of the hierarchy.

Interactive Genetic Algorithm. The steps of the proposed IGA are the following:

1. The DM should establish the criteria relative importance by using a series of pairwise comparisons, which are made though the scale given in Table 1. The criteria are pairwise compared against the goal for importance. AHP weights are extracted.

Table 1. Proposed algorithm parameter values

Parameter	Value	Tested values
Number of initial generations	200	50, 100, 200, 500
Population size	500	100, 200, 500, 1000
Crossover probability	0.7	0.5, 0.7, 0.9
Mutation probability	0.3	0.1, 0.2, 0.3
Tournament size	2	2, 3, 4
No. interactive generations	100	25, 50, 100
No. clusters	9	7, 9
Fuzziness in c-means	1.025	1.025, 1.05, 1.1

2. An initial random population of N individuals is generated.
3. The process of clustering is applied over the initial population, grouping the individuals into c categories. This allow to evaluate a large population of solutions with just c evaluations of the DM.
4. A representative element of each cluster is displayed to the DM.
5. If the DM is satisfied with the algorithm result, then, the process ends. Otherwise, the system takes the subjective evaluations from the DM about the c representative solutions of the population.
6. If one or more solutions are judged interesting to the DM, he/she can keep them as favourites. They will be visible to the DM during the entire process. This assures an improvement. In this way, none of the solutions that the DM considers interesting will be lost in the IGA evolution.
7. Considering the marks given by the DM to the c representative elements of the clusters, the AHP importance weights for the criteria and the membership grade to each cluster, the fitness for each individual is computed.
8. The selection method (here the Tournament Selection) is applied to select the individuals that will be involved in the evolutionary operations. Additionally, the Elitism method is included to preserve a percentage of the best individuals.
9. Crossover and Mutation operators are applied to the individuals with a probability given by the DM.
10. The new population is created. Go to step 3.

Crossover Operator. The crossover operator is applied depending on the chromosome segment. In the first segment, the Partial Mapped Crossover (PMX) (Eiben (2003)) has been implemented. In the second one, the recombination method used is N-Point Crossover (Holland 1992).

Mutation Operator. In the facility sequence segment, positions are randomly chosen and their content is switched. In the segment of bay divisions, a random position is selected and its value is changed to its opposite.

3 Test Example

3.1 Methodology

Explained that the methodology will be test our approach taking in to account the quantitative results given in the example and also the qualitative ones that we described into our previous work. Then we will compared them for extract the results of applying our new approach.

The novel proposed approach has parameters that must be set up. Table 1 shows the values tested for each parameter and the value chosen to be used levels.

An ovine slaughterhouse projected in Córdoba (Spain) raised this facility layout design problem. Additionally, between facilities that made up this real plant, there are material flow movements, distance requirements and adjacency relationships. They are detailed in Salas-Morera et al. (1996). Additionally, Decision Maker has some qualitative preferences which have been extracted from García-Hernández et al. (2015) in order to compare with the results obtained by them. Moreover, it is necessary to establish the criteria priorities by means of the pairwise comparisons. In this sense, designer gives the importance to one criterion face to the remaining, as shown in Table 2.

Table 2. Importance pairwise comparisons between factors.

Weight	Pairwise comparisons
9	Designer versus flow
9	Designer versus distance
9	Designer versus adjacency
3	Flow versus distance
1/5	Flow versus adjacency
1/7	Distance versus adjacency

3.2 Results

Running the novel approach on the ovine slaughterhouse problem returned satisfactory solutions. The best one reached is shown in the Fig. 1.

This solution fully satisfies all DM preferences that were the factor considered as preferred and more important face to remaining ones by DM. The following factor considered very important for designer was adjacency, in this sense, all adjacency requirements are also fully satisfied. In addition, all facilities that made up the plant layout accomplished aspect ratio constraint. If we focus on distance requirements, solution reached by our proposed approach fulfills most of distance requests with the exception of request of facility J must be far to facility B.

Referring to material flow, this solution presents a value of '5280.48'. In terms of material flow, this value is in the range of better values obtained by García-Hernández et al. (2015). In this work, an IGA without AHP and Enea's algorithm presented as

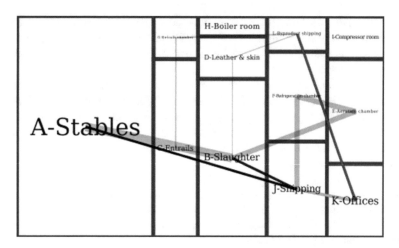

Fig. 1. The best solution reached by our proposed approach.

good solutions those having values of material flow that oscillate between: '4578' and '8456'. Unfortunately, in their approach they only considered material flow and the designer opinion (by a designer mark). Moreover, they presented solutions that their algorithm reached as 'good solutions' even when the preferences of the DM were not satisfied.

Thus, by means of the approach we suggested, we are able to reach solutions that take into account aspects considered in García-Hernández et al. (2015) and also, improve the quality solutions by the consideration of adjacency requests, distance requirements and aspect ratio constrains. Moreover, our approach is able to assign more importance to one of the factors face to the remaining ones depending on the different facility layout problem. This fact is very interesting for obtaining good solutions that could be adapted to each situation.

4   Conclusions

In this work, an IGA that combines both DM's expert knowledge and AHP method is used to address the UA-FLP. This way, is it possible to consider both quantitative and qualitative aspects into UA-FLP. Our approach allows the DM to interact with the algorithm, guiding the search process by means of his/her evaluations. In order not to overburden the DM with excessive evaluations, these evaluations were realized exclusively over a subset of representative solutions of the total population in each IGA iteration, which are sufficiently different and are chosen using the c-Means clustering method.

From the empirical study, it is shown that the suggested IGA is able of capturing the preferences that the DM would like in the solution and also, incorporating some quantitative aspects (as material flow movements, adjacency requests, distance requirements, aspect ratio constraint) without losing the efficient adaptation of our IGA to the interests that the DM would like in the final design. Moreover, these solutions are obtained in a reasonable number of generations. These numbers of generations depended on the randomness of the initial population and on the complexity of the DM's preferences.

Additionally, our approach is able to assign more importance over a factor versus the remaining ones depending on the different facility layout problem. This fact is very interesting for obtaining good solutions that could be adapted to design problem.

As future work, a promising line of research could be to study alternative methods of offering visual information to the DM with the aim of transmitting this information in a more ergonomic way.

Acknowledgements. The authors would like to thank the program to promote research of the University of Córdoba (Spain) for funding this research under the Research Project XXII PP. Mod. 4.1 in Plan Propio de Investigación de la Universidad de Córdoba y del Programa Operativo de fondos FEDER Andalucía 6.

References

Aiello, G., Enea, M., Galante, G.: A multi-objective approach to facility layout problem by genetic search algorithm and electre method. Robot. Comput. Integr. Manuf. **22**, 447–455 (2006)

Aiello, G., Scalia, G.L., Enea, M.: A multi objective genetic algorithm for the facility layout problem based upon slicing structure encoding. Expert Syst. Appl. **39**, 10352–10358 (2012)

Armour, G.C., Buffa, E.S.: A heuristic algorithm and simulation approach to relative location of facilities. Manage. Sci. **9**, 294–309 (1963)

Babbar-Sebens, M., Minsker, B.S.: Interactive Genetic Algorithm with Mixed Initiative Interaction for multi-criteria ground water monitoring design. Appl. Soft Comput. **12**(1), 182–195 (2012). http://dx.doi.org/10.1016/j.asoc.2011.08.054

Bhushan, N., Rai, K.: Strategic Decision Making: Applying the Analytic Hierarchy Process. Springer, London (2004)

Brintup, A.M., Ramsden, J., Tiwari, A.: An interactive genetic algorithm-based framework for handling qualitative criteria in design optimization. Comput. Ind. **58**, 279–291 (2007)

Drira, A., Pierreval, H., Hajri-Gabouj, S.: Facility layout problems: a survey. Ann. Rev. Control **31**, 255–267 (2007)

Eiben, A., Smith, J.: Introduction to Evolutionary Computing. Springer (2003)

García-Hernández, L., Arauzo-Azofra, A., Salas-Morera, L., Pierreval, H., Corchado, E.: Facility layout design using a multi-objective interactive genetic algorithm to support the DM. Expert Syst. **32**, 94–107 (2015)

García-Hernández, L., Pierreval, H., Salas-Morera, L., Arauzo-Azofra, A.: Handling qualitative aspects in unequal area facility layout problem: an interactive genetic algorithm. Appl. Soft Comput. **13**, 1718–1727 (2013)

Gomez, A., Fernández, Q., la Fuente García, D.D., García, P.: Using genetic algorithms to resolve layout problems in facilities where there are aisles. Int. J. Prod. Econ. **84**, 271–282 (2003)

Holland, J.H.: Adaptation in Natural and Artificial Systems. MIT Press, Cambridge (1992)

Kouvelis, P., Kurawarwala, A.A., Gutierrez, G.J.: Algorithms for robust single and multiple period layout planning for manufacturing systems. Eur. J. Oper. Res. **63**, 287–303 (1992). European Journal of Operational Research, 223, 614–628

Saaty, T.: The Analytic Hierarchy Process, Planning, Priority Setting, Resource Allocation. McGraw-Hill, New York (1980)

Salas-Morera, L., Cubero-Atienza, A.J., Ayuso-Munoz, R.: Computer-aided plant layout. Informacion Tecnologica **7**, 39–46 (1996)

Steiguer, J.E., Duberstein, J., Lopes, V.L.: The Analytic Hierarchy Process as a Means for Integrated Watershed Management (2003)

Tompkins, J., White, J., Bozer, Y., Tanchoco, J.: Facilities Planning, 4th edn. Wiley, New York (2010)

Tong, X.: SECOT: A Sequential Construction Technique For Facility Design. University of Pittsburg: Doctoral Dissertation. (1991)

Ulutas, B.H., Kulturel-Konak, S.: An artificial immune system based algorithm to solve unequal area facility layout problem. Expert Syst. Appl. **39**, 5384–5395 (2012)

Vaidya, O.S., Kumar, S.: Analytic hierarchy process: an overview of applications. Eur. J. Oper. Res. **169**, 1–29 (2006)

Wong, K.-Y., Komarudin: Solving facility layout problems using flexible bay structure representation and ant system algorithm. Expert Syst. Appl. **37**, 5523–5527 (2010)

Robust Noisy Speech Recognition Using Deep Neural Support Vector Machines

Rimah Amami[1(✉)] and Dorra Ben Ayed[2]

[1] Deanship of Preparatory Year and Supporting Studies,
Imam Abdulrahman Bin Faisal University, Dammam, Kingdom of Saudi Arabia
`raamami@iau.edu.sa`
[2] Department of Electrical Engineering, National School of Engineering of Tunis,
Tunis, Tunisia
`dorra.BAyed@gmail.com`

Abstract. This paper aims to classify noisy sound samples in several daily indoor and outdoor acoustic scenes using an optimized deep neural networks (DNNs). The advantage of a traditional DNNs lies in using at the top layer a softmax activation function which is a logistic regression in order to learn the output label in a multi-class recognition problem. In this paper, we optimize the DNNs by replacing the softmax activation function by a linear support vector machine.

In this paper, a novel deep neural networks (DN) using Support Vector Machines (SVM) instead of the multinomial logistic regression is proposed. We have verified the effectiveness of this new method using speech samples from Aurora speech database recorded in noisy conditions. The experimental results obtained with the method DN-SVM demonstrates a significant improvement of the performance with noisy sound samples classification.

Keywords: Deep learning · SVM · Speech recognition
Neural networks · MFCC

1 Introduction

In recent years, deep learning achieved great success in pattern recognition domain and it is very widely used in the learning machine applications. The deep learning, known as a multi-layer neural network, is composed of several computational layers in order to learn the datasets in an hierarchical architecture [2]. The Deep learning improved the performance in many domains such as speech recognition, image recognition, detection problems [1,2].

Deep Neural Networks (DNNs) is a widely used approach for supervised machine learning and classification. Though DNNs are based in a powerful mechanism, some apparent failure were detected. For example, the performance of the DNNs decrease when they are applied to tasks with small amount of training data. They are, also, known to be complicated models which depend on

© Springer International Publishing AG, part of Springer Nature 2019
F. De La Prieta et al. (Eds.): DCAI 2018, AISC 800, pp. 300–307, 2019.
https://doi.org/10.1007/978-3-319-94649-8_36

robust computational facilities for the training stage. On the other hand, Support Vector Machines (SVMs) has been a very successful supervised algorithm for solving multi-class recognition problems. The SVMs is an approach based on the Vapnic-Chervonenkis (VC) theory and the principle of Structural Risk Minimization (SRM). Furthermore, the robust learning performance of this approach is related to the fact that SVMs apply a linear algorithm to the data in a high dimensional space. This work aims to replace the softmax active function at the above layer, in the traditional DNNs for classification, by linear SVM. and we optimized the primal problem of the standard SVM. The main purpose of this study is to investigate and evaluate the performance of the method DN-SVM applied to recognize noisy speech samples. To show its effectiveness, we support our strategy with empirical evaluation using distorted speech from Aurora speech database.

The rest of this paper is organized as follows: in the next section, we introduce a brief overview of the deep neural networks. Section 3 describes our proposed method. Experimental setup and results are given in Sect. 4. The paper concludes in Sect. 5.

2   Related Works

In the last decades, deep neural networks [8] were the main subject of various studies and were included in a wide range of domains such as speech recognition [3,5,20] and visual object recognition [6,7]. It must be pointed out, that all the mentioned studies used the traditional deep neural networks.

Indeed, the deep neural networks have shown that they outperforms powerful machine learning approaches such Support Vector Machines [9–11]. On the other hand, several studies combined the deep neural networks method with different method in order to ensure an efficient performance. The authors Li et al. [12] proposes a deep neural networks combined to Hidden Markov Model (HMM) for the speech emotion recognition. They proved that the proposed method DNN-HMMs outperform the GMM-HMMs method by 11.67%. The method proposed by Yu et al. [13] is based, also, on the hybrid system DNN-HMM which outperforms significantly the standard GMM-HMM applied on a vocabulary continuous speech recognition problem. Their proposed method combine the powerful mechanism of the deep neural learning and the ability of HMM to model efficiently. Furthermore, Zhang et al. present in [14] a novel method for speech recognition based on deep neural learning using SVM. TIMIT corpus was used to verify the effectiveness of the proposed method applied to continuous speech recognition task. The proposed model using the deep neural and SVM reduce the error rate compared to the standard deep neural network by 8%.

Tang [15] presents, also, an deep learning using the linear SVM for the task of facial expression recognition. The proposed method have shown that combining deep neural and SVM is significantly beneficial when applied to common deep learning datasets such as MNIST and CIFAR-10.

In this context, we propose a new approach which handle the problem of combining deep neural networks (DNNs) and Support Vector Machins (SVMs) for the recognition of a noisy speech samples from Aurora speech database.

3   Deep Neural SVM

For the classification, the traditional deep neural networks use at the top layer a softmax active function, called also logistic regression.

In a multi-class classification problem, the softmax layer has k nodes given by a discrete probability distribution p_i where $i = 1....k$. The equation p_i is given by:

$$p_i = \frac{\exp(h_k W_{ki})}{\sum_1^k \exp(h_k W_{ki})} \tag{1}$$

The output of softmax layer can be expressed as

$$\sum_k h_k W_{ki} \tag{2}$$

Where W_{ki} are the weights connecting the last layer to the softmax layer ki and h is the output vector of the top layer in a deep neural networks.

In a traditional DNNs, the classification of a speech sample is given by the following equation:

$$\arg\max_i p_i \tag{3}$$

On the other hand, the Support Vector Machines are based on an implementation of the SRM induction principle [10].

Thus, SVMs propose to maximize the margin which is the distance from a separating hyperplane to the nearest sample. The optimal hyperplane separating samples can be expressed as follow:

$$\frac{1}{2}\|w^{ij}\|^2 + C \sum_{i=1}^m \xi^{ij} \tag{4}$$

Where C is a regularization parameter and ξ is a slack variable. It should be noted that the minimization of the margin in the multiclass problem is defined by:

$$\frac{1}{2}\|w^{ij}\|^2 + C \sum_{i=1}^m \xi^{ij} \tag{5}$$

Moreover, the dual lagrangian objective function is optimized as follows:

$$L_d = \max_{\alpha_i} \sum_{i=1}^m \alpha_i - \sum_{i=1}^m \sum_{j=1}^m \alpha_i \alpha_j y_i y_j K(x_i, x_j). \tag{6}$$

where $K(x_i, x_j)$ is the kernel function and the coefficients α_i are the lagrange multipliers [9,10]. As mentioned above, we use, in this paper, a linear SVMs instead of a logistic regression known as softmax activation function.

Hence, in the proposed method, the optimization problem given by Eq. 5 will be replaced in order to train deep neural for the speech recognition. Thus, in the new objective function $a(w)$ the input sample x in Eq. 5 is replaced with h which is the output vector of the top hidden layer in traditional DNN:

$$\frac{\partial(a(w))}{\partial(h_n)} = 2C[1 + w_{\bar{s}_t}^T h_t - w_{s_t}^T h_t] + (w_{\bar{s}_t} - w_{s_t}) \tag{7}$$

Where $\bar{s}_t = arg\ max_{\bar{s}_t} w_{\bar{s}_t}^T h_t$ and s_t is the output state [14].

Following this update in the Eq. 7, the backpropagation algorithm is the same principle as the traditional DNNs.

The architecture of proposed method used to recognize the noisy speech is illustrated in Fig. 1. Consequently, the result of applying a deep neural Support Vector machines approach on noisy speech recognition task have the ability to guarantee better classification results since the proposed system is robust to the noisy objects.

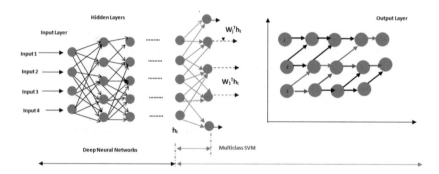

Fig. 1. The architecture of the proposed method deep neural support vector machines.

4   Experimental Analysis

We have develop a noisy speech recognition system in MATLAB for the evaluation of our proposed Deep Neural SVM method. To prove its effectiveness and robustness, we verified our method with an empirical evaluation using distorted speech samples included in Aurora speech database. The following subsection describes the database used in this study.

4.1   Noisy Speech System Description

Nowadays, the speech recognition applications become more attractive and widely used in various real-world problems.

In this paper, we study continuous speech recognition using the database Aurora 2 which consist of connected digits task spoken by Native American English speakers [16,17].

Furthermore, Aurora database is designed to verify the performance of speech recognition systems in noisy conditions. Thus, the database includes a selection of many real-world background noise. For the speech features extraction, the Mel-Frequency Cepstral Coefficients (MFCC) are used in order to extract information from input noisy speech analog signal samples.

The MFCC are the cepstral coefficients calculated from the mel-frequency warped Fourier transform representation of the log magnitude spectrum.

Moreover, we include, in this paper, the temporal cepstral derivative in order to enhance the performance of speech recognition system. In the state of art, those coefficients are known to capture efficiently the transitional characteristics of the speech signal [18, 19].

In the next subsection, the experiments results will be verified. The impact of using a combined deep neural SVM for the recognition of distorted samples is discussed. We will, also, make a comparison of the performance of the proposed system and some different learning known algorithms.

4.2   Results and Discussion

As mentioned before, for our experimentation we used the Aurora 2 database. It includes speech samples pronounced by English American adults (male and female) speaking isolated digits. At a rate of 8 kHz, each speaker record up to 7 digits.

At the feature extraction stage, 39-dimensional MFCC feature vectors are extracted from the speech signals.

The following list present the noisy environment that have been added to the recorded speech samples:

- Airport
- babble
- Suburban train
- Street
- Restaurant
- Train station
- Exhibition hall
- Car

At this stage, we present the recognition results using the proposed method Deep Neural SVM, the traditional SVMs and the traditional DNNs.

The Table 1 represents the Noisy TIdigits recognition accuracy for speech samples with noisy background.

The results in Table 1 demonstrated that the recognition system based on the standard SVMs and the standard DNNs showed moderate performance in all noises backgrounds. Meanwhile, we noticed that the use of deep neural SVM improves the overall performance of the TIdigits recognition in a noisy environment.

When the DN-SVM method was employed, the recognition error was reduced a 91.74% in comparison to the misclassification rate obtained with the standard

SVM and the standard DNN with the car noise background, which represents a decrease of the recognition error of respectively 07.53% and 10.63%.

Moreover, when the babble noise is added to the recorded speech samples, the recognition error was decreased by 13% with DN-SVM method in comparison to the error rate obtained with the standard SVM and by 14% in comparison to the error rate obtained with the standard DNN.

Table 1. % TIdigit recognition accuracy for test set with noisy training data based on standard DNNs, standard SVMs and the proposed DN-SVM

Methods	Subway	Restaurant	Babble	Street	Car	Airport	Exhibition	Train-station
DNN	84.16	75.81	72.17	70.95.44	81.11	86.27	73.14	76.89
SVM	81.27	73.28	70.38	72.37	84.21	81.16	71.83	74.72
DN-SVM	90.12	84.33	85.65	85.18	91.74	90.32	87.51	88.35

Looking at the results for the test set shown in the Table 1, it may be deduced that an improvements of the Noisy TIdigits recognition is produced after applying a deep neural learning based on SVMs. Indeed, the proposed method show its robustness for the speech recognition in a noisy environment. Moreover, the use of deep neural SVMs method allows to enhance the performance of the Noisy TIdigits recognition by 15% compared to the traditional SVM and the traditional DNNs.

In all experimentation set, the recognition results show significant improvements for all the recording conditions.

For our proposed method, the best performance is produced with the car noise within 91%. Those recognition rates with the database Aurora 2 shows the effectiveness of the method DN-SVM for the recognition of distorted speech samples.

5   Conclusion and Future Works

In this paper, we have proposed an updated version of the traditional DNNs. We present a deep neural learning based on SVMs. Our new method aims to generate a robust performance while dealing with distorted speech from Aurora database.

Based on the experimentation set, the proposed system based on Deep Neural SVM lead to better results in terms of recognition accuracy.

Furthermore, the perspective of this work intend to explore this new instantiation of the proposed method by adding the soft computing principal in uncertain context in order to verify the efficiency of our deep neural SVMs method.

References

1. Hemsoth, N.: The next wave of deep learning applications. Next Platform, September 2016
2. Sze, V., Yu-Hsin Chen, Y.H., Yang, T.J., Fellow, J.E.: Efficient processing of deep neural networks: a tutorial and survey. Proc. IEEE **105**(12), 2295–2329 (2017)
3. Mohamed, A., Dahl, G.E., Hinton, G.E.: Deep belief networks for phone recognition. In: NIPS Workshop on Deep Learning for Speech Recognition and Related Applications (2009)
4. Quoc, L., Ngiam, J., Chen, Z., Chia, D., Koh, P.W., Ng, A.: Tiled convolutional neural networks. In: NIPS 23 (2010)
5. Dhal, G.E., Ranzato, M., Mohamed, A., Hinton, G.E.: Phone recognition with the mean-covariance restricted Boltzmann machine. In: Neural Information Processing Systems Conference 23 (2010)
6. Jarrett, K., Kavukcuoglu, K., Ranzato, M., LeCun, Y.: What is the best multi-stage architecture for object recognition. In: Proceedings International Conference on Computer Vision (ICCV09). IEEE (2009)
7. Krizhevsky, A., Sutskever, I., Hinton, G.E.: Imagenet classification with deep convolutional neural networks. In: NIPS, pp. 1106–1114 (2012)
8. Goodfellow, I., Bengio, Y., Courville, A.: Deep Learning. MIT Press (2016)
9. Vapnik, V.: The Nature of Statistical Learning Theory. Springer, New York (1995). 8(6), 188
10. Cortes, C., Vapnik, V.: Support-vector networks. Mach. Learn. **20**(3), 273–297 (1995)
11. Schölkopf, B., Burges, C., Vapnik, V.: Extracting support data for a given task. In: Conference on Knowledge Discovery and Data Mining (1995)
12. Li, L., Zhao, Y., Jiang, D., Zhang, Y., Wang, F., Gonzalez, I., Valentin, E., Sahli, H.: Hybrid deep neural network-hidden markov model (DNN-HMM) based speech emotion recognition. In: Proceedings of the 2013 Humaine Association Conference on Affective Computing and Intelligent Interaction, pp. 312–317 (2013)
13. Yu, D., Deng, L.: Deep neural network-hidden markov model hybrid systems. In: Automatic Speech Recognition. Signals and Communication Technology. Springer, London (2015)
14. Zhang, S.X., Liu, C., Yao, K., Gong, Y.: Deep neural support vector machines for speech recognition. In: IEEE International Conference on Acoustics, Speech and Signal Processing (ICASSP) (2015)
15. Tang, Y.: Deep learning using linear support vector machines. In: International Conference on Machine Learning Challenges in Representation Learning Workshop, Atlanta, Georgia, USA (2013)
16. Zhu, Q., Iseli, M., Cui, X., Alwan, A.: Noise robust feature extraction for asr using the aurora 2 database. In: 7th European Conference on Speech Communication and Technology, 2nd INTERSPEECH Event, Aalborg, Denmark, 3–7 September 2001
17. Hirsch, H.G., Pearce, D.: The AURORA experimental framework for the performance evaluations of speech recognition systems under noisy condition. In: ISCA ITRW ASR 2000 Automatic Speech Recognition: Challenges for the Next Millennium, France (2000)

18. Amami, R., Ben Ayed, D., Ellouze, N.: Phoneme recognition using support vector machine and different features representations. In: The 9th International Conference Distributed Computing and Artificial Intelligence (DCAI). Advances in Intelligent and Soft Computing, vol. 151, pp. 587–595, Salamanca, Spain. Springer, Heidelberg (2012)
19. Davis, S.B., Mermelstein, P.: Comparison of parametric representations for monosyllabic word recognition in continuously spoken sentences. Acoust. Speech Sig. Process. **28**(4), 357–366 (1980)
20. Deng, L., Li, J., Huang, J.-T., Yao, K., Yu, D., Seide, F., Seltzer, M., Zweig, G., He, X., Williams, J., et al.: Recent advances in deep learning for speech research at Microsoft. In: ICASSP (2013)

Preliminary Study of Mobile Device-Based Speech Enhancement System Using Lip-Reading

Yuta Matsunaga[1], Kenji Matsui[1(✉)], Yoshihisa Nakatoh[2],
Yumiko O. Kato[3], Daniel Lopez-Sanchez[4], Sara Rodriguez[4],
and Juan Manuel Corchado[4]

[1] Osaka Institute of Technology, Osaka 535-8585, Japan
mlm17r22@st.oit.ac.jp, kenji.matsui@oit.ac.jp
[2] Kyushu Institute of Technology, Kitakyushu 804-8550, Japan
[3] St. Marianna University School of Medicine, Kawasaki 216-8511, Japan
[4] BISITE Research Group, University of Salamanca, Salamanca, Spain

Abstract. Inconspicuous speech enhancement system for laryngectomies using lip-reading is proposed to improve the usability and the speech quality. The proposed system uses a tiny camera on mobile phone and recognize the vowel sequences using lip-reading function. Three types of Japanese vowel recognition algorithms using MLP, CNN, and MobileNets, were investigated. 3,000 image datasets for training and testing were prepared from five persons while uttering discrete vowels. Our preliminary experimental result shows that the MobileNets is appropriate for embedding mobile devices in consideration of a performance both recognition accuracy and calculation cost.

Keywords: Lip-reading · Neural network · Speech recognition

1 Introduction

People who have had laryngectomies have several options for restoration of speech, but currently available devices are not satisfactory.

The electrolarynx (EL), typically a hand-held device which introduces a source vibration into the vocal tract by vibrating the external walls, has been used for decades by laryngectomees for speech communication. It is easy to master with relatively short-term practice period regardless of the post-operative changes in the neck. However, it has a couple of disadvantages. Firstly, it does not produce airflow, so the intelligibility of consonants is diminished and the speech is very mechanical tone that does not sound natural. Secondly, it is far from normal appearance. Alternatively, esophageal speech does not require any special equipment, but requires speakers to insufflate, or inject air into the esophagus. It takes long time to master the speech, especially, elderly laryngectomees face difficulty in mastering the speech or keep using esophageal speech because of the waning strength.

To understand the user needs more precisely, questionnaires were used to understand implicit user needs with 121 laryngectomees (87% male, 13% female), including

65% esophageal talkers, 12% EL users, 7% both, and 21% used writing messages to communicate.

We extracted primary needs of laryngectomees from the result as shown in Fig. 1. Then, this time, we focused on the three of them, i.e. "Use of existing devices", "Ordinary-looking", "Easy to Use".

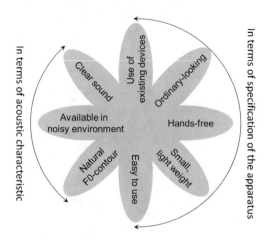

Fig. 1. The primary needs of laryngectomees

(1) "Use of existing devices": Mobile phone, especially smartphone is becoming very popular and it has a lot of computational power. Therefore, we use the smartphone as the central unit of speech enhancement system.

(2) "Ordinary-looking": Again, mobile devices are widely used and no one thinks it is strange even if you are talking to your mobile phone. We plan to develop a system which can recognize your lip motion and generate the corresponding synthesized speech.

(3) "Easy to use": By combining lip-reading and speech synthesis, people can communicate without using either electrolarynx or esophageal speech. That makes users much easier to communicate.

Recently, image processing technique (lip-reading) for handicapped people has been developed by several researchers [1] such as visual only speech recognition (VSR). However, most of those technologies are developed for ordinary people and implemented on PC. The present study aims to develop the speech enhancement tool using technique of lip-reading on mobile device for laryngectomies to meet the essential user needs.

2   Speech Enhancement System

Figure 2 shows a concept image of the proposed system. If uses want to talk, they just need to move their mouth. That is like silently talking to the smart phone. Then the system captures the lip images via the tiny camera, and recognizes each phoneme using lip-reading function. Then, the speech synthesis application converts unrestricted phoneme sequence into speech signal, and send it to the audio output of the mobile device or wireless loud speaker. We can expect to obtain relatively higher lip-reading accuracy by utilizing the feedback from the display.

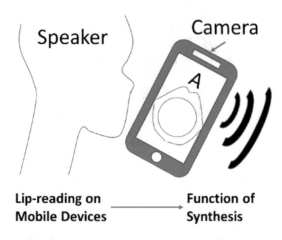

Fig. 2.   A concept image of the proposed system

3   Vowel Recognition

Figure 3 shows the screen shot of the real-time monitoring program. The green colored symbol is the recognition result displayed on top of the screen. The experimental system is running on PC while small web camera is capturing the lip image.

3.1   Block Diagram of Vowel Recognition System

The block diagram of the presented approach is shown in Fig. 4. Firstly, in order to extract the face images, we used HOG (Histogram of Oriented Gradients) detector and SVM based algorithm [3]. Next, the face image inputs a GBDT (Gradient Boosting Decision Tree) based algorithm to extract lip images [4]. Finally, the mouth region of interest inputs a neural network and classify the Japanese vowel.

3.2   Experiments Using Multilayer Perceptron

Based on our experimental system design, we took multilayer-perceptron (MLP) for the neural part at first. As for the performance evaluation, we prepared 3,000 image

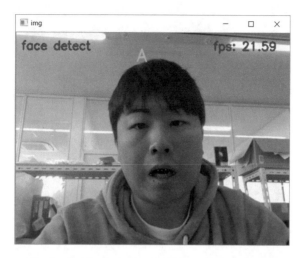

Fig. 3. Screen shot of real-time monitoring program

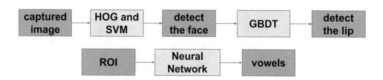

Fig. 4. A block diagram of proposed system

datasets for training and testing. The datasets are from five young male adults captured by a web camera while uttering six phonemes ("a", "i", "u", "e", "o", "silence"), and 600 images each phonemes. A structure of MLP is shown in Fig. 5. The number of dimensions is 1,024 (32 × 32 pixels). In addition, 2,100 images were used for the training and 900 images were used for the testing. 128 neurons are in the hidden layer. The experimental system was implemented in a PC with Intel Core-i5 2.2 GHz processor, 4 GB RAM. The experimental result was 87.1% accuracy.

3.3 Experiments Using Convolutional Neural Network

To make the result more accurate and stable, we used convolutional neural network (CNN). The experimental environment and the datasets (training data, test data) are the same as the previous evaluation. The structure of CNN is shown in Fig. 6. The CNN has two convolution layers and archived 96.9% accuracy. That result was higher than the MLP-based system. However, the computational cost was much higher than the previous one. Since we need to develop very fast and accurate performance using mobile device computational power, we need to reduce the computational cost while improving the processing time.

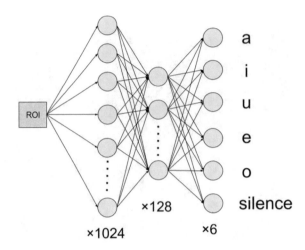

Fig. 5. Structure of the tested multilayer-perceptron

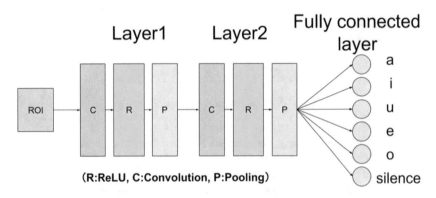

Fig. 6. Structure of the tested convolutional neural network

3.4 Experiments Using MobileNets

MobileNets are based on a streamlined architecture that uses depthwise separable convolutions [5] that is suitable for mobile. Figure 7 shows the standard convolutional filters. The standard convolutional filter combines features based on the convolutional kernels to reduce new representation from image data.

The MobileNets separates convolutional filter into two steps that called depthwise separable convolutions. The method is made up two layers: depthwise convolutions and pointwise convolutions. Figures 8 and 9 shows description of those filters.

The filter has two features. (1) A size of network is small. (2) We are able to reduce the computational cost. In addition, the number of parameters and number of multiply-add can be modified by using the 'alpha' parameter, which increases or decreases the number of filters in each layer. In this section, we selected 0.5 of 'alpha' parameter and describe the structure of MobileNets in Fig. 10. Also, the experimental environment

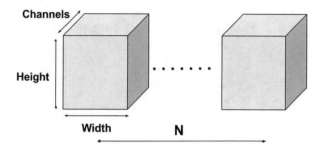

Fig. 7. A standard convolutional filter

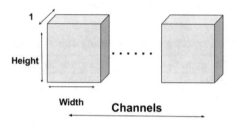

Fig. 8. A depthwise convolution filter

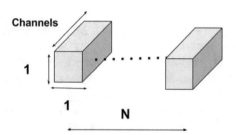

Fig. 9. A pointwise convolution filter

and the datasets are the same as the last one. In this case, the system archived 92.4% accuracy.

3.5 Evaluation of Accuracy and Computing Speed

We evaluated three types neural network that should be suitable for the mobile system. Table 1 shows the summary of evaluation results. We evaluated the processing time of all 900-testing data.

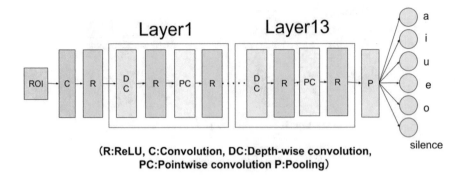

(R:ReLU, C:Convolution, DC:Depth-wise convolution,
PC:Pointwise convolution P:Pooling)

Fig. 10. A structure of MobileNets

Table 1. An accuracy and processing time of all testing data

Algorithm s	Accuracy (%)	Processing time (s)
MLP	87.1	0.014
CNN	96.9	8.321
MobileNets	92.4	2.333

4   Discussion

Based on the preliminary experimental results, MLP type neural network was obviously the fastest one. However, the accuracy is lower than 90%. MobileNets is appropriate for embedding mobile devices in consideration of a performance between accuracy and speed. CNN is promising. If you look at the computational power trend of mobile device, we might be able to reduce the delay time significantly. From the user interface point of view, Fig. 11 shows a concept of detection method for lips that has two steps. (1) Use lip detecting system (in Fig. 4) only once when the user taps a screen. (2) Use optical flow for tracking the lip image. We might be able to reduce the computational cost by using such method for detecting lips. There are a lot of potentials

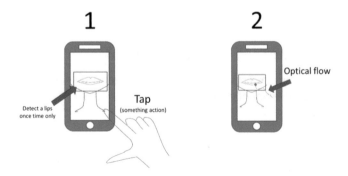

Fig. 11. A concept image of detecting lips on real time

in terms of designing very effective UI for such mobile based system. In addition, we need to work on developing algorithms for predicting phoneme, word and sentence. As part of such work, using local binary pattern in three orthogonal planes (LBP-TOP) [6] to extract dynamic features is investigated.

5 Conclusions

We performed preliminary comparison study between three types of vowel recognition system which converts lip-image data into vowel sequences. Although our target system is mobile phone, we utilized PC and webcam this time. We aim to apply this approach on mobile devices as our next step. Also, we need to develop continuous recognition system including both vowels and consonants. To make the system fast enough, more effective feature parameters need to be extracted from the image.

Acknowledgment. This work was supported by JSPS KAKENHI Grant-in-Aid for Scientific Research(C) Grant Number 15K01487.

References

1. Lin, B.-S., Yao, Y.-H., Liu, C.-F., Lienand, C.-F., Lin, B.-S.: Development of Novel Lip-Reading Recognition Algorithm, pp. 2169–3536. IEEE, 6 March 2017
2. Kimura, K., et al.: Development of wearable speech enhancement system for laryngectomees. In: NCSP 2016, pp. 339–342, March 2016
3. King, D.E.: Max-Margin Object Detection, arXiv:1502.00046v1 [cs.CV], 31 January 2015
4. Kazemi, V., Sullivan, J.: One millisecond face alignment with an ensemble of regression trees. In: IEEE Conference on Computer Vision and Pattern Recognition, pp. 1867–1874 (2014)
5. Howard, A.G., Zhu, M., Chen, B., Kalenichenko, D., Wang, W., Weyand, T., Andreetto, M., Adam, H.: MobileNets: Efficient Convolutional Neural Networks for Mobile Vision Applications, arXiv:1704.04861v1 [cs.CV], 17 April 2017
6. Rathee, N.: Investigating back propagation neural network for lip reading. In: ICCCA, pp. 373–376 (2016)

Social Services Diagnosis by Deep Learning

Emilio Serrano$^{(\boxtimes)}$ and Pedro del Pozo-Jiménez

Ontology Engineering Group, Universidad Politécnica de Madrid,
Madrid, Spain
emilioserra@fi.upm.es, pedrodel.pozo.jimenez@alumnos.upm.es

Abstract. Machine learning and Deep Learning are revolutionizing the field of medicine. These Artificial Intelligence technologies also have the potential to profoundly impact social services. This paper experiments with various architectures of deep artificial neural networks to diagnose cases of chronic social exclusion. The results improve on several metrics previous predictive models based on other machine learning paradigms such as logistic regression or random forests. These models, far from replacing social workers, allow them to be more responsive, efficient, and proactive.

Keywords: Social exclusion · Social services · Deep Learning
Data analysis · Machine learning · Data mining

1 Introduction

Deep Learning (DL) is a subfield of machine learning based on the use of different artificial neural network architectures that, through a hierarchy of layers with non-linear processing units, learn high-level abstractions for data. These representations facilitate the resolution of certain tasks in various fields among which stand out: artificial vision, natural language processing, speech recognition, and reinforcement learning. DL is replacing other machine learning paradigms more and more because of its great predictive power. Furthermore, building DL based systems is relatively easy and cheap thanks to open-source tools such as TensorFlow or Spark and massive amounts of computation power through cloud providers such as Amazon Web Services and Google Cloud.

Medicine is one of the fields most benefited by the DL revolution. DL models can: recognize cancerous tissue at a level comparable to trained physicians [5], predict drug response in cancer [16], and even diagnose rare genetic diseases from a simple face photo [2].

Social services, also called welfare services or social work, include publicly or privately provided services intended to aid disadvantaged, distressed, or vulnerable persons or groups. The economic crisis is undermining the sustainability of social protection systems in the EU [1]: 24% of all the EU population (over 120 million people) are at risk of poverty or social exclusion.

F. De La Prieta et al. (Eds.): DCAI 2018, AISC 800, pp. 316–323, 2019.
https://doi.org/10.1007/978-3-319-94649-8_38

As in the field of medicine, DL has the potential to profoundly impact social services. This article continues the research line of previous works [10,11] for the use of artificial intelligence technologies to achieve more responsive, efficient, and proactive social services. More specifically, the paper contributes with a DL model capable of predicting chronic social exclusion with an accuracy of 74.0% and a recall of 84.4%. The model has been trained with social services data of Castilla y León (CyL), which is the largest region in Spain and counts with around two and a half million inhabitants.

The paper outline is as follows. After revising some of the most relevant related works in Sect. 2, the process used to analyze the data is explained in Sect. 3. Section 4 reports the outcomes of the experiments conducted. Section 5 explains, analyzes, and compares the results. Finally, Sect. 6 concludes and offers future works.

2 Related Works

Risk prediction models are widely used in insurance companies to allow customers to estimate their policies cost. Manulife Philippines [3] offers a number of online tools to calculate the likelihood of disability, critical illness, or death before the age of 65; based on age, gender, and smoking status. Health is another application field where risk estimations are undertaken for preventive purposes. More specifically, the risk of heart disease can be estimated at different websites such as at the Mayo clinic web [4].

There are a number of data analysis works in social exclusion that are detailed enough to extrapolate some of their methods to the research presented here. Ramos and Valera [9] use the *logistic regression* (LR) model to study social exclusion in 384 cases labeled by social workers through a heuristic procedure. Lafuente-Lechuga and Faura-Martínez [8] undertake an analysis of 31 predictors based on segmentation methods and LR. Haron [7] studies the social exclusion in Israel and proposes the *linear regression* as a better alternative to the LR. Suh et al. [15] analyze over 35 K cases of 34 European countries using LR. Although these works are significant contributions to the social exclusion problem; the use of linear classifiers exclusively such as LR may hinder models from achieving a better predictive power.

3 Methodology

The methodology employed for this data analysis research is the widely used *Knowledge Discovery in Databases* (KDD) process described by Fayyad et al. [6]. This includes the following steps: Selection, Preprocessing, Transformation, Data Mining, and Evaluation and/or Interpretation. Although the KDD is an iterative and incremental process, some of the decision made in the different steps are presented unlooped here for the shake of clarity.

3.1 Selection and Preprocessing

Eleven databases (DBs) with social services information were available to select relevant data. More specifically, the DBs were implemented with the Oracle object-relational database management system. After several meetings with the social workers experts, 63 relevant variables from those DBs were selected to further study and preprocess. More information about the selection and preprocessing is described in [11].

3.2 Transformation

In this phase, data are transformed into forms appropriate for mining. This includes, among others: (1) standardization of numeric variables; (2) transforming internal numerical codes into interpretative nominal values; (3) aggregation for the multi-instance learning (where each example in the data comprises several different instances, such as persons with not one but a number of values for a specific variable); and, (5) dealing with the imbalanced classification problem.

Deep Learning methods also require extra transformations to represent inputs as vectors or real values. For the categorical predictors, each variable with x possible values is encoded in x new binary variables, so that in a sample only one of the variables can take the value of one, i.e. samples contain a one-hot vector with x dimensions. A real number is used when there is a clear order between the values of the predictor. The use of embedding layers or columns to code categorical values as dense vectors has not been considered in this work.

The original dataset contemplated in this research [11] counted with 63 predictors and 16535 instances: 4205 of the positive class and 12330 of the negative class. After the new encoding for DL networks, the dataset includes 5573 variables. Regarding the different amount of positive and negative case, this situation is known as imbalanced classification. As a result, a high accuracy is achieved by just predicting always the negative class. This problem is also typical in the application of machine learning to medical domains.

Some approaches to cope with this situation include: (1) penalized models; (2) undersampling the over represented class (negative); (3) oversampling the underrepresented class (positive); and, (4) generating synthetic samples. Section 4 shows several experiments based on oversampling the underrepresented class which is the transformation that leads to the most challenging evaluation [11].

3.3 Data Mining

In this phase, machine learning and data mining paradigms are applied to create a hypothesis that explains the observations [13]. *Deep feedforward networks* (DFN) have been used in the data mining phase of the KDD for the classification problem. Some of the main hyperparameters to define different networks architectures are: number of layers; number of neurons per layer; neuron activation function; dropout layers, method, and parameter; regularization method, and parameter; learning rate; momentum; batch size; loss function; and, number of epochs.

Most of these hyperparameters have well known effects in underfitting and overfitting situations. For instance, an overfitting state can be improved by increasing the regularization parameter, increasing the dropout layers, or decreasing the number of epochs.

The *Rectified Linear Unit* (ReLU) activation function is used in the neurons, a SoftMax layer is employed as output, the cross entropy is the loss function considered, and the batch size is of 128 samples. The experiments with these DFNs are compared with previous results [11] training *random forest* (RF) and LR.

3.4 Evaluation

For the evaluation of the models, the cross validation is typically contemplated when the performance allows it. Using 10 folds involves rebuilding the machine learning model for the data 11 times. The training in DL networks has a considerable computational cost, making the cross validation too expensive. More importantly, when oversampling methods are applied, the cross validation method leads to overoptimistic results since the validation fold considers instances that are also present in the training folds.

Thus, the classic partition between training and validation has been undertaken ensuring: (1) the splitting (80/20% is considered) preserves the overall class distribution of the data (25% of positive cases); (2) and, the oversampling is performed after this splitting both in the training and the validation data.

A third partition for testing is not considered given that: (1) the limited positive cases with regard to the negative examples; and, (2) the results with DL are compared with the ones achieved by RF and LR obtained in [11]. In previous work, default values for the hyperparameters of the learning algorithms were used, i.e. validation results were not employed to optimize hyperparameters. Moreover, since there is hyperparameters optimization, the accuracy has to be reported not only for the validation set but also for the training. The difference between these values indicates a possible underfitting or overfitting situation and the strategies to improve them.

4 Results

Experiments were performed in a computer with 8 GB RAM memory with a QuadCore Intel i7-4710HQ 3300 MHz process and a nvidia geforce GTX 850M graphic video card with 2 GB dedicated. The experimenters lasted between 10 min and 2 hours depending on the neural network architecture. Keras[1] has been used in the experiments. Keras is an open source neural network library written in Python capable of running on top of TensorFlow, CNTK, or Theano. TensorFlow[2] has been used as backend scientific computing library.

As explained in Sect. 3.2, all experiments use oversampled data on the positive class with random sampling replacement before splitting the validation data.

---

[1] https://keras.io/.

[2] https://www.tensorflow.org/.

Table 1 details the hyperparameters used to define different artificial neural networks architectures, see Sect. 3.3. For each configuration, the accuracy is reported both for the training and the validation data. Precision and recall for the validation data are also reported, see Sect. 3.4. Moreover, to compare with previously used methods [11], the results with LR and RF are also reported.

Table 1. Experiments results. Columns description: Id.: Experiment identification. L: Number of Layers. Neurons: Neurons in each layer (range). RL1: L1 regularization parameter. Drop: Dropout rate (range or fixed). The dropout configuration can include with the range: n, representing normal dropout instead of the alpha dropout; h to indicate dropout in half the layers. LR: Learning Rate. E: Epochs. Acc.Tr: Accuracy in training. Acc: Accuracy in validation. P: Precision in validation. R: Recall in validation.

Id.	L	Neurons	RL1	Drop	LR	E	Acc.Tr	Acc	P	R
1 - RL	–	–	–	–	–	–	–	55.5%	78.0%	15.4%
2 - RF	–	–	–	–	–	–	–	**67.8%**	**88.6%**	**40.9%**
3	1	512	0.01	–	0.001	50	96.4%	70.4%	80.3%	54.1%
4	2	128–256	0.01	0.3–0.4 n	0.001	50	97.0%	69.4%	78.7%	53.2%
5	3	64–256	0.01	0.2–0.4 n	0.001	50	96.1%	70.6%	80.5%	54.3%
6	3	256–512	0.001	0.3–0.4 n	0.0001	80	98.4%	70.3%	79.7%	54.5%
7	10	1024–5096	0.01	–	0.001	50	99.2%	68.6%	79.4%	50.2%
8	10	512–2048	0.01	0.2	0.0001	10	92.4%	70.7%	78.7%	56.8%
9	10	512–2048	0.001	0.2	0.0001	20	98.3%	69.2%	78.9%	52.3%
10	20	512–2048	0.01	0.2	0.001	20	95.3%	70.0%	78.2%	55.6%
11	20	512–2048	0.01	0.2	0.001	70	97.4%	68.8%	77.6%	52.9%
12	20	512–2048	0.001	0.2	0.0001	100	99.5%	67.2%	83.7%	42.8%
13	20	512–5096	0.01	0.1	0.001	100	91.7%	57.1%	54.2%	92.5%
14	20	512–5096	0.01	0.1	0.0001	20	77.7%	**74.0%**	69.9%	**84.4%**
15	20	512–5096	0.01	0.1	0.0001	100	86.4%	**72.5%**	**81.5%**	58.3%
16	40	512–2048	0.001	0.2 h	0.0001	100	99.9%	68.9%	78.5%	51.9%

5 Discussion

The experiments id3 to id13 and id16 show a clear overfitting, with more than a 20% difference between the accuracy for training and validation data. Even with this clear limitation, an accuracy equal to or greater than RF is obtained except in experiments id12 and id13. The recall of these experiments is also markedly superior to RF: more than 10% in most experiments. The recall is particularly exceptional, 92.5% in experiment id13, but at the expense of a very low accuracy, which leads to a model of type "everything is a case of social exclusion". Finally, precision is lower than the RF in all the experiments carried out, with experiment id12 being the one closest to it with 4.9% less precision.

Finally, experiments id14 and id15 are those that offer less overfitting. In addition, they also give the best balance between accuracy and recall (experiment id14); and, between accuracy and precision (experiment id15). Both selected experiments id14 and id15 have similar neural network architecture but the number of epochs is different: 20 vs 100. Figure 1 shows the evolution of the loss function and accuracy with respect to the number of epochs for these two selected experiments. The figure shows that increasing the number of epochs could reduce the error in experiment id14 while it does not seem beneficial for experiment id15. Therefore, it could be interesting to choose a number of epochs between the two cases. The slow evolution of the loss function displayed in the figures is motivated by the low learning rate, which also have obtained the best results.

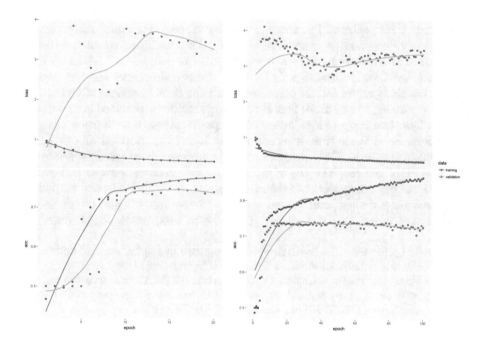

Fig. 1. Training and validation accuracy evolution per epochs, experiment id 14 (on the left) and id 15 (on the right).

Although experiment id12 is the one that offers results more similar to RF, RF is still suggested as a prediction model to achieve the best precision or positive predictive value, which is a quality measure in prediction, i.e. most of the predictions of cases as socially excluded will be correct. Furthermore, experiment id14 offers a DL model that improves substantially the recall or probability of detection of a social exclusion condition, which is a quantity measure, i.e. most of the social exclusion cases will be detected. In addition, model id14 offers the best accuracy when compared with a numerous experiments that include those described in Table 1, those presented in previous works [11], and in a large number of alternative experiments not presented because of its inferior results.

6   Conclusion and Future Works

This paper introduces a model to predict the risk of suffering chronic social exclusion with Deep Learning (DL). The deployed model obtains an 84.4% recall and a 74.0% accuracy. These results outperform previously models based on the logistic regression and random forests (RFs) [11]. On the other hand, the RF model gets better results in precision, 88.6% vs 69.9%, and therefore it is still useful in combination with the DL model.

These artificial intelligence powered tools are an important asset to generate automatic alerts of possible social exclusion cases. Then social workers can further study their patients. The recovery process is accelerated noticeably because of the early detection as in medical diseases. The machine learning predictive models are fed with data from a whole Spanish region: eleven databases from the social services of Castilla y León. Both the trained model and the dataset can be obtained under formal agreement with the Social Services of CyL.

As explained, the artificial neural networks architectures tested are relatively simple: less than forty hidden layers and experiments with a duration of up to two hours. Even these simplistic approaches have outperformed all our previous models accuracy. Therefore, our main future work is to test new and more complex architectures. The use of different transformations such as embedding columns to code categorical variables as dense vectors will also be contemplated. In a more general sense, multi-agent system architectures will also be considered to study reputation [12] and privacy [14] issues in these prediction services.

Acknowledgements. This research work is supported by the Universidad Politécnica de Madrid under the education innovation project "Aprendizaje basado en retos para la Biología Computacional y la Ciencia de Datos", code IE1718.1003; and by the Spanish Ministry of Economy, Indystry and Competitiveness under the R&D project Datos 4.0: Retos y soluciones (TIN2016-78011-C4-4-R, AEI/FEDER, UE).

References

1. European Commission's DG for Employment, Social Affairs & Inclusion. http://ec.europa.eu/social/main.jsp?catId=751. Accessed Feb 2018
2. Face Time: How AI Can Diagnose Rare Genetic Diseases Faster. https://blogs.nvidia.com/blog/2017/03/14/ai-diagnose-rare-genetic-diseases/. Accessed Feb 2018
3. Manulife Philippines. Calculate your risk, your partner's risk or both. http://www.insureright.ca/what-is-your-risk. Accessed Feb 2017
4. Mayo Clinic. Heart Disease Risk Calculator. http://www.mayoclinic.org/diseases-conditions/heart-disease/in-depth/heart-disease-risk/itt-20084942. Accessed Feb 2017

5. Esteva, A., Kuprel, B., Novoa, R.A., Ko, J., Swetter, S.M., Blau, H.M., Thrun, S.: Dermatologist-level classification of skin cancer with deep neural networks. Nature **542**, 115–118 (2017)
6. Fayyad, U.M., Piatetsky-Shapiro, G., Smyth, P.: Advances in knowledge discovery and data mining. In: Data Mining to Knowledge Discovery: An Overview, pp. 1–34. American Association for Artificial Intelligence, Menlo Park (1996)
7. Haron, N.: On social exclusion and income poverty in israel: findings from the european social survey, pp. 247–269. Springer, Boston (2013)
8. Lafuente-Lechuga, M., Faura-Martínez, U.: Análisis de los individuos vulnerables a la exclusión social en españa en 2009. Anales de ASEPUMA (21) (2013)
9. Ramos, J., Varela, A.: Beyond the margins: analyzing social exclusion with a homeless client dataset. Social Work Society **14**(2) (2016)
10. Serrano, E., Bajo, J.: Towards social care prediction services aided by multi-agent systems. In: Montagna, S., Abreu, P.H., Giroux, S., Schumacher, M.I. (eds.) Agents and Multi-Agent Systems for Health Care - 10th International Workshop, A2HC 2017, São Paulo, Brazil, May 8, 2017, and International Workshop, AHEALTH 2017, Porto, Portugal, 21 June 2017, Revised and Extended Selected Papers, vol. 10685, Lecture Notes in Computer Science, pp. 119–130. Springer (2017)
11. Serrano, E., del Pozo-Jiménez, P., Suárez-Figueroa, M.C., González-Pachón, J., Bajo, J., Gómez-Pérez, A.: Predicting the risk of suffering chronic social exclusion with machine learning. In: Omatu, S., Rodríguez, S., Villarrubia, G., Faria, P., Sitek, P., Prieto, J. (eds.) 14th International Conference Distributed Computing and Artificial Intelligence, Advances in Intelligent Systems and Computing, DCAI 2017, vol. 620, Porto, Portugal, 21–23 June 2017, pp. 132–139. Springer (2017)
12. Serrano, E., Rovatsos, M., Botía, J.A.: A qualitative reputation system for multi-agent systems with protocol-based communication. In: van der Hoek, W., Padgham, L., Conitzer, V., Winikoff, M. (eds.) International Conference on Autonomous Agents and Multiagent Systems, AAMAS 2012, vol. 3, Valencia, Spain, 4–8 June 2012, pp. 307–314. IFAAMAS (2012)
13. Serrano, E., Rovatsos, M., Botía, J.A.: Data mining agent conversations: a qualitative approach to multiagent systems analysis. Inf. Sci. **230**, 132–146 (2013)
14. Serrano, E., Such, J.M., Botía, J.A., García-Fornes, A.: Strategies for avoiding preference profiling in agent-based e-commerce environments. Appl. Intell. **40**(1), 127–142 (2014)
15. Suh, E., TiffanyVizard, P., AsgharBurchardt, T.: Quality of life in Europe: social inequalities. 3rd European Quality of Life Survey (2013)
16. Vougas, K., Krochmal, M., Jackson, T., Polyzos, A., Aggelopoulos, A., Pateras, I.S., Liontos, M., Varvarigou, A., Johnson, E.O., Georgoulias, V., Vlahou, A., Townsend, P., Thanos, D., Bartek, J., Gorgoulis, V.G.: Deep learning and association rule mining for predicting drug response in cancer. A personalised medicine approach. bioRxiv (2017)

Combining Image and Non-image Clinical Data: An Infrastructure that Allows Machine Learning Studies in a Hospital Environment

Raphael Espanha[1,2,3], Frank Thiele[1,2], Georgy Shakirin[1,2],
Jens Roggenfelder[1], Sascha Zeiter[1], Pantelis Stavrinou[1],
Victor Alves[4(✉)], and Michael Perkuhn[1,2]

[1] University Hospital Cologne, Cologne, Germany
r.a.espanha@gmail.com, {frank.thiele,georgy.shakirin,
jens.roggenfelder,sascha.zeiter,pantelis.stavrinou,
michael.perkuhn}@uk-koeln.de
[2] Philips Research Laboratories Aachen, Aachen, Germany
[3] Department of Informatics, University of Minho, Braga, Portugal
[4] Centro Algoritmi, University of Minho, Braga, Portugal
valves@di.uminho.pt

Abstract. Over the past years Machine Learning and Deep Learning techniques are showing their huge potential in medical research. However, this research is mainly done by using public or private datasets that were created for study purposes. Despite ensuring reproducibility, these datasets need to be constantly updated.

In this paper we present an infrastructure that transfers, processes and stores medical image and non-image data in an organized and secure workflow. This infrastructure concept has been tested at a university hospital. XNAT, an extensible open-source imaging informatics software platform was extended to store the non-image data and later feed the Machine Learning models. The resulting infrastructure allowed an easy implementation of a Deep Learning approach for brain tumor segmentation with potential for other medical image research scenarios.

Keywords: Medical image research · Clinical environment
Machine learning workflow · Brain tumor segmentation
Deep learning · Convolution neural networks · Docker · XNAT

1 Introduction

The industrial and scientific world have witnessed the revolutionary changes in recent years brought about by the development of information technology. These changes have been modernizing many disciplines and industries, and biomedicine is no exception. The importance of biomedical information technology and its application has expanded beyond the boundary of health services, leading to the discovery of new knowledge in life sciences and medicine. Many emerging areas have recently been

© Springer International Publishing AG, part of Springer Nature 2019
F. De La Prieta et al. (Eds.): DCAI 2018, AISC 800, pp. 324–331, 2019.
https://doi.org/10.1007/978-3-319-94649-8_39

developed, including e.g., health informatics, bioinformatics, medical imaging informatics, neuroinformatics, medical biometrics [1].

The increasing ability to obtain digital information in medical and biological neuroimaging research has led to a vast increase of scientific data from across a variety of spatial and temporal scales [2]. Moreover, neuroimaging laboratories are faced with several data management challenges. Within the laboratory, data must be passed through a series of capture, quality control, processing, and utilization steps. Additionally, neuroimaging studies routinely incorporate measures from a range of other experimental methods, e.g. genetic, clinical and neuropsychological, which must be integrated, in a secure way, with the imaging measures into a unified dataset [3]. This means that, apart from the subjects, the imaging process is intrinsically a digital electronic enterprise: image acquisition, processing, databasing, and sharing are all accomplished in the digital domain. Each step of this process, therefore, affords the opportunity to capture all the pertinent information that characterizes the steps [4].

Hence, many organizations, over the past years, started to develop several tools to facilitate collaborative research and data sharing in neuroimaging. Different technologies and approaches were used such as ontologies [5], data format exchange [6] as well as databases, and data management systems including the Human Imaging Database (HID) [7], LONI (Laboratory of Neuro Imaging) IDA (Image Data Archive) [2] and the eXtensible Neuroimaging Archive Toolkit (XNAT) [3, 8]. Amongst those technologies, XNAT is one of the most commonly used and with an increasing adoption in the community [8, 9].

In clinical environment, the workflow is predefined and optimized for diagnosis and therapy decisions in a well-known clinical question, while in medical research the aim is to gain knowledge, not only about individuals, but about cohorts resulting in a heterogeneous environment of toolsets and data where new algorithms and workflows still have to be validated. With the exponential increase in available biomedical data repositories it is important for the studies to be close to real world in its size and scope [10].

In medical image research it is common to use public datasets from repositories like The Cancer Imaging Archive (TCIA) [11] or Multimodal Brain Tumor Segmentation (BraTS) challenge [12] which offer a static and already known datasets. Although it allows reproducibility of the results between several studies, it does not fully address the problem of a real environment where the data is constantly growing and updating which might lead to different study results over time. Additionally, these studies are done in a close environment, where the dataset is transferred locally inside the hospital network after pseudonymization or anonymization with all the requirements for the study and the pipeline is run locally.

1.1 Extensible Neuroimaging Archive Toolkit (XNAT)

XNAT is an open source imaging informatics software platform developed by the Neuroinformatics Research Group at Washington University and it was designed to facilitate common management and handling of neuroimaging and associated data [3]. XNAT's core functions manage importing, archiving, processing and securely distributing imaging and related study data. It captures data from multiple sources, to maintain the data in a secure repository, and to distribute the data to approved users (Fig. 1).

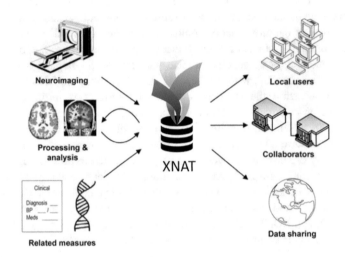

Fig. 1. XNAT captures data from multiple sources and distributes it to a variety of end users. Based on [3].

Therefore, XNAT implements a workflow to support the quality, integrity, and security of data from acquisition, and storage of analysis and public sharing. Imaging data from the scanners enter the workflow using mechanisms like the Digital Imaging and Communications in Medicine (DICOM) "pushes", Secure File Transfer Protocol (SFTP), or portable hard media. Non-imaging data such as clinical assessments, subject demographics and genetic measures are passed via web-based forms, spreadsheet uploads, or XML (eXtensible Markup Language) [7]. Moreover, it allows the development of custom data types which can easily be added to a XNAT instance.

1.2   Advanced Image Viewing and Processing

For the overall infrastructure, an application for advanced image viewing as well as image processing is required. We use a commercial research solution that is available as zero-client server application within the hospital's private network (IntelliSpace Discovery, Philips Healthcare, Best, The Netherlands). Accessibility from a standard web browser promotes collaboration between different medical departments. The application specifically offers functionality for image pre-processing (bias field correction, registration, organ mask segmentation). Furthermore, Machine Learning (ML) results like tumor segmentation can be visually inspected and further processed, e.g. as input for radiomics analysis.

1.3   Containerization Platform

Docker is a way to isolate a process from the system on which it is running. It isolates the code written to define an application and the resources required to run the application from the hardware on which it runs [13]. To completely encapsulate software applications, Docker uses Linux Containers (LXC) which are an operating system level virtualization technology that creates a completely sandboxed virtual environment in Linux without the overhead of a full-fledged Virtual Machine (VM) [14]. All steps of the scientific process, from data collection and processing, to analyses, visualizations and conclusions depend ever more on computation and algorithms which makes computational reproducibility to receive an increasing level of attention throughout the scientific community [13]. Docker solves this problem via containerization ensuring compatibility across platforms and, therefore, eliminating the "works on my machine" common issues in research and software developing.

2   Infrastructure Development

Developing an infrastructure using state of the art models in a hospital environment requires the first priority on data security and a good balance between scalability and extensibility. With that in mind, magnetic resonance images (MRI) (and other medical images) necessary for a research study are transferred to the image processing server (ISD, Philips IntelliSpace Discovery) in a hospital intranet, from PACS (Picture Archiving and Communication System) (Fig. 2-A). Here, radiologists and researchers can visualize and create tumor segmentations on registered MRI images (Fig. 2-B). Both Volumes of Interest (VOIs) and registered MRI are transferred to XNAT Server using ISD-XNAT Interface (Fig. 2-C). From hospital HIS (Hospital Information System), clinical information for a study, such age or blood test results, could be collected and transferred to XNAT Server (Fig. 2-D). For the current study, we did not implement an interface to HIS and clinical information was obtained from the Neurosurgery department as a study table. Both image and non-image data stored in XNAT are pulled from Docker System using PyXNAT, a Python module that interacts with XNAT by exposing the XNAT Web Services and unify their features (Fig. 2-E) [8]. Machine learning workflow can now use both image and non-image data (Fig. 2-F). Any image result predicted from the models can then return to ISD to be visualized.

2.1   Extending XNAT Data Types

Despite being able to deal with non-image data types, XNAT needs to be extended to the user's needs if new and specific data types are to be created. Creating a new data type in XNAT consists of creating a XML Schema Document that defines the datatype itself [9]. Using the new version of XNAT 1.7, custom data types and new functionality can be added via plugin, that is, in the form of a fully compiled, self-contained package that is separated from XNAT server but that, once enabled, runs in the same process space with XNAT as a fully integrated extension to XNAT server core. The plugin is written in JAVA and, after compiling it, the generated JAR (Java ARchive) file must be

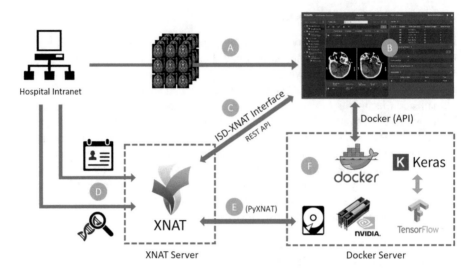

Fig. 2. Machine Learning Infrastructure at the hospital.

placed in Plugins folder inside XNAT files system. After restarting the Tomcat server, the new data type is ready to be used in XNAT web interface as well as being accessible from its REST API.

3   Combining Image and Non-image Data for Brain Tumor Segmentation

In order to test the extended infrastructure, we studied a DL model that combined MRI image and non-image clinical information for brain tumor segmentation. The image data consisted of FLAIR sequences of 103 patients diagnosed with Glioblastoma tumor, along with the corresponding manually segmented edemas as VOIs. The patient's age, the presence of MGMT and the volume information of tumor left after the surgery were taken as clinical information parameters.

After pseudonymization the data was combined in a single CSV (Comma Separated Values) file where each row refers to a subject with the correspondent MRI sequence and masks paths along with the non-image values of each parameter, already cleaned and normalized. In this way, train/test splits were easier and faster to apply since the image data was not loaded. From 103 subjects, 61 were used for training, 21 for validation and 21 for testing the model.

MRI images are altered by the bias field distortion making the intensity of the same tissues to vary across the image. For that reason, the scan images were normalized to White Matter (WM) average using masks, that is, dividing the image voxel values by the mean values of intersected voxels. To balance the training classes, all tumor voxels were kept along with randomly selected non-tumor voxels on a similar amount inside the brain mask. This brain mask was also used to remove the black background and cranial voxels. In this study we took a 2D patch approach [15, 16]. From the image data of previous step, 9×9 patches were created with a stride of 1 which greatly increased the amount of data. The central voxel of each patch defined the label value (tumor or non-tumor) as it was intersecting, or not, the VOI mask. All patches' voxels were then normalized to their mean value and unit variance.

The model which combined image and non-image data (Fig. 3) takes two input layers. The image input is followed by convolution and pooling layers which create and learn from feature maps. On the other side, regular Density Connected Layers (DCL) were used to learn from clinical inputs. After flattening the final layers of both branches with the same shape, a merge layer applies a dot product creating a single and merged layer.

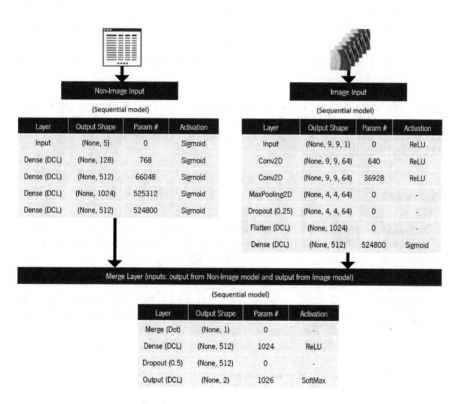

Fig. 3. Summary of combined model.

A similar model was created without the non-image input for comparison. Each model was trained in each GPU using different containers from the same docker image. This parallelized both training processes allowing a quick comparison of results. The DICE Similarity Coefficient (DSC) was used to compare both combined and image-only models [17].

4 Conclusions

A ML infrastructure in the hospital was extended to allow an easy development of DL approach combining medical images and non-image clinical information. XNAT customization was essential to store heterogenous medical data types. Docker allowed to consume these data in a secure way with scalability potential. A DL model was developed in this infrastructure which combined MRI image and non-image clinical information in brain tumor segmentation scenario. The combined model achieved a slightly better DSC values (0.894 ± 0.025) over image only model (0.882 ± 0.025). This approach also resulted in an end-to-end classifier where convolution layers perform the feature extraction in MRI input without the need of this additional step. Furthermore, the developed infrastructure supports the continuation of this kind of studies at the hospital. It offers a dynamic approach for medical image studies where acquired data (image and non-image) can immediately enter the workflow and improve/update ML models.

Acknowledgements. This work was supported by University Hospital Cologne, Department of Radiology and Philips Research Aachen. It was also supported by COMPETE: POCI-01-0145-FEDER-007043 and FCT – Fundação para a Ciência e Tecnologia within the Project Scope: UID/CEC/00319/2013.

References

1. Feng, D.D.: Biomedical Information Technology. Academic Press, Burlington (2011)
2. Van Horn, J.D., Toga, A.W.: Is it time to re-prioritize neuroimaging databases and digital repositories? Neuroimage **47**(4), 1720–1734 (2009)
3. Marcus, D.S., Olsen, T.R., Ramaratnam, M., Buckner, R.L.: The extensible neuroimaging archive toolkit. Neuroinformatics **5**(1), 11–33 (2007)
4. Poline, J.-B., et al.: Data sharing in neuroimaging research. Front. Neuroinform. **6**, 9 (2012)
5. Bug, W.J., et al.: The NIFSTD and BIRNLex vocabularies: building comprehensive ontologies for neuroscience. Neuroinformatics **6**(3), 175–194 (2008)
6. Gadde, S., et al.: XCEDE: an extensible schema for biomedical data. Neuroinformatics **10**(1), 19–32 (2012)
7. Keator, D.B., et al.: A national human neuroimaging collaboratory enabled by the Biomedical Informatics Research Network (BIRN). IEEE Trans. Inf Technol. Biomed. **12**(2), 162–172 (2008)
8. Schwartz, Y., et al.: PyXNAT: XNAT in python. Front. Neuroinform. **6**, 12 (2012)
9. HHMI Washington University School of Medicine, Harvard University, "XNAT Documentation." https://wiki.xnat.org/documentation. Accessed: 29 Oct 2016
10. de Herrera, A.G.S., Demner-Fushman, D., Bedrick, S.: Evaluating performance of biomedical image retrieval systems—an overview of the medical image retrieval task at ImageCLEF 2004–2013. Comput. Med. Imaging Graph. **39**, 55–61 (2015)
11. Clark, K., et al.: The cancer imaging archive (TCIA): Maintaining and operating a public information repository. J. Digit. Imaging **26**(6), 1045–1057 (2013)
12. Menze, B.H., et al.: The Multimodal Brain Tumor Image Segmentation Benchmark (BRATS). IEEE Trans. Med. Imaging **34**(10), 1993–2024 (2015)

13. Boettiger, C.: An introduction to Docker for reproducible research. ACM SIGOPS Oper. Syst. Rev. **49**(1), 71–79 (2015)
14. Cook, J.: "Docker", in Docker for Data Science, pp. 29–47. Apress, Berkeley (2017)
15. Cui, Z., Yang, J., Qiao, Y.: Brain MRI segmentation with patch-based CNN approach. In: Chinese Control Conference, CCC, 2016 August, pp. 7026–7031 (2016)
16. Pereira, S., Pinto, A., Alves, V., Silva, C.A.: Brain tumor segmentation using convolutional neural networks in MRI Images. IEEE Trans. Med. Imaging **35**(5), 1240–1251 (2016)
17. Zou, K.H., et al.: Statistical validation of image segmentation quality based on a spatial overlap index1. Acad. Radiol. **11**(2), 178–189 (2004)

GarbMAS: Simulation of the Application of Gamification Techniques to Increase the Amount of Recycled Waste Through a Multi-agent System

Alfonso González-Briones[1(✉)], Diego Valdeolmillos[1],
Roberto Casado-Vara[1], Pablo Chamoso[1], José A. García Coria[1],
Enrique Herrera-Viedma[2], and Juan M. Corchado[1,3,4]

[1] BISITE Digital Innovation Hub, University of Salamanca,
Edificio Multiusos I+D+i, 37007 Salamanca, Spain
{alfonsogb,dval,rober,chamoso,
jalberto,corchado}@usal.es
[2] Department of Computer Science and Artificial Intelligence,
University of Granada, 18071 Granada, Spain
viedma@decsai.ugr.es
[3] Department of Electronics, Information and Communication,
Faculty of Engineering, Osaka Institute of Technology, Osaka 535-8585, Japan
[4] Pusat Komputeran dan Informatik, Universiti Malaysia Kelantan,
Karung Berkunci 36, Pengkaan Chepa, Kota Bharu, Kelantan, Malaysia

Abstract. The increase in population is increasing the growth of the number of residues. A large amount of this waste can be recycled so that it does not remain in uncontrolled landfills, pollute air, land or water. Although many campaigns and policies of recycling have been developed all-way there is not a total awareness about this problem and a large number of waste does not end up in the right place to be recycled. It is necessary to increase the amount of recycled waste and for this citizen participation is key. It is vital to involve the population in a more active way to recycle through some kind of social benefit. For this reason, GarbMAS proposes a system that generates a motivation for citizen participation by reducing the garbage tax applied by its local government. Thus, the increase in the amount of waste collected to be recycled is expected. GarbMAS employs a multi-agent system that simulates data collection and efficient waste management in cities through gamification techniques to produce the change motivation of citizens and therefore increase participation and recycled quantity. The case study in which the simulation was carried out showed an increase in citizen participation by 34.2% and an increase of 29.4% in the amount of waste collected.

Keywords: Behavioral change · Simulation · Multi-agent system
Context-awareness · Social computing

J. A. García Coria—IEEE.

1 Introduction

The increase in population has led to an increase in the number of wastes. A large amount of this waste can be recycled so that it is not left in uncontrolled landfills, polluting the air, soil or water. Even though many recycling campaigns and policies have been developed, there is still not complete awareness of this problem and a large number of wastes do not end up in the right place to be recycled. It is a big problem that affects especially in large cities, as well as in the whole population of the planet, due to the increase in population, the development of modern human activities and high consumerism, which have greatly increased the amount of garbage that occur.

One of the solutions that were implemented to solve the problem of what to do with garbage, was to adopt measures that would allow recycling these wastes. In this sense, many countries began to build urban waste treatment plants to manage the waste generated in the metropolises. At the beginning of its implementation these plants only allowed the recycling of some waste such as paper and cardboard, glass or some plastic components allowing the recycling of other elements as the knowledge and technology in this field has evolved. However, for these plans to perform their recycling function it was necessary to develop a waste collection network. In the European Union there is no common method in waste management. Some countries have deployed a series of colored containers in which the user introduces the waste according to their typology, in other countries it is in supermarkets where the consumer is allowed to deliver a series of waste and containers for recycling. The use of these measures has allowed the recovery of numerous tons of garbage that were previously completely discarded in landfills, obtaining numerous benefits for both humans and the environment.

Although these measures have been widely accepted, there are still many tons of recyclable materials that end up in landfills without any reuse. The glass recycling rate in the European Union reached 73% in 2013 according to The European Container Glass Federation (FEVE) [15]. The recycling rate of Paper/cardboard is around 83%, plastic around 34.3% and wood around 37.7% [3, 18]. This shows that the recycling model of glass and paper/cardboard works properly, but it is necessary to achieve a greater percentage of recycling of the rest of materials.

One of the ways that would help increase the recycling rate would be a more active participation of citizens in the recycling chain. One of the ways to achieve greater involvement of citizens consisted in granting benefits to citizens who participate more actively in the recycling chain. The concept of gamification is a technique that allows the process of recycling to be stimulated and carried out more dynamically so that certain results are achieved. Therefore, it is necessary to develop a system that allows increasing the rate of recycling of all materials through greater citizen participation. The objective of the paper presented in this paper is to provide a new multi-agent system for the development of context-aware that, through the granting of economic benefits, encourages citizen participation by increasing the number of waste collected in the recycling chain.

The development of this system is carried out using CAFCLA (Context-Aware Framework for Collaborative Learning Applications) as a basis for its technical and social implementation [5, 6]. The system will learn from the actions carried out by citizens to adapt to them and provide new solutions that allow increasing the rate of recycling and citizen participation. For this purpose, the system is developed using the mul-ti-agent (MAS) systems paradigm for learning the actions of users and making decisions.

This article is organized as follows: Sect. 2 describes the state of the art of multi-agent systems and gamification, Sect. 3 describes the GarbMAS system, Sect. 4 presents the case-study and Sect. 5 conclusions.

2   Background

This section shows the different methods currently implemented for the recycling of waste in the European Union, detailing its characteristics, advantages and disadvantages. It also explains the current state of the techniques used by the proposed system for its adoption within the problem of waste recycling, and how these techniques will help us increase citizen participation.

2.1   European Union Recycling Methods

To prevent large quantities of materials from being deposited in landfills and becoming garbage, it was necessary to develop mechanisms that would allow selective collection that would allow subsequent recycling. Within the European Union each country was developing different methods to deal with the problem of garbage collection. In Spain, the first container for glass collection was installed in 1982. It was in that same year, when a model of cooperation between Autonomous Communities, local corporations and manufacturers of glass containers was established to manage the recycling of this element. In 1994, Directive 94/62/EC dictates the framework on which all the legislation of each of the countries of the European Community will be developed. Within this European legislative framework, two models of predominant recycling were developed in which the majority of the countries that integrate the EU are grouped.

The first of these lies in the deployment of containers identified by colors according to the container for which they are focused. Within this model there are small differences according to each country, such as the color of the container for each container. In Spain, only three containers are deployed: blue for paper and cardboard, yellow for plastic waste and green for glass containers, apart from the urban waste and oil container. In France, this system has five containers, with yellow being used for the collection of paper and cardboard, blue for collecting plastic and aluminum, and green for collecting glass and two additional containers for domestic and organic waste. The collection of waste is governed by an annual plan that sets the days to throw away the garbage (usually two days per week). The garbage must be deposited the day before the day set in the planning, being able to receive a fine if they do not recycle. In Italy, the system also employs the deployment of color containers, which vary with respect to Spain and France. However, it is also governed by a municipal collection plan and may also receive a fine if it does not recycle. In Norway the system is slightly different, there is a container for paper and cardboard and another for organic waste. The collection of plastic is done by introducing these containers in a special bag that is deposited next to the paper and cardboard container. The glass is deposited in special containers next to the supermarkets.

The second method is Deposit Refund System (DRS), which is being implemented in Germany, Sweden and Denmark. In this system the containers are engraved with a recycling tax which is paid at the time of purchase, once the container is deposited in perfect condition in the machines destined for collection, a ticket is received in which it is obtained € 0.25 for each container. This method in some countries such as Germany is complemented with the method of containers by colors.

Both methods are widely accepted, however, they require an evolution to bring about an improvement in the recycling system. The two methods previously presented have several drawbacks. The recycling system by means of colored containers only allows the collection of some materials and the DRS model only serves to recover the containers of water, soft drinks and cereals, whether they are plastic or metal containers. Specifically, it manages 8% of the containers, while, in the colored container model, it handles 80% of the containers. The operation of the DRS model is that its operation is linked to machines installed in supermarkets, so that citizens only deposit their containers at set times. Another big drawback presented by the DRS model is that the containers are engraved with a tax which the citizen has to pay when buying a product. This amount is not reimbursed in case of damage or dent, so there is no real guarantee to recover the money.

It has been detected that in the system of recycling through the system by containers, it is done by imposition (possibility of being fined). This makes this process to be carried out with reluctance and without paying attention to the correct recycling. On the contrary, the DRS system simply recovers the money that was paid for each container when they were purchased.

2.2 Change in User Behaviour by Means of Gamification Techniques

Within the educational sphere, the games have been part of a set of activities that have been conducive to learning in a playful way. These types of games have used the typical features of entertainment games to address learning tasks, understanding or social impact, addressing both cognitive and affective dimensions. The dynamic and playable concepts that make participation and interaction a stimulus in the learning process are used for this purpose.

The use of game features for a purpose beyond entertainment and more focused on learning is gamification. The gambling in the recycling field allows to increase the citizen participation in these tasks, thanks to the obtaining of rewards when they carry out correct actions and receiving penalties in the opposite case.

The use of gamification will increase the level of participation while increasing citizen commitment to the environment. This technique has recently been used in fields such as energy efficiency in the home with very satisfactory results [4, 6]. However, it has been detected that it has not been used in the field of recycling, so it is interesting to know if this technique produces satisfactory results in this field as well.

2.3   Multi-agent Systems

Multi-agent systems (MAS) have been widely accepted by the scientific community due to their intrinsic characteristics that allow them to be used in a wide variety of contexts. They are widely used to model behaviors, simulate situations or solve difficult or impossible problems for a monolithic system.

The agents of a MAS communicate, coordinate, interact and cooperate for the performance of different actions and for these reasons that have been applied in work with objectives as varied as obtaining genes with behavioral patterns associated with a given disease [8, 9], detection of drivers under the influence of drugs [1], classification of facial images according to gender and age [10], optimization of energy consumption [11, 12] and all of them with good results.

The agents of a MAS have their own autonomy which allows them to interact with each other without any action on the part of the users, the ability to perceive changes in the environment and react to them makes it a very propitious methodology for the data capture, learning behavior patterns and decision making in the face of changes. That is why the multi-agent systems have been used in different proposals within the field of recycling, from behavioral simulation, learning behaviors associated with recycling, efficient management of waste within supply chains or learning of actions of user behavior.

Next, it highlights some outstanding related work. Meng *et al.* They presented a platform that performs simulations of actions to recycle solid domestic waste together with social surveys [13]. The platform is based on a set of agents that simulate behavior in domestic scenarios, making decisions in relation to simulations. This study was carried out under two scenarios whose conditions are fixed. The system is based on a multi-agent system in which three agents simulate the behavior and decision-making under two scenarios of fixed conditions. In this work, a multi-agent system is used that simulates behavior under various conditions, but does not encourage citizen participation in recycling tasks. Yang *et al.* developed a system for evaluating the economic sustainability of a waste recovery system using agent-based simulation [19]. Sustainability metrics are used to evaluate and compare two recovery processes, which allows to know the economic sustainability, recycling rate in the process and economic efficiency. To make automatic judgments about the effectiveness of effective waste recycling Mishra *et al.* they developed a multi-agent system whose agents coordinate efficiently to perform different tasks such as waste categorization, transport, waste recycling, waste management and reusable product allocation [14]. The agents allow to manage all the activities that take place in the supply chain, the management of these complex tasks is carried out efficiently thanks to the cooperation and communication of the agents that model the supply chain. Each new experience in the supply chain allows agents to learn to make new judgments.

Therefore, the proposed system will contribute to the efficiency management in garbage collection chains and efficient collection decision-making while encouraging the participation of citizens. As can be seen from the previous works, the use of multi-agent systems provides a series of advantages such as the possibility of simulating this type of problems and learning from the actions taken to make decisions. It is necessary to develop an architecture based on agents that adapts to the needs raised in our work (integration of different devices, distributed services, applications). For this reason, deploying a multi-agent system allows the management of context-independent environments, which makes it possible to manage waste collection services distributed within a Smart city. In addition, it allows rewarding users who participate in the recycling chain obtaining benefits. The use of agents allows to establish lines of communication with smartphones so that each user can know the amount of waste deposited to receive and the benefits obtained. For the design of the architecture, the model proposed by Rodríguez et al. [15], whose contribution is the ability to perform dynamic and adaptive planning to distribute tasks among agents (container status, transport management, gamification process or user management, among others).

3 Proposed System

In this section, we detail the technical development of the multi-agent system responsible for the collection and transmission of data, the gamification system and the participation of users.

3.1 Infrastructure Needed for Multi-agent System Management

In order for the multi-agent system to have access to the amount of waste that each citizen recycles, it is necessary to deploy an infrastructure that allows knowing the quantity, the type of waste, the user, the filling status of the containers, the degree of occupation of the nearest waste treatment plant. This infrastructure does not propose a change in the traditional business model of recycling, although it is necessary to incorporate a series of devices (i) a QR code reader to identify the user. (ii) GPS locator, indicates the coordinates in which the container is located. (iii) Weight sensor, weighs the amount of waste introduced in the container. (iv) Volumetric sensor, allows to know the state of filling of the container. (v) NarrowBand IOT (NB-IoT) communication technology, for the transmission of data to the system. The use of NB-IoT responds to being able to transmit data over long distances, up to 50 km in rural areas and between 5 and 10 km in urban areas. (vi) Solar panel, which feeds the different sensors deployed in the container so that the system is independent at the energy level.

When a citizen inserts a waste bag in the corresponding container, a data structure is generated with the following information: user id, type of waste, quantity of waste, container id, location of the container, state of filling the container. This data structure is sent through the MQTT protocol using the NB-IoT network to a local station. The local station will act as MQTT Broker sending the data in JSON form through REST services. A communication line is needed between each container and the waste recycling plant that allows sending a collection truck to the container site.

3.2 GarbMAS Architecture

As the GarbMAS requires an update of an associated cloud that may undergo changes such as increasing the number of containers, changing the location of the containers, etc. It is necessary that the management system is a highly dynamic platform that uses self-adapting capabilities at the time of execution. This allows the behavior of a maritime agent to be determined by the goals you want to achieve (amount of waste on sale, percentage of participation, etc.), while taking into account the goals of other agents and any changes that may arise in the environment. The core of the architecture is a group of deliberative agents that act as controllers and administrators of all applications and services. The functionalities of the agents are not within their structure, but modeled as services. This approach offers a greater capacity for error recovery and greater flexibility to change the behavior of agents at the time of execution. The use of a multi-agent system, as has been observed in the state of the art, allows us to adapt to a changing context such as the increase or change in the location of the containers. A MAS allows modeling a highly dynamic platform that uses self-adapting capabilities at runtime.

JADE framework has been chosen for the development of the GarbMAS. This Java framework provide mechanisms to incorporates agents that manage security aspects in order to measures so that the information collected is really the one that is transmitted and analyzed [7]. The architecture has been developed for the incorporation of multiple heterogeneous sensors, so that functionality can grow in the future for the incorporation of new services [16]. The system is specifically designed to analyze user behaviour, sensor data, include data from third parties and decision-making.

GarbMAS is organized in different layers according to the functionality of the agents, as shown in Fig. 1.

- **Manage Layer.** Agent that is in charge of the deployment of the rest of the main agents of each layer. This agent also coordinates the rest of the agents to attend to all requests of the system.
- **Smart City Layer.** This layer receives the user's recycling action data structures, the status of the containers is checked in order to make decisions. Big Data Engine Agents analyzes the data received through Complex Event Processing for pattern recognition. GarbMAS has an Open Data agent to acquire relevant data for the analysis, such as meteorological data.
- **Data Layer.** The agents in this layer are responsible for capturing data from the infrastructure deployed in the Smart City. IoT agents communicate with the agents deployed in the Infrastructure layer to build the data structure that collects the data about the waste deposit in a container, which is sent in JSON format through REST services. IoT Broker is responsible for the NGSI-to-NGSI (Next Generation Services Interface) conversion between IoT agents and the Context Broker agents, and the data is collected by Data Communication agents, which transmits them for analysis to Smart City Intelligence Layer.
- **Smart Government Layer.** In this layer, the results obtained by Smart City layer for the collection of waste from the containers or a greater deployment of containers in a certain area are managed if the demand for waste collection or a larger truck shipment is not met. This layer is responsible for the process and gamification with the user, subsidizing or penalizing users with a decrease or increase in the rate of garbage imposed by the local government.
- **Infrastructure Layer.** The agents in this layer act as Middleware between the users and the multi-agent system. Each container is associated with an agent that is

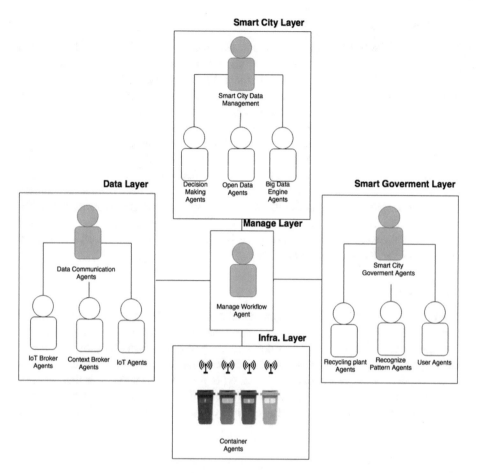

Fig. 1. GarbMAS architectures.

responsible for generating a data structure (quantity, type of waste deposited, user that recycles, state of filling of the container, degree of occupation of the nearest waste treatment plant). These agents through LPWAN send the information to the Data Layer agents.

4   Case-Study

In this section we will detail the characteristics of the case study in which GarbMAS has been tested, and the results obtained.

4.1 Smart City Experimental Set-Up

To evaluate if GarbMAS represents an advance in citizen participation in the recycling process. The system has been evaluated with data from an urbanization outside the urban core of Zaragoza with an estimated population of 2200 inhabitants, in which the deployment of thirty containers with the required infrastructure has been simulated (ten blue containers for paper and cardboard, ten containers yellow plastic waste, ten green containers for glass and ten red containers of urban waste not monitored), for this it has counted on the collaboration of the company of urban waste collection of the city.

The experiment has been divided into two phases and took place between the months of August and November 2017. The first phase of the experiment was carried out during the months of July and August, in this phase it has been counted the number number of people who recycle and the amount of waste collected in each container. The second phase was carried out during the months of September and October, in this stage the data were acquired exclusively by GarbMAS. In this second phase, the behavior of the users has been simulated.

4.2 Results

The process consists in that each agent simulates a user who log in by reading the QR code that identifies the container and identifies the user in a way that enables the opening of the container for depositing and weighing the waste, as shown in Fig. 2. Once the waste has been introduced, the container has the amount of waste, sends the generated data structure (amount of waste, type of waste deposited, user that recycles, state of filling of the container, degree of occupation of the waste treatment plant). Nearest waste through the local relay antenna deployed using LPWAN.

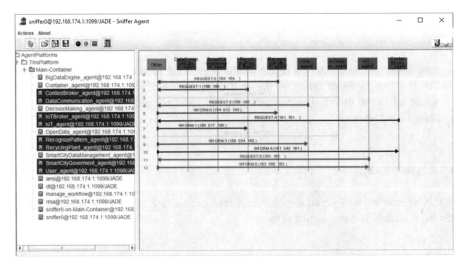

Fig. 2. Communication between the agents that are part of the Data Layer and Smart Government Layer

Table 1. Quantity (Kg) of waste collected from each container

	Before system	After system
Blue container	2,412.23	3,295.98
Yellow container	1,951.42	2,701.26
Green container	1,659.92	2,501.12
Total	6,023.57	8498,36

The information reaches the system and the agents of the different layers are responsible for making decisions such as sending a truck to collect the waste of a specific container if it is full, or updating the bonus or penalty in the user's profile. Each agent allows to visualize the achievements of the user accumulated during the current month (each month the profile is restarted, the achievements are not cumulative for the next month). If the user reaches the goals proposed by the city council, in the case of the study, it was proposed to increase the amount of waste by 25%. The users who obtained it obtained a reduction of € 5 on the monthly rate of garbage (€ 50). At the end of the experiment, the amount of waste deposited according to the simulated behavior was measured in order to measure the efficiency of the system, as can be seen in Table 1. In which an increase of 29.4% is observed.

5 Conclusions

This work presents GarbMAS an innovative approach that allows a multi-function system whose main objective is to increase the recycling of urban waste through gamification techniques. For this, GarbMAS adapts the recycling process to the methodology of the games so that benefits are achieved if users deposit a greater amount of waste. In order for the system to allow this efficient management, it is necessary to deploy an infrastructure in each container that allows capturing the information about the user, the amount of waste introduced, the type of waste, the degree of occupancy of the customer and the nearest waste treatment plant.

The case study, has allowed us to simulate different behaviors of the users in a way that shows us how our proposal provides some results that improve the current recycling rates. The results that the case study confirms this assertion, since the increase that has occurred in citizen participation has been of 34.2% and in terms of the number of residues it has increased by 29.4% with respect to the data obtained if GarbMAS is displayed in the location of the case study. As future work lines it is intended to carry out the real deployment of the system. This will allow new studies to validate the results in a real scenario. At a social level, GarbMAS will identify the least participative people and develop awareness campaigns. It is proposed in the next version of this work to make partial achievements, so that the benefit of each user depends on the amount of waste deposited. The functionality of the system allows the system to be used as a tool to perform measurements of social scope that allows knowing the behavior patterns of the users (average amount of waste deposited, frequency with which it goes to the containers, time, days of the week, etc.).

Acknowledgements. This research has been partially supported by the European Regional Development Fund (FEDER) within the framework of the Interreg program V-A Spain-Portugal 2014-2020 (PocTep) under the IOTEC project grant 0123_IOTEC_3_E and by FEDER funds under the projects TIN2016-75850-R. The research of Alfonso González-Briones has been co-financed by the European Social Fund (Operational Programme 2014-2020 for Castilla y León, EDU/310/2015 BOCYL).

References

1. Briones, A.G., González, J.R., de Paz Santana, J.F.: A drug identification system for intoxicated drivers based on a systematic review. ADCAIJ Adv. Distrib. Comput. Artif. Intell. J. **4**(4), 83–101 (2015)
2. Chamoso, P., De la Prieta, F., De Paz, F., Corchado, J.M.: Swarm agent-based architecture suitable for internet of things and smartcities. In: 12th International Conference on Distributed Computing and Artificial Intelligence, pp. 21–29. Springer, Cham (2015)
3. Eurostat, Statistics Explained. Waste statistics (2018). http://ec.europa.eu/eurostat/statistics-explained/index.php/Waste_statistics. Accessed 19 Dec 2017
4. García, Ó., Alonso, R.S., Prieto, J., Corchado, J.M.: Energy efficiency in public buildings through context-aware social computing. Sensors **17**(4), 826 (2017)
5. García, Ó., Alonso, R.S., Tapia, D.I., Corchado, J.M.: CAFCLA: a framework to design, develop, and deploy AMI-based collaborative learning applications. In: Recent Advances in Ambient Intelligence and Context-Aware Computing, pp. 187–209. IGI Global (2015)
6. García, O., Chamoso, P., Prieto, J., Rodríguez, S., de la Prieta, F.: A serious game to reduce consumption in smart buildings. In: International Conference on Practical Applications of Agents and Multi-Agent Systems, pp. 481–493. Springer, Cham (2017)
7. González Briones, A., Chamoso, P., Barriuso, A.L.: Review of the Main Security Problems with Multi-Agent Systems used in E-commerce Applications (2016)
8. González, A., Ramos, J., De Paz, J.F., Corchado, J.M.: Obtaining relevant genes by analysis of expression arrays with a multi-agent system. In: 9th International Conference on Practical Applications of Computational Biology and Bioinformatics, pp. 137–146. Springer, Cham (2015)
9. González-Briones, A., Ramos, J., De Paz, J.F., Corchado, J.M.: Multi-agent system for obtaining relevant genes in expression analysis between young and older women with triple negative breast cancer. J. Integr. Bioinform. **12**(4), 1–14 (2015)
10. González-Briones, A., Villarrubia, G., De Paz, J.F., Corchado, J.M.: A multi-agent system for the classification of gender and age from images. Comput. Vis. Image Underst. (2018). https://doi.org/10.1016/j.cviu.2018.01.012
11. González-Briones, A., Chamoso, P., Yoe, H., Corchado, J.M.: GreenVMAS: virtual organization based platform for heating greenhouses using waste energy from power plants. Sensors **18**, 861 (2018)
12. González-Briones, A., Prieto, J., De La Prieta, F., Herrera-Viedma, E., Corchado, J.M.: Energy optimization using a case-based reasoning strategy. Sensors **18**, 865 (2018)
13. Meng, X., Wen, Z., Qian, Y.: Multi-agent based simulation for household solid waste recycling behavior. Resour. Conserv. Recycl. **128**, 535–545 (2018)
14. Mishra, N., Kumar, V., Chan, F.T.: A multi-agent architecture for reverse logistics in a green supply chain. Int. J. Prod. Res. **50**(9), 2396–2406 (2012)
15. Rodríguez, S., de Paz, Y., Bajo, J., Corchado, J.M.: Social-based planning model for multiagent systems. Expert Syst. Appl. **38**(10), 13005–13023 (2011)

16. Tapia, D.I., Alonso, R.S., De Paz, J.F., Corchado, J.M.: Introducing a distributed architecture for heterogeneous wireless sensor networks. In: International Work-Conference on Artificial Neural Networks, pp. 116–123. Springer, Heidelberg (2009)
17. The European Container Glass Federation. FEVE PR recycling 2013 (2017). http://feve.org/wp-content/uploads/2016/04/Press-Release-EU.pdf. Accessed 19 Dec 2017
18. Wood woRking INdustry RecycliNG. Waste statistics (2017). https://ec.europa.eu/growth/tools-databases/eip-raw-materials/en/content/wood-working-industry-recycling. Accessed 19 Dec 2017
19. Yang, Q.Z., Sheng, Y.Z., Shen, Z.Q.: Agent-based simulation of economic sustainability in waste-to-material recovery. In: 2011 IEEE International Conference on Industrial Engineering and Engineering Management (IEEM), pp. 1150–1154. IEEE (2011)

A Computational Analysis of Psychopathy Based on a Network-Oriented Modeling Approach

Freke W. van Dijk and Jan Treur[✉]

Behavioural Informatics Group, Vrije Universiteit Amsterdam,
Amsterdam, Netherlands
frekevd@gmail.com, j.treur@vu.nl

Abstract. In this paper psychopathy is analysed computationally by creating a temporal-causal network model. The network model was designed using knowledge from Cognitive and Social Neuroscience and simulates the internal neural circuit for moral decision making. Among others, empathy and fear are considered to affect the decision making. This model provides a basis for a virtual agent for simulation-based training or a support application for medical purposes.

Keywords: Psychopathy · Network-Oriented Modeling
Computational analysis

1   Introduction

Modeling is the key to simplifying complex real-world processes in science which allows one to study these processes in tangible portions. The field of Artificial Intelligence has addressed modeling of cognitive processes behind intelligence since the beginning, and the influence of emotion in cognitive models is growing. These human-like models are necessary for engineering virtual agents and emotions are now a broader accepted influence of the decision making processes in the brain; e.g., (Loewenstein and Lerner 2003, p. 619). The mental disorder psychopathy is the target domain of this computational analysis. This disorder is characterized by brain deviations and differences in behavior compared to healthy individuals. Research has been done on how this can be mapped out and simulated. As there are not many computational models yet for the specific domain of Psychopathy, the research presented in this paper forms a possible basis for more future research in this area.

A temporal-causal network model (Treur 2016) for psychopathy was created and tested in a situation that draws out psychopathic behavior in people with this disorder. The setting is a room in the public area, for example, a waiting room at a train station. One person walks out of the room and accidentally leaves behind his or her wallet. Another person might pick up the wallet and give it back, or might choose to steal it, despite that there are other people present. Our model will computationally address how this (mostly unconscious) decision making process works and how it differs between healthy individuals and persons with psychopathy. The main aspects found

© Springer International Publishing AG, part of Springer Nature 2019
F. De La Prieta et al. (Eds.): DCAI 2018, AISC 800, pp. 344–356, 2019.
https://doi.org/10.1007/978-3-319-94649-8_41

that play an important role in this are empathy and fear, which would both inhibit one's propensity to grab the wallet. In contrast, a motivator to grab the wallet would be gain, which represents how taking the wallet would improve the situation of this person.

This computational analysis is mainly targeted at the Cognitive Agent Modeling field of Artificial Intelligence and integrating the fields Neuroscience, Criminology and Artificial Intelligence in a manner that benefits all. It provides a basis for a virtual agent that addresses psychopathy. This virtual agent model could be a great advantage in empirical research or in an educational setting. A model closest to a computational model for psychopathy is a more general agent-based model for criminal behaviour (Bosse et al. 2009). In the model introduced in the current paper the relevant cognitive and neural processes are addressed in more detail.

In Sect. 2, the (neurological) background of this research and related literature are discussed. Section 3 covers the modeling segment. The computational network model, which is based on Sect. 2, is introduced. Section 4 addresses the experimental work and some of the results that were achieved. Finally, Sect. 5 is a discussion.

2 Background on Psychopathy

Psychopathy falls under Antisocial Personality Disorder, or ASPD. Whereas Conduct Disorder is the childhood version of ASPD and psychopathy, adults either have ASPD or ASPD with psychopathy. ASPD is characterized by a lack of remorse or empathy for others, cruelty, poor behavioral controls, recurring difficulty with the law, promiscuity, inability to tolerate boredom, irresponsibility and the tendency to violate the boundaries of others (Vervaeke 2016). On top of these characteristics, people with ASPD and psychopathy also are often manipulative, highly intelligent, have a grandiose sense of self-worth, superficial charm and they are not good at detecting emotions, especially fear, in others as well as themselves. Psychopathy is considered to form the dark triad together with narcissism and Machiavellianism. Narcissists often also express psychopathic tendencies (Muris et al. 2017; Ronningstam 2010) but the research in this paper will focus on people with ASPD and psychopathy.

How do these phenotypical characteristics look in the brain? Neuroimaging studies found that the dorsal and ventral regions of the prefrontal cortex (PFC), the amygdala, hippocampus, angular gyrus and posterior cingulate are functionally or structurally deviating in persons with psychopathy. These overlap somewhat with the regions that are activated in moral judgement tasks, such as whether to steal a wallet or not (Raine and Yang 2006). These regions are the medial PFC, ventral PFC, amygdala, and the regions less impaired in psychopaths; the angular gyrus and posterior cingulate.

Rule-breaking behavior is partly caused by impairments in the structures dorsal and ventral PFC, amygdala, and angular gyrus, which serve moral cognition and emotion (Vervaeke 2016). Similar findings were described by Blair and James (2013). They found that several functions were compromised. These regions are the amygdala, ventromedial PFC (vmPFC), the connectivity from amygdala to vmPFC (especially when making moral judgements), and the uncinate fasciculus (a route of white matter in the brain that connects, among others, the amygdala with the frontal brain structures). They also found that amygdala responsiveness to emotional cues was reduced in psychopaths, which is associated with a lack of empathy (callousness) and a lack of guilt. No impairment in cognitive empathy was observed. Structurally, the amygdala volume is found to be reduced as well as the structural integrity of the vmPFC. Combined with the reduced functional connectivity from amygdala to vmPFC in individuals with psychopathic traits, this means in particular that the amygdala, vmPFC, and the upward connection from amygdala to vmPFC is impaired in psychopaths. These regions are associated with emotion and emotion processing (Salzman and Fusi 2010).

Empathy was defined by Keysers and Valeria Gazzola (2014) as "feeling what we would feel in another's stead". Using this definition, Keysers makes the point that there are many ways to measure empathy, and these all acknowledge different facets to empathy. A distinction between these facets could be between cognitive or emotional empathy, or between fantasizing, perspective taking, personal distress or empathic concern. Another common distinction is between actions, emotions and sensations. Keysers' argues that a different dissociation, one between ability and propensity of empathy for different facets of empathy, could help neuroscience. Shirtcliff et al. (2009) has a definition of empathy that is very close to Keysers'. It is described as "the recognition and sharing of another's emotional state", which touches some more on emotions than Keysers did. Their focus is mainly on emotional empathy, but they also acknowledge cognitive empathy, which is closer to the Theory of Mind (Blair et al. 1996; Bosse et al. 2011). Theory of Mind is the ability to represent someone else's mental sates such as thoughts, desires, hopes and feelings. This cognitive empathy or Theory of Mind is shown to be present in psychopaths, who perform even better in Theory of Mind than most highly able autistic adults (Blair and James 2013). From this it can be concluded that psychopaths do not, per definition, have a deficit in cognitive empathy. The ability of cognitive empathy is present (Blair et al. 1996). Experiments have also shown that psychopathic individuals can have the ability to empathize in situations that encourage empathy. However, individuals with psychopathy have a different propensity for this ability than healthy individuals do (Keysers and Valeria Gazzola 2014). Shirtcliff et al. (2009) found something similar, namely that empathy is not completely automatic. Some of the processes behind this ability to empathize depend on attention or motivation. It is likely that motivating situations encourage psychopaths to empathize with others, while other situations do not encourage them when helping is costly. Empathizing with someone can increase that person's trust in you, giving you access to their resources. A psychopath would be more likely to empathize in such a scenario (Keysers and Valeria Gazzola 2014). From a neuroscientific perspective, psychopaths have shown reduced amygdala activation and fear, although this normalizes when attention is focused on emotion of the stimuli (Marsh et al. 2013; Keysers and Valeria Gazzola 2014). Lesions in amygdala, associated with psychopathy, reduce fear attribution unless this person's gaze is directed to the eyes of the other person. This, combined with the dependency of empathy on attention and motivation, indicates some very powerful regulating circuit for empathy.

Similar to empathy, fear has several definitions. While it is sometimes described as "a feeling of great worry or anxiety caused by the knowledge of danger" (Kosson et al. 2016) in research about psychopaths it is defined with "anxiety as emotions to describe unpleasant state or tension". Physiological arousal is associated with these emotions, including increased heart rate, respiration and sweating (Kosson et al. 2016). *Earlier,* psychopathy was associated with the absence of nervousness and reduced major affective reactions. Psychopaths were considered to be fearless of danger and punishment. They were thought to be capable of experiencing momentary discomfort or fear of an immediate danger, but not the kind of worrying about future consequences, which ultimately means no fear to be arrested when committing a crime (Kosson et al. 2016). More recently, youths with psychopathic traits reported that they experience less frequent and less intense fear than non-psychopathic youths did. This might be evidence for the disruptions in fear processing in psychopaths that psychopathy is known for (Kosson et al. 2016). The structural and functional impairments of the amygdala discussed above are also observed by Schultz et al. (2016). This structure is associated with fear and emotion processing which lead to the support of a low-fear model of psychopathy. Schultz argues for a distinction between two types of psychopaths who express fear in different ways. He found that primary psychopaths, characterized by constitutional fearlessness, show normal fear expression in a fear conditioning study. In the same study, Secondary psychopaths, characterized by problematic behavior motivated by other factors than fearlessness, expressed a pattern consistent with fear inhibition. He argues that the low fear that is associated with psychopaths might be specific to secondary psychopaths. While psychopathy can be split up in two types, the primary and secondary type, fear can be as well, Hoppenbrouwers et al. (2016) claim. Fear has been used generically, while differences between threat detection, responsivity and the conscious experience of fear are observed. This differentiation might be the cause of the dissension that can be observed in research into fear in psychopaths. Hoppenbrouwers and Bulten make a distinction between threat responsivity and detection versus the conscious experience of fear. Their findings suggest that psychopaths have behavioral and neurobiological deficiencies in threat detection and responsivity, but seem to be able to consciously experience fear as an emotion. Kosson et al. (2016) state that it appears that psychopaths are not generally under-responsive to aversive stimuli. They seem to have a conscious or unconscious coping style that protects them from the negative emotional effects of aversive stimuli when they are prepared for those stimuli. This might indicate that psychopaths also have some sort of fear regulation system that is more developed than in normal persons. This, in a sense more advanced coping style in threat detection, might form an explanation why psychopaths do not avoid the aversive situations that healthy individuals would avoid. It was also found that attention plays an important role in the relationship between psychopathy and the startle reactions of fear, although this relationship is complex (Newman and Baskin-Sommers 2011). Research shows that the deficits in fear and anxiety in psychopaths can be attributed to lack of attention, similar to the findings on empathy. When psychopaths are less focused on emotion, their brain systems for emotion are less activated (Kosson et al. 2016). Psychopaths seem to tune out threatening stimuli by their form of fear regulation.

3 Modeling Psychopathy

The adopted modelling approach is the Network-Oriented Modeling approach descri-bed in (Treur 2016). This is an approach that addresses the complexity of human and social processes and uses temporal-causal networks as a vehicle, in which causes and consequences are set out over time. Treur (2016) shows that any smooth dynamical system can be modeled by a temporal-causal network, especially from the Cognitive, Affective and Social domains. In a temporal-causal network model the states are assumed to have (activation) levels that vary over time. As not all causal relations are equally strong, some notion of *strength of a connection* is used. Furthermore, when more than one causal relation affects a state, some way to *aggregate multiple causal impacts* on a state is used. Moreover, a notion of *speed of change* of a state is used for timing of the processes. These three notions are covered by elements in the Network-Oriented Modelling approach based on temporal-causal networks, and form the defining part of a conceptual representation of a specific temporal-causal network model:

- **Strength of a connection** $\omega_{X,Y}$ Each connection from a state X to a state Y has a *connection weight value* $\omega_{X,Y}$, often between 0 and 1, sometimes below 0 (negative effect) or above 1.
- **Combining multiple impacts on a state** $c_Y(..)$ For each state (a reference to) a *combination function* $c_Y(..)$ is chosen to combine the causal impacts of other states on state Y.
- **Speed of change of a state** η_Y For each state Y a *speed factor* η_Y is used to represent how fast a state is changing upon causal impact.

Combination functions can have different forms, as there are many different approaches possible to address the issue of combining multiple impacts. The appli-cability of a specific combination rule for this may depend much on the type of application addressed, and even on the type of states within an application. Therefore, the Network-Oriented Modelling approach based on temporal-causal networks incor-porates for each state, as a kind of label or parameter, a way to specify how multiple causal impacts on this state are aggregated. For this aggregation a number of standard combination functions are made available as options and a number of desirable properties of such combination functions have been identified, some of which are the identity function and the advanced logistic sum function (Treur 2016), Chap. 2, Sec-tions 2.6 and 2.7:

$$\mathbf{id}(V) = V$$

$$\mathbf{alogistic}_{\sigma,\tau}(V_1, \ldots, V_k) = [(1/(1 + e^{-\sigma(V_1 + .. + V_k - \tau)})) - (1/(1 + e^{\sigma\tau}))](1 + e^{-\sigma\tau})$$

with $\sigma, \tau \geq 0$

Here σ describes the *steepness* of the function when the sum $V_1 + \ldots + V_k$ is around the *threshold* value τ. A main scenario for the model for psychopathy intro-duced here is based on a situation where two persons that do not know each other are in a room, for example a waiting room at a train station. Person X walks out and unknowingly leaves his or her wallet behind. Person Y is left behind in the room with the wallet of person X and needs to decide to either take the wallet or not to take it, risking other people in the waiting room witnessing this. This decision is based on three main components that contribute to valuing of the decision options: gain, empathy and fear. The theories covered in Sect. 2 are the basis for the temporal-causal network model. Thereby the theories on the different systems that a psychopath's decision is based on, that is empathy and fear, are simplified to their essence in this model. The most relevant theories that were considered in designing this model are the following:

- The ability versus propensity of empathy in psychopaths.
- The dissociation between threat processing, automatic physiological fear processing and the conscious experience of fear in psychopaths.
- The amygdala is the key region in fear and emotion processing as well as empathy. This is both functionally and structurally deviating in psychopaths.
- The vmPFC and amygdala-vmPFC connection is essential in empathy. These structures are reduced in integrity and function in psychopaths.

In addition to these, the aspect of gain was addressed. Gain in this model comes purely from agent Y's prediction of how their situation will improve when they execute the action of grabbing the wallet for themselves. The states that are used are shown in Table 1. Except for ws_s, every state is representing a state of agent Y. The weights $\omega_{X,Y}$ represent with how much strength the first state X has impact on Y. Figure 1 shows a graphical conceptual representation of the network model for psychopathy. The simulation starts when the wallet is left, which reinforces itself because it does not change until the final action is executed. If the wallet is left behind in the room with person Y, the assumption is that person Y then observes the wallet. However, this connection can be adapted in the computational model. What follows is that observing the wallet strengthens the preparation of action a, which is to grab the wallet. The state of preparing to grab the wallet influences three representation states (gain, empathy and fear) as a way of valuing the action. In return these three states influence the preparation of action a. Each connection can be either a reinforcement or an inhibition. Finally, if the preparation state of action A reaches its threshold, it will trigger the execution of action a, which is to grab the wallet for themselves.

Additionally, fear is impacted by context, in a very risky situation fear will not be inhibited. Section 2 shows that empathy is possible in psychopaths, but it is most likely to be used in situations where empathy is estimated as useful by the psychopath. This is represented in the conceptual representation by the state "empathy regulation".

When empathy is useful, this reinforces the level of empathy in the psychopath. When empathy is not estimated as useful, empathy is inhibited. This is activated by person Y's sensory representation of gain. Meanwhile, the empathy state is monitored by the empathy regulation state.

The introduced model is connected to the background of neuroscience of psychopathy in the following ways. The regulation states, empathy regulation and fear regulation, and emotion states, empathy and fear, are each represented in different brain structures. The regulation states of the emotions fear and empathy take place in the vmPFC and the emotion states, empathy and fear, are represented in the amygdala. The connections between these states also represent the white matter connection between

Table 1. Overview of states and their explanations

State	Full name state	Explanation
ws_c	World State for context c	The context of the situation
ws_s	World State for stimulus s	Person X leaves wallet behind
ss_c	Sensor State for context c	Observing the context of the situation
ss_s	Sensor State stimulus s	Observing wallet
ps_a	Preparation State of action a	Preparation of grabbing wallet
srs_g	Sensory Representation State for g	Gain
srs_e	Sensory Representation State for e	Empathy, based in the amygdala
srs_f	Sensory Representation State for f	Fear, based in the amygdala
frs	Fear Regulation State	Fear regulation, based in vmPFC
ers	Empathy Regulation State	Empathy regulation, based in vmPFC
es_a	Execution State of action a	Execution of grabbing wallet
ws_w	World State stimulus w	The wallet being gone

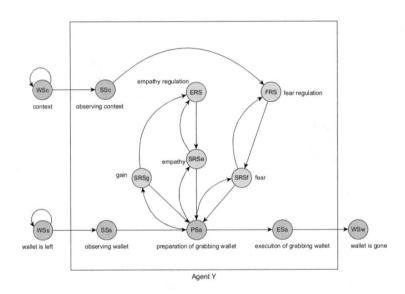

Fig. 1. Graphical conceptual representation of the network model

these brain structures. The regulation states, or control states model mechanisms that detect and adapt undesired levels of emotion by inhibiting connections. The more negative these connections, the stronger the suppression.

To run simulations, a numerical representation of above model of psychopathy is created. For simplicity, during the computational simulations only the in-agent portion of our model was used, where agent Y represents the person described in our scenarios. This is the person who needs to decide whether to take the wallet or not to take it. The simulations of these scenarios were ran using numerical software. In this software the conceptual representation of the model shown above is transformed into a numerical representation in an automated manner as follows; see Ch 2 in (Treur 2016):

- For each time t every state of the Y model has a real value in the interval $[0, 1]$
- For each time t every condition X connected to state Y has an effect on the state Y defined as $\text{impact}_{X,Y}(t) = \omega_{X,Y}Y(t)$
- The aggregated impact of state X_i on state Y at time t is determined by the combination function $\mathbf{c}_Y(..) : \mathbf{aggimpact}_Y(t) = \mathbf{c}_Y(\text{impact}_{X_1,Y}(t), \ldots, \text{impact}_{X_k,Y}(t))$
$$= \mathbf{c}_Y(\omega_{X_1,Y}X_1(t), \ldots, \omega_{X_k,Y}X_k(t))$$

- The effect of $\mathbf{aggimpact}_Y(t)$ on Y is exerted over time gradually, depending on the speed factor $\mathbf{\eta}_Y$: $Y(t + \Delta t) = Y(t) + \mathbf{\eta}_Y [\mathbf{aggimpact}_Y(t) - Y(t)] \Delta t$

$$\mathbf{d}Y(t)/\mathbf{d}t = \mathbf{\eta}_Y[\mathbf{aggimpact}_Y(t) - Y(t)]$$

- This generates a difference and differential equation for Y

$$Y(t + \Delta t) = Y(t) + \mathbf{\eta}_Y [\mathbf{c}_Y(\omega_{X_1,Y}X_1(t), \ldots, \omega_{X_k,Y}X_k(t)) - Y(t))] \Delta t$$
$$\mathbf{d}Y(t)/\mathbf{d}t = \mathbf{\eta}_Y [\mathbf{c}_Y(\omega_{X_1,Y}X_1(t), \ldots, \omega_{X_k,Y}X_k(t)) - Y(t)]$$

As an example, the obtained difference equation for state ps_a is

$$\text{ps}_a(t + \Delta t) = \text{ps}_a(t)$$
$$+ \mathbf{\eta}_{\text{ps}_a} [\mathbf{c}_{\text{ps}_a}(\omega_{\text{ss}_s,\text{ps}_a} \text{ss}_s(t), \omega_{\text{srs}_g,\text{ps}_a} \text{srs}_g(t), \omega_{\text{srs}_e,\text{ps}_a} \text{srs}_e(t),$$
$$\omega_{\text{srs}_f,\text{ps}_a} \text{srs}_f(t)) - \text{ps}_a(t))] \Delta t$$
$$\text{with } \mathbf{c}_{\text{ps}_a}(..) = \mathbf{alogistic}_{\sigma,\tau}(\ldots..)$$

Different software templates have been developed that automatically perform the above transformation and can be used to do simulation experiments, e.g., in Matlab, Python.

4 Simulation Experiments

A number of scenarios were created to use for running simulations and testing the model. Two of them will be discussed here.

- First, a basic scenario as starting point is the following. Person X leaves their wallet behind and person Y is a general person without psychopathy. This situation is called the 'Regular Situation'.
- The second scenario is the situation in which a psychopath steals the wallet. This situation is called the 'Psychopath'.

The expected results are the following for each scenario. In the basic scenario the goal is that person Y does not steal the wallet of person X. This person Y, as a regular person, would have healthy fear regulation. They would be apt to cooperate, which means that empathy is estimated as useful. The second scenario Psychopath addresses a case in which Y steals the wallet. This person Y would have strong fear regulation, which causes low fear. This is expressed by increased inhibiting connections from fear regulation to fear and increased support from gain to empathy regulation since gain plays an important part in the decision-making process. Empathy is not estimated as useful for this case, which is characterized by a strong inhibiting connection from empathy regulation to empathy.

The connections that are assumed to be impaired in individuals with psychopathy are: (ss$_c$, frs), (frs, srs$_f$), (srs$_f$, fr), (ers, srs$_e$), (srs$_e$, ers), (srs$_g$, ers) and (srs$_g$, ps$_a$). These are the connections that differ in weights and thresholds between the test scenarios. The in-agent portion of our domain model was used in simulations for simplification purposes. This is possible because all connections from outside to inside agent Y and from inside agent Y to outside are weighted 1. They are single impacts which means that values are passed through directly. Many simulations were ran to find values that would create a scenario very close to the domain model to test whether this model comes close to our interpretation of the neurological truth. These simulations also show a portion of what is possible to simulate with our model. In the graph for each scenario the colors are used with their representations as presented in Table 2. Furthermore, the weights that remain constant through all four tested scenarios are all 1 except for (srs$_e$, ps$_a$), (srs$_f$, ps$_a$), (frs, srs$_f$) and (ers, srs$_e$) which are -1. These are the connections that are not impaired in healthy individuals nor in individuals with psychopathy.

Table 2. Combination functions used

Color	Name	Description	Function
Turquoise	X1	Observing wallet	id(.)
Light Blue	X2	Observing Context (observing risk of scenario)	id(.)
Green	X3	Preparation of action (preparation of grabbing the wallet)	alogistic$_{\sigma,\tau}$(..)
Pink	X4	Gain	id(.)
Red	X5	Empathy	alogistic$_{\sigma,\tau}$(..)
Orange	X6	Fear	alogistic$_{\sigma,\tau}$(..)
Blue	X7	Fear Regulation	alogistic$_{\sigma,\tau}$(..)
Purple	X8	Empathy Regulation	alogistic$_{\sigma,\tau}$(..)
Lilac	X9	Execution of action	id(.)

The simulations all used speed factor 0.2. Initial values were 0 for all states except X1 and X2 (observing the wallet and observing the context). These states have initial values of 1 to initiate the simulation and to keep these states constant. In this section two figures are inserted as graphical representation of our scenarios. In these graphs the x-axis represents time and the y-axis represents activation levels ranging from 0 to 1.

Scenario 1 is the scenario where our agent Y is a non-criminal individual without mental illness. The graph in Fig. 2 represents the course of the simulation. In this graph, and in the other figure in this section, some similar characteristics can be observed. The states X1 and X2 are constant through all scenarios. For state X1, the observation of the wallet, and state X2, the observation of the context, this applies because these are constant states that are kept active. Note that some of the nine lines are not visible because they are covered by other lines. For example, gain is not visible because it is covered by the execution state.

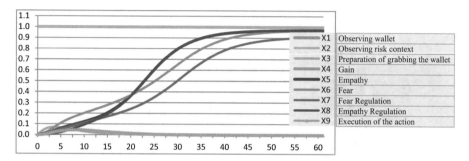

Fig. 2. A simulation of scenario 1: a regular person who does not take the wallet (time on horizontal axis, activation level on vertical axis)

Figure 2 shows state X3, that represents the preparation of action, and state X9, that represents the execution of the action. Both never rise higher than an activation of approximately 0.1. In this figure empathy (X1) works without suppression from empathy regulation (X8), which leads to a line that rises above activation of 0.9. Fear (X6) is also without suppression from fear regulation (X7), because the context of this situation is risky, considering that other people might witness the theft of the wallet. In this scenario the presence of fear and empathy outweigh gain in such a manner that the action of stealing the wallet is not executed. The steepness and threshold values that were used in this simulation are presented in Table 3 below. The connection weight for (ss_c, frs) was 0.3, and for (srs_g, ers) −0.1, the others were 1.

Table 3. Steepness and threshold values in scenario 1

State	X1	X2	X3	X4	X5	X6	X7	X8	X9
Steepness σ	5	5	15	5	5	5	5	5	200
Threshold τ	–	–	1.1	–	0.2	0.1	0.8	0.2	3.4

Scenario 2 is the main target scenario, when agent Y is an individual with psychopathy who steals the wallet due to fear and empathy being inhibited by their corresponding regulation states. Here the connection weights and thresholds are shown in Tables 4 and 5. They were determined based on the expected pattern.

Figure 3 shows, as the only scenario that does so, that preparation of action (X3) comes far of the ground and execution of action (X9) follows. Fear (X6) is inhibited by the fear regulation state (X7). Empathy (X5) is inhibited as well. The action of stealing the wallet is executed (X9 rises to an activation of 1) because the sensory representation of gain (X4) outweighs the inhibited fear and empathy.

Table 4. Connection weights in scenario 2

Connection	(ss_c, frs)	(srs_g, ps_a)	(srs_g, er)	(srs_e, er)	(ers, sr_e)	(srs_f, fr)	(frs, srs_f)
Weight	0.2	2	1	0.2	−1	0.8	−1

Table 5. Steepness and threshold values in scenario 2

State	X1	X2	X3	X4	X5	X6	X7	X8	X9
Steepness σ	5	5	15	5	5	5	5	5	200
Threshold τ	–	–	1.1	–	1	1.2	0.8	0.3	3.4

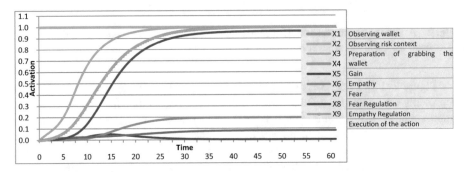

Fig. 3. A simulation of scenario 2: a psychopath who takes the wallet (time on horizontal axis, activation level on vertical axis)

5   Discussion

In this paper a temporal-causal network model was introduced to perform simulation experiments for internal processes and behaviour of a psychopath. It was found that psychopaths do not, by definition, have a deficit in cognitive empathy. It seems that at least for them empathy is not completely automatic, but that it is dependent on their attention or motivation. Similar findings exist on fear. Although it was found that there are differences in experience of fear between primary and secondary types of psychopaths, psychopaths seem to have a coping style for fear that protects them from these negative emotions, more than an average person. These findings indicate that there are some strong circuits active in psychopaths that regulate fear and empathy. This designed model was tested on four scenarios (two of which were presented in the paper) using numerical software, which have led to findings that approximate the expected outcomes of the test scenarios.

The parameters of the model such as thresholds and connection weights in this model can be altered to different types of simulation scenarios. The higher the threshold, the less easily the state of this threshold will be activated. The results from the scenarios that were tested come close to expected outcomes of the scenarios. To create realistic simulations, these weights were adapted to the values that are presented in our findings, but still a finer tuning of the model would be possible. Although the model was compared to qualitative empirical information through the scenarios, it has not been compared to real numerical empirical data yet.

In order to further refine this model, several options are possible. First, the model would benefit from further tuning, as mentioned above. Furthermore, literature about fear in psychopaths also touches upon problems in fear conditioning in psychopathic individuals. (Schultz et al. 2016; Hoppenbrouwers et al. 2016; Lopez et al. 2012). The amygdala is where associative learning of threat takes place, and it is a compromised structure in psychopathy. This fear conditioning problem could be an interesting addition to this model in the future; it would add learning capabilities to the model and provide an adaptive temporal-causal network model. Secondly, as empathy is represented by one state with one regulating state in our model, some refinement could be achieved here too, for example by adopting some elements from Treur (2016, Chap. 9).

Some longer term applications of the further developed model are the following. A virtual patient model that can be used for educational purposes. Such a virtual patient model enables students to see what the effect is of certain aspects of and changes in a patient without having to study a real patient. These virtual patient models could be personalized by changing the values of weights and thresholds. Another way such a virtual patient model could be useful is in the field of Criminology. Criminologists could use this model on offenders instead of on patients in order to study their behavior and their psyche.

References

Blair, R., Sellars, C., Strickland, I., Clark, F., Williams, A., Smith, M., Jones, L.: Theory of mind in the psychopath. J. Forensic Psychiatry **7**, 15–25 (1996)

Blair, R., James, J.: The neurobiology of psychopathic traits in youths. Nat. Rev. **14**, 786–799 (2013)

Bosse, T., Gerritsen, C., Treur, J.: Towards integration of biological, psychological and social aspects in agent-based simulation of violent offenders. Simulation **85**, 635–660 (2009)

Bosse, T., Memon, Z., Treur, J.: A recursive BDI agent model for theory of mind and its applications. Appl. Artif. Intell. **25**, 1–44 (2011)

Kosson, D.S., Vitacco, M.J., Swogger, M.T., Steuerwald, B.L.: Emotional experiences of the psychopath. In: Gaconoco, C.B. (ed.) The Clinical and Forensic Assessment of Psychopathy: A Practitioner's Guide, 2nd edn., pp. 73–96 (2016)

Hoppenbrouwers, S., Bulten, E., Brazil, I.: Parsing fear: a reassessment of the evidence for fear deficits in psychopathy. Psychol. Bull. **142**, 573–600 (2016)

IBM via Google. Temporal Causal Models. (https://www.ibm.com/support/knowledgecenter/en/SS3RA7_17.0.0/components/tcm/tcm_intro.html). Accessed 28 Dec 2017

Keysers, C., Valeria Gazzola, V.: Dissociating the ability and propensity for empathy. Trends Cognit. Sci. **18**(4), 163–166 (2014)

Loewenstein, G., Lerner, J.: The role of emotion in decision making. In: Davidson, R.J., Goldsmith, H.H., Scherer, K.R. (eds.) The Handbook of Affective Science, pp. 619–642. Oxford University Press, Oxford (2003)

Lopez, R., Poy, R., Patrick, C., Moltó, J.: Deficient fear conditioning and self-reported psychopathy: the role of fearless dominance. Psychophysiology **50**, 210–218 (2012)

Marsh, A., Finger, E.C., Fowler, K.A., Adalio, C.J., Jurkowitz, I.T.N., Schechter, J.C., Pine, D. S., Decety, J., Blair, R.J.R.: Empathic responsiveness in amygdala and anterior cingulate cortex in youths with psychopathic traits. J. Child Psychol. Psychiatry **54**(8), 900–910 (2013)

Muris, P., Merckelbach, H., Otgaar, H., Meijer, E.: The malevolent side of human nature: a meta-analysis and critical review of the literature on the dark triad (narcissism, machiavellianism, and psychopathy). Perspect. Psychol. Sci. **12**, 183–204 (2017)

Newman, J.P., Baskin-Sommers, A.: Early selective attention abnormalities in psychopathy: implications for self-regulation. In: Posner, M. (ed.) Cognitive Neuroscience of Attention, pp. 421–439. Guilford Press, New York (2011)

Raine, A., Yang, Y.: Neural Foundations to Moral Reasoning and Antisocial Behavior. Oxford University Press, Oxford (2006)

Ronningstam, E.: Narcissistic personality disorder: a current review. Curr. Psychiatry Rep. **12**, 68–75 (2010)

Salzman, C.D., Fusi, S.: Emotion, cognition, and mental state representation in amygdala and prefrontal cortex. Ann. Rev. Neurosci. **33**, 173–202 (2010)

Schultz, D.H., Balderston, N.L., Baskin-Sommers, A.R., Larson, C.L., Helmstetter, F.J.: Psychopaths show enhanced amygdala activation during fear conditioning. Front. Psychol. **7**, 348 (2016)

Shirtcliff, E., Vitacco, M.J., Graf, A.R., Gostisha, A.J., Merz, J.L., Zahn-Waxler, C.: Neurobiology of empathy and callousness: implications for the development of antisocial behavior. Behav. Sci. Law **27**(2), 137–171 (2009)

Treur, J.: Network-Oriented Modeling: Addressing Complexity of Cognitive, Affective and Social Interactions. Springer, Cham (2016)

Vervaeke, H.: Neurobiology of morality and antisocial behavior. Lectures at Vrije Universiteit Amsterdam (2016)

An Adaptive Temporal-Causal Network for Representing Changing Opinions on Music Releases

Sarah van Gerwen, Aram van Meurs, and Jan Treur[✉]

Behavioural Informatics Group, Vrije Universiteit Amsterdam, Amsterdam, Netherlands
sarahvangerwen@hotmail.com, aramvanmeurs@hotmail.com, j.treur@vu.nl

Abstract. In this paper a temporal-causal network model is introduced representing a shift of opinion about an artist after an album release. Simulation experiments are presented to illustrate the model. Furthermore, mathematical analysis has been done to verify the simulated model and validation by means of an empirical data set and parameter tuning has been addressed as well.

1 Introduction

For many people, music is an important part of daily life. Music can be useful in many situations. While entertainment immediately comes to mind as a primary function of music, it can also provide relief, distraction or emotional engagement (Lundqvist et al. 2008).

Because of the potential impact an artist, album or even a single song can have on someone, it is not surprising that strong fandoms can arise even from a single release. A fan of a specific artist has likely been influenced significantly by one or more releases in the artist's discography. As a result of this, fans do not only tend to monitor the artist's activity in the music business more closely, but they are also likely to have higher expectations of any new music releases that an artist may release in the future.

However, an artist may not always be able to live up to these expectations. They may 'lose their touch' over time, or release new material that is not in line with the expectations set by the artist's core fan base. In such a situation, part of the fan base may publicly declare their dislike for the new released material. In turn, they are likely to influence other members of the fan base as well. In the current era of easy digital communication through social media platforms such as Twitter, Facebook or Youtube, it is very easy to declare an opinion that is public and can be read by a potentially large amount of people. In this paper, we will research this phenomenon in social networks through an analysis and simulation of such situations.

The modeling approach used is the Network-Oriented Modeling approach described in (Treur 2016; 2017). This approach makes use of temporal-causal networks as a vehicle, enabling to model dynamic and adaptive aspects of networks. The network model presented here incorporates both social contagion and network evolution in the sense that network connections change over time by a homophily principle (Byrne

© Springer International Publishing AG, part of Springer Nature 2019
F. De La Prieta et al. (Eds.): DCAI 2018, AISC 800, pp. 357–367, 2019.
https://doi.org/10.1007/978-3-319-94649-8_42

1986; McPherson et al. 2001), which makes connections stronger when persons are more alike and weaker when they are less alike; see also (Mislove et al. 2010; Steglich et al. 2010; Sharpanskykh and Treur 2014; Macy et al. 2003).

In the paper, first in Sect. 2 the Network-Oriented Modeling approach is briefly summarized. Section 3 introduces the temporal-causal network that was designed. Section 4 discusses some simulation results. In Sect. 5 it is discussed how the model was verified by Mathematical Analysis. Section 6 shows how the model was validated by empirical data and parameter tuning. Finally, Sect. 7 is a discussion.

2   The Network-Oriented Modeling Approach Used

In this section the Network-Oriented Modeling approach used described in (Treur 2016), is briefly explained. This Network-Oriented Modeling approach is based on adaptive temporal-causal networks. Causal modeling, causal reasoning and causal simulation have a long tradition in AI; e.g., (Kuipers and Kassirer 1983; Kuipers 1984; Pearl 2000). The Network-Oriented Modeling approach described in (Treur 2016) can be viewed both as part of this causal modeling tradition, and from the perspective on mental states and their causal relations in Philosophy of Mind (e.g., (Kim 1996)). It is a widely usable generic dynamic AI modeling approach that distinguishes itself by incorporating a dynamic and adaptive temporal perspective, both on states and on causal relations. This dynamic perspective enables modeling of cyclic and adaptive networks, and also of timing of causal effects. This enables modelling by adaptive causal networks for connected mental states and for evolving social interaction.

As discussed in detail in (Treur 2016), Chap. 2, temporal-causal network models can be represented at two levels: by a conceptual representation and by a numerical representation. These model representations can be used not only to display graphical network pictures, but also for numerical simulation. Furthermore, they can be analyzed mathematically and validated by comparing their simulation results to empirical data. They usually include a number of parameters for domain, person, or social context-specific characteristics. A conceptual representation of a temporal-causal network model in the first place involves representing in a declarative manner states and connections between them that represent (causal) impacts of states on each other, as assumed to hold for the application domain addressed. The states are assumed to have (activation) levels that vary over time. In reality not all causal relations are equally strong, so some notion of *strength of a connection* is used. Furthermore, when more than one causal relation affects a state, some way to *aggregate multiple causal impacts* on a state is used. Moreover, a notion of *speed of change* of a state is used for timing of processes. These three notions are covered by elements in the Network-Oriented Modelling approach based on temporal-causal networks, and are part of a conceptual representation of a temporal-causal network model:

- **Strength of a connection** $\omega_{X,Y}$ Each connection from a state X to a state Y has a *connection weight value* $\omega_{X,Y}$ representing the strength of the connection, often between 0 and 1, but sometimes also below 0 (negative effect).

- **Combining multiple impacts on a state $c_Y(..)$** For each state (a reference to) a *combination function* $c_Y(..)$ is chosen to combine the causal impacts of other states on state Y.
- **Speed of change of a state η_Y** For each state Y a *speed factor* η_Y is used to represent how fast a state is changing upon causal impact.

Combination functions can have different forms, as there are many different approaches possible to address the issue of combining multiple impacts. The applicability of a specific combination rule for this may depend much on the type of application addressed, and even on the type of states within an application. Therefore the Network-Oriented Modelling approach based on temporal-causal networks incorporates for each state, as a kind of parameter, a way to specify how multiple causal impacts on this state are aggregated. For this aggregation a number of standard combination functions are made available as options and a number of desirable properties of such combination functions have been identified (see Treur 2016, Chap. 2, Sects. 2.6 and 2.7), for example, the scaled sum function with scaling factor $\lambda > 0$:

$$\mathbf{ssum}_\lambda(V_1 + \ldots + V_k) = (V_1 + \ldots + V_k)/\lambda$$

A conceptual representation of temporal-causal network model can be transformed in a systematic or even automated manner into a numerical representation of the model as follows (Treur 2016, Chap. 2):

- at each time point t each state Y in the model has a real number value in the interval $[0, 1]$, denoted by $Y(t)$
- at each time point t each state X connected to state Y has an impact on Y defined as **impact**$_{X,Y}(t) = \omega_{X,Y} X(t)$ where $\omega_{X,Y}$ is the weight of the connection from X to Y
- The *aggregated impact* of multiple states X_i on Y at t is determined using a *combination function* $c_Y(..)$:

$$\begin{aligned}\mathbf{aggimpact}_Y(t) &= c_Y(\mathbf{impact}_{X_1,Y}(t), \ldots, \mathbf{impact}_{X_k,Y}(t)) \\ &= c_Y(\omega_{X_1,Y}X_1(t), \ldots, \omega_{X_k,Y}X_k(t))\end{aligned}$$

 where X_i are the states with outgoing connections to state Y
- The effect of **aggimpact**$_Y(t)$ on Y is exerted over time gradually, depending on speed factor η_Y:

$$Y(t + \Delta t) = Y(t) + \eta_Y[\mathbf{aggimpact}_Y(t) - Y(t)]\Delta t$$

$$\text{or} \quad \mathbf{d}Y(t)/\mathbf{d}t = \eta_Y[\mathbf{aggimpact}_Y(t)) - Y(t)]$$

- Thus, the following *difference* and *differential equation* for Y are obtained:

$$Y(t + \Delta t) = Y(t) + \eta_Y[c_Y(\omega_{X_1,Y}X_1(t), \ldots, \omega_{X_k,Y}X_k(t)) - Y(t)]\Delta t$$
$$\mathbf{d}Y(t)/\mathbf{d}t = \eta_Y[c_Y(\omega_{X_1,Y}X_1(t), \ldots, \omega_{X_k,Y}X_k(t)) - Y(t)]$$

The above three concepts (connection weight, speed factor, combination function) can be considered as parameters representing characteristics in a network model. In a non-adaptive network model these parameters are fixed over time. But to model processes by adaptive networks, not only the state levels, but also the values of these parameters can change over time.

3   The Temporal-Causal Network Model

In the current network we will model opinions and influences according to two principles. The first principle is social contagion, which can be explained as the principle that the opinions of people who are strongly connected will become more alike over time; this is a well-known principle. The second, a bit less known principle is called homophily (Byrne 1986; McPherson et al. 2001). Homophily can be explained as the principle that people whose opinions are more alike will become more strongly connected over time; in this way the network becomes adaptive. When a network follows both these principles, it creates a circular effect.

For the sake of simplicity, all persons are initially randomly connected in our network. Furthermore, two nodes will be added that are connected to all other nodes but are not affected by the homophily principle. One node will represent the overall opinion by summing the values of all nodes. The other node symbolizes the release of the album. This node will represent a sentiment, one can see it as the definition of the way the album is received i.e. if the album is received negatively the value of this node will be low, if the album is received positively the value of this node will be high. The value of this node stays constant over time for simplicity sake. See Fig. 1 for the conceptual representation.

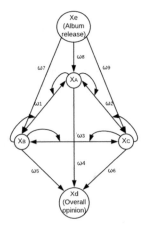

Fig. 1. Conceptual representation of the temporal-causal network model

A numerical representation of the social contagion part of the model, using the scaled sum function as combination function to calculate the aggregated impact, is as follows.

A. Modeling *social contagion* between the states X_B:

- Difference equation:

$$X_B(t + \Delta t) = X_B(t) + \eta_{X_B}[\mathbf{c}_{X_B}(\omega_{X_{A_1},X_B}X_{A_1}(t), \ldots, \omega_{X_{A_k},X_B}X_{A_k}(t)) - X_B(t)]\Delta t$$

- Choosing combination function $\mathbf{ssum}_\lambda(..)$ with scaling factor $\lambda = \omega_{XB} = \omega_{X_{A_1},X_B} + \ldots + \omega_{X_{A_k},X_B}$ provides the following difference equation for $X_B(t)$:

$$\begin{aligned} X_B(t + \Delta t) &= X_B(t) + \eta_{X_B}[\mathbf{ssum}\lambda(\omega_{X_{A_1},X_B}X_{A_1}(t), \ldots, \omega_{X_{A_k},X_B}X_{A_k}(t)) - X_B(t)]\Delta t \\ &= X_B(t) + \eta_{X_B}[(\omega_{X_{A_1},X_B}X_{A_1}(t) + \ldots + \omega_{X_{A_k},XB}X_{A_k}(t))/\omega_{X_B} - X_B(t)]\Delta t \end{aligned}$$

- The corresponding differential equation is:

$$dX_B(t)/dt = \eta_{X_B}[(\omega_{X_{A_1},X_B}X_{A_1}(t) + \ldots + \omega_{X_{A_k},X_B}X_{A_k}(t))/\omega_{X_B} - X_B(t)]$$

B. Modeling the *dynamics of the adaptive connection weights* ω, from X_A to X_B:

- Difference equation:

$$\begin{aligned} \omega_{X_A,X_B}(t + \Delta t) = &\omega_{X_A,X_B}(t) \\ &+ \eta_{\omega_{X_A,X_B}}[\mathbf{c}_{\omega_{X_A,X_B}}(X_A(t), X_B(t), \omega_{X_A,X_B}(t)) - \omega_{X_A,X_B}(t)]\Delta t \end{aligned}$$

- Choosing combination function $\mathbf{c}_{\omega_{X_A,X_B}}(..) = \mathbf{slhom}_{\alpha,\tau}(..)$:

$$\mathbf{slhom}_{\alpha,\tau}(V_1, V_2, W) = W + \alpha W(1 - W)(\tau - |V_1 - V_2|)$$

provides the following difference equation for $\omega_{X_A,X_B}(t)$:

$$\begin{aligned} \omega_{X_A,X_B}(t + \Delta t) = &\omega_{X_A,X_B}(t) \\ &+ \eta_{\omega_{X_A,X_B}}[\mathbf{slhom}_{\alpha,\tau}(X_A(t), X_B(t), \omega_{X_A,X_B}(t)) - \omega_{X_A,X_B}(t)]\Delta t \\ \omega_{X_A,X_B}(t + \Delta t) = &\omega_{X_A,X_B}(t) \\ &+ \eta_{\omega_{X_A,X_B}}[\alpha\,\omega_{X_A,X_B}(t)(1 - \omega_{X_A,X_B}(t))(\tau - |X_A(t) - X_B(t)|)]\Delta t \end{aligned}$$

- The corresponding differential equation is:

$$d\omega_{X_A,X_B}(t)/dt = \eta_{\omega_{X_A,X_B}}[\alpha\,\omega_{X_A,X_B}(t)(1 - \omega_{X_A,X_B}(t))(\tau - |X_A(t) - X_B(t)|)]$$

4   Simulation Results

A first example scenario that was simulated in Python using the introduced model is as follows. Here, the scenario is that an album was received negatively but the overall opinions of people were not harshly affected by this. For this simulation, the model was set up for 100 persons, all interconnected with each other with values for the connection weights randomly assigned. Parameter settings are presented in Fig. 2. Initial values were set randomly. See Fig. 2 for a graph of the outcomes for the state values (For parameter values, see Table 1).

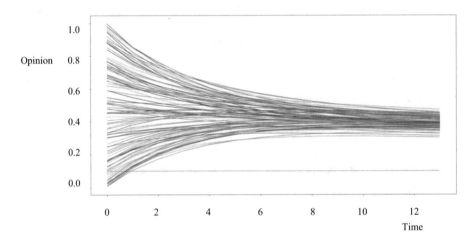

Fig. 2. Graphical representation of an example simulated scenario

Table 1. Parameter settings

Parameters	α	η_X	η_ω	τ	time	Δt
Values	0.5	0.5	0.5	0.05	14	0.5

The final network was also analyzed. Because all 102 nodes (except the overall opinion node and the album node) were interconnected in both directions, only modularity was examined. It was expected that subgroups would be found due to the homophily principle and social contagion principle. When people were more alike their connections became stronger, and the connections became weaker when they were different thus giving the opportunity for groups to arise. But the expectation was that modularity would be low due to the fact that all nodes were initially interconnected randomly and initial state values were set randomly. Modularity was 0.042 and four communities were found. These findings were in accordance with the expectations.

5 Verification of the Network Model by Mathematical Analysis

The behaviour of both parts of the model was mathematically verified, by analysing stationary points. A state Y has a stationary point at t if $dY(t)/dt = 0$. The following criterion is useful; for more details, see (Treur 2016, Chap. 12; Treur 2017).

Criterion In a temporal-causal network model a state Y has a stationary point at t if and only if $c_Y(\omega_{X_1,Y}X_1(t), \ldots, \omega_{X_k,Y}X_k(t)) = Y(t)$.

First, the social contagion part of the model was verified by using the above criterion. The state that has been verified is the overall-opinion node. See Table 2 for the overview for this state and Table 3 for the connections. The second row shows the time point considered, the third row the simulated state value at that time point (right hand side of the above criterion). The third row shows the outcome of the aggregated impact of the other states on this state (left hand side of the above criterion). The last row shows the difference between the two rows above it (the difference between left hand side and right hand side of the above criterion). The stationary point was accurate because the deviation was less than 0.01.

Table 2. Mathematical analysis of stationary point for the overall-opinion node

	Overall-opinion node
Time point	75
State value	0.25
Aggregated impact	0.24409
Deviation	0.00591

Secondly, in a similar manner the homophily part of the model was verified by again looking at stationary points. Here, according to the above criterion a connection $\omega_{XA,XB}$ is stationary at a point t in time if and only if $\omega_{X_A,X_B}(1 - \omega_{X_A,X_B})(\tau - |X_A(t) - X_B(t)|) = 0$. For every cluster of the simulated model, as presented in Fig. 2, one connection within each group has been verified. Furthermore, for every group, connections between groups have been verified. See Table 3 for the results. All stationary points were accurate (absolute deviation < 0.01).

Next, it is shown how the equilibrium values for the connection weights were determined by solving the equilibrium equations. Recall that the combination function for chosen homophily was $\mathbf{slhom}_{\alpha,\tau}(..)$. Therefore the criterion for a stationary point of ω_{X_A,X_B} is

$$\alpha\,\omega_{X_A,X_B}(t)\,(1 - \omega_{X_A,X_B}(t))\,(\tau - |X_A(t) - X_B(t)|) = 0$$
$$\omega_{X_A,X_B}(t) = 0 \text{ or } \omega_{X_A,X_B}(t) = 1 \text{ or } |X_A(t) - X_B(t)| = \tau$$

It was indeed found in the simulation experiments that in an equilibrium state all connection weights have one of these three values. Actually, only the values 0 and 1 occurred. The third option does not occur in simulations; it turns out non-attracting.

Table 3. Mathematical analysis of stationary points for connections in different groups

	ωs037, s053	ωs051, s002	ωs008, s060	ωs004, s027
Group	Green	Red	Blue	Purple
Time Point	199	147	199	148
τ	0.05	0.05	0.05	0.05
A	0.7533396	0.41653445752	0.732069432	0.37019925
B	0.7462730	0.4665799073	0.753700675	0.3170979
$\omega_{XA,XB}$	0.9735395	0.9947619214	0.99904304	0.3170979
Deviation	0.00006	-0.00000002	0.00002	-0.00002
Green	ω_s001, s003		ω_s001, s075	ω_s001, s017
Group	Red		Blue	Purple
Time point	199		199	199
τ	0.05		0.05	0.05
A	0.7550216029699692		0.7550216029699692	0.7550216029699692
B	0.0706582564182544		0.31859763865295626	0.37415177835490465
$\omega_{XA,XB}$	0.0013878034662381867		0.0071999214810507407	0.0057620295145630971
Deviation	−0.0006		−0.0013	−0.008
Red	ω_s003, s096		ω_s003, s009	ω_s003, s011
Group	Blue		Green	Purple
Time point	199		199	199
τ	0.05		0.05	0.05
A	0.0706582564182544		0.0706582564182544	0.0706582564182544
B	0.7339988291681084		0.7525078027695329	0.7655270352677902
$\omega_{XA,XB}$	0.00061083816794847766		0.003810955419157624	0.0049627422806713751
Deviation	−0.0002		−0.0018	−0.0024
Blue	ω_s021, s004		ω_s021, s100	ω_s021, s043
Group	Purple		Red	Green
Time point	185		152	151
τ	0.05		0.05	0.05
A	0.5643423754280926		0.40710342094874463	0.40240624936064046
B	0.4985616546776945		0.45795017406714145	0.4529622564974337
$\omega_{XA,XB}$	0.98564508572479936		0.99694175836884025	0.93834989764832599
Deviation	−0.00003		−0.0000003	−0.000003
Purple	ω_s027, s022		ω_s027, s065	ω_s027, s015
Group	Green		Red	Blue
Time point	199		199	199
τ	0.05		0.05	0.05
A	0.3406215358493736		0.3406215358493736	0.3406215358493736
B	0.7123216865241855		0.7431349070323119	0.7538605497167643
$\omega_{XA,XB}$	0.066607495201250019		0.025797442526141176	0.027427080454125591
Deviation	−0.0084		−0.004	−0.0045

6 Validation Using Empirical Data and Parameter Tuning

To validate the introduced model, a real life example was sought after. This example was the release of Eminem's new album, which is called "Revival". The focus was on the appreciation of this album. The release of this album was accompanied by very negative opinions and reviews, especially compared to Eminem's previous album release. Since the opinions about this new album were so vastly different from the preceding general opinions about Eminem, this seemed to be a good use case for validating the model. Data about the opinions on this album was collected by using the Twitter API. Twitter was used as the preferred social network, since it is a platform where people can very easily express their opinions about specific subjects. Since hashtags are very commonly used by Twitter users, it is easy for anyone to chime in and say something about any subject they desire. Additionally, these opinions can then be read by anyone who is interested in that hashtag. This makes Twitter an ideal platform for expressing opinions.

A Python script was written in order to interact with Twitter's API to collect relevant tweets about Eminem's 'Revival' album. To do this, the Python script was ordered to search for any tweets containing the hashtag '#Revival' during two different time spans. We wanted to measure opinions about the album both before and after the album was released on 15-12-2017. Therefore, the following time spans were used: before release (12-12-2017 to 14-12-2017), and after release (16-12-2017 to 18-12-2017). Ideally, longer time spans would be used. However, this was not possible due to the Twitter API's 7-day search limit. The search returned 175 results before the album's release, and 464 results after the release, all labeled with their sender. These tweets were then analysed using NLTK's sentiment analysis module. The results of this analysis are displayed in Table 4. The difference between the positive tweets and negative tweets from before the album release and after was not very large. Still, a small difference was found. Therefore, this data was still used to validate the network model by parameter tuning. Due to data-retrieval-constraints, connections between the different senders could not be identified. Therefore, all nodes were randomly connected with each other as they were in the previous simulation. Initial values were set to represent the attitudes before the release, 8 people received a value below the 0.5 while the others received a value above the 0.5, again randomly assigned. The album release node represented the negative impact. The value assigned to represent this negative sentiment was 0.35.

Using exhaustive search (with grain size 0.05), the optimal value of the speed factor for the contagion part of the model was identified, which was 0.6. As a result, the graphical representation presented in Fig. 3 was obtained.

Table 4. Data collected from Twitter API search and sentiment analysis

	Before release	After release
Positive tweets	160	410
Negative tweets	15	54
Negativity percentage	8.6%	11.6%

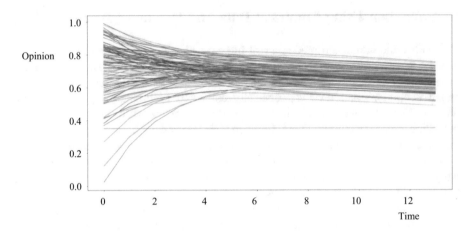

Fig. 3. Graphical representation of simulation of the final model

7   Discussion

In this paper a temporal causal network has been introduced representing a shift of opinion after an album release. One initial simulation experiment has been done. Furthermore, mathematical analysis has been done to verify the simulated model and validation by means of an empirical data set and parameter tuning has been addressed. The data set had some limitations as only data about states over time were available and no data about connections over time.

The Network-Oriented Modeling approach used is based on adaptive temporal-causal networks. Causal modeling has a long tradition in AI; e.g., (Kuipers and Kassirer 1983; Kuipers 1984; Pearl 2000). The Network-Oriented Modeling approach based on temporal-causal networks described in (Treur 2016) can be viewed as part of this causal modeling tradition. In (Treur 2017) it is shown that it is a widely usable generic dynamic modeling approach that distinguishes itself by incorporating a dynamic and adaptive temporal perspective, both on states and on causal relations. This dynamic perspective enables modeling of cyclic and adaptive networks, and also of timing of causal effects. This enables modelling by adaptive causal networks for connected mental states and for evolving social interaction. The model presented in the current paper shows these dynamics and adaptivity. In contrast to other literature such as (Mislove et al. 2010; Steglich et al. 2010; Sharpanskykh and Treur 2014; Macy et al. 2003) the work presented in the current paper addresses application to music appreciation and social media. Note that in the model the album-release node has a constant value for simplifying reasons though it could very well be imaginable that this sentiment shifts according to the overall opinion. Future research could therefore add a relation between these two nodes.

References

Byrne, D.: The attraction hypothesis: Do similar attitudes affect anything? J. Pers. Soc. Psychol. **51**, 1167–1170 (1986)

Kim, J.: Philosophy of Mind. Westview Press, Boulder (1996)

Kuipers, B.J.: Commonsense reasoning about causality: deriving behavior from structure. Artif. Intell. **24**, 169–203 (1984)

Kuipers, B.J., Kassirer, J.P.: How to discover a knowledge representation for causal reasoning by studying an expert physician. In: Proceedings of the 8th International Joint Conference on Artificial Intelligence, IJCAI 1983, Karlsruhe. William Kaufman, Los Altos, CA, pp. 49–56 (1983)

Lundqvist, L.O., Carlsson, F., Hilmersson, P., Juslin, P.N.: Emotional responses to music: experience, expression, and physiology. Psychol. Music **37**, 61–90 (2008)

Macy, M., Kitts, J.A., Flache, A.: Polarization in dynamic networks: a hopfield model of emergent structure. In: Dynamic Social Network Modeling and Analysis, pp. 162–173. National Academies Press, Washington, DC (2003)

McPherson, M., Smith-Lovin, L., Cook, J.M.: Birds of a feather: homophily in social networks. Annu. Rev. Sociol. **27**(1), 415–444 (2001)

Mislove, A., Viswanath, B., Gummadi, K.P., Druschel, P.: You are who you know: inferring user profiles in online social networks. In: Proceedings of the WSDM 2010, New York City, New York, USA, pp. 251–260 (2010)

Pearl, J.: Causality. Cambridge University Press, Cambridge (2000)

Sharpanskykh, A., Treur, J.: Modelling and analysis of social contagion in dynamic networks. Neurocomputing **146**, 140–150 (2014)

Steglich, C.E.G., Snijders, T.A.B., Pearson, M.: Dynamic networks and behavior: separating selection from influence. Sociol. Methodol. **40**, 329–393 (2010)

Treur, J.: Network-Oriented Modeling: Addressing Complexity of Cognitive, Affective and Social Interactions. Springer Publishers, Cham (2016). https://doi.org/10.1007/978-3-319-45213-5

Treur, J.: On the applicability of network-oriented modeling based on temporal-causal networks: why network models do not just model networks. J. Inf. Telecommun. **1**(1), 23–40 (2017)